THE OXFORD AUTHORS

General Editor: Frank Kermode

GERARD MANLEY HOPKINS was born in 1844 at Stratford in Essex. After attending Highgate School, he entered Balliol College, Oxford, as an exhibitioner in 1863. Drawn into the religious controversy still active there, he was converted to Catholicism and upon graduation took a post as teacher in Newman's school, the Oratory, near Birmingham. The following year he decided to enter the Society of Jesus and burned copies of the poems he had written in symbolic dedication of himself to his new vocation. Two largely enjoyable years at the Novitiate in London were followed by three spent in further study amidst the bleaker beauty of Lancashire. It was some years later, while Hopkins was studying theology in Wales, that he returned to poetic composition, writing in less than two years 'The Wreck of the Deutschland' and more than a dozen good short poems. Between the end of his study of theology in 1877 and the respite of his tertianship in 1881 he held seven different posts up and down the country from London and Oxford to Liverpool and Glasgow. The final five years of his life were spent in Ireland as Professor of Greek and Latin Literature in the newly formed Catholic University College, Dublin. Here he was beset by overwork, distress at Irish hatred of England, and most of all, despite writing a number of poems now considered among his very best, by the growing conviction that he could not accomplish any literary work of recognized value. He died of typhoid in June 1889.

CATHERINE PHILLIPS is Fellow and Director of Studies in English at Downing College, Cambridge. She has prepared an edition of Hopkins's letters for the Oxford Selected Letters series (1990, paperback 1991) and has written a biography of Robert Bridges (Oxford University Press 1992). An edition of W. B. Yeats's play, *The Hour-Glass*, in the Cornell Yeats series, is forthcoming, and she is currently writing a book on Hopkins and the Arts.

FRANK KERMODE, retired King Edward VII Professor of English Literature at Cambridge is the author of many books, including *Romantic Image, The Sense of an Ending, The Classic, The Genesis of Secrecy, Forms of Attention,* and *History and Value*; he is also co-editor with John Hollander of *The Oxford Anthology of English Literature.*

THE OXFORD AUTHORS

GERARD MANLEY HOPKINS

EDITED BY

CATHERINE PHILLIPS

Oxford   New York

OXFORD UNIVERSITY PRESS

Oxford University Press, Walton Street, Oxford OX2 6DP

Oxford New York Toronto
Delhi Bombay Calcutta Madras Karachi
Petaling Jaya Singapore Hong Kong Tokyo
Nairobi Dar es Salaam Cape Town
Melbourne Auckland
and associated companies in
Berlin Ibadan

Oxford is a trade mark of Oxford University Press

First published 1986 as an Oxford University Press paperback
and simultaneously in a hardback edition
Paperback reprinted 1989, 1990 (with corrections), 1991, 1992

British Library Cataloguing in Publication Data
Hopkins, Gerard Manley
Gerard Manley Hopkins.—(The Oxford authors)
I. Title II. Phillips, Catherine
821'.8 PR4803.H44
ISBN 0-19-281386-2 Pbk

Printed in Great Britain by
Biddles Ltd.
Guildford and King's Lynn

CONTENTS

PROSE

xii   CONTENTS

ABBREVIATIONS

MS *A*, *B*, *C* I, *C* II, *C*, *D*, *F*, *H*. For detailed descriptions of these see p. xl.

Bodl.   The Bodleian Library, Oxford, in which most of the above manuscripts are housed.

CH   Campion Hall, the private Jesuit hall associated with the University of Oxford, which possesses a number of G. M. H.'s manuscripts.

BL   The British Library, London.

G. M. H. Gerard Manley Hopkins.

R. B.   Robert Bridges, see p. xix.

Dixon   Richard Watson Dixon, see pp. xxix and 394.

Bacon   Father Francis Bacon, who transcribed several of G. M. H.'s poems.

H. H.   Humphry House, see *J*, below.

W. H. G. W. H. Gardner. References to his statements are, unless otherwise stated, to *The Poems of Gerard Manley Hopkins*, 4th edn., which he edited with N. H. MacKenzie (Oxford, 1967, 1982).

N. H. M. N. H. MacKenzie, co-editor of *Poems*, 4th edn. and author of *A Reader's Guide*. The abbreviation is also used to indicate my indebtedness to him for information that will be published in his Oxford English Texts edition.

C. L. P. C. L. Phillips.

L I   *The Letters of G. M. H. to R. B.*, edited by C. C. Abbott (1935, 1955).

L II   *The Correspondence of G. M. H. and R. W. Dixon*, edited by C. C. Abbott (1935, 1955).

L III   *The Further Letters of G. M. H.*, edited by C. C. Abbott, 2nd edn. (1956).

J   *The Journals and Papers of G. M. H.*, edited by Humphry House and Graham Storey (1959).

S   *The Sermons and Devotional Writings of G. M. H.*, edited by Christopher Devlin, SJ (1959).

For details of critical books mentioned, see Further Reading, pp. 401–3.

INTRODUCTION

GERARD MANLEY HOPKINS once wrote about a family holiday at the seaside: 'bathing delightful, horses and boats to be obtained . . . sketches charming, walking tours and excursions . . . My brothers and cousin catch us shrimps, prawns and lobsters, and keep aquariums' (*L* III, p. 201). As the description reminds us, Gerard grew up in a family in many ways typical among well-to-do Victorians. That they were, however, among the more pious of moderate High Anglicans is suggested by the story of Gerard's daily reading of the Bible while at school despite initially being mocked by the other boys, and by the fact that he and his sister Milicent both entered celibate religious orders.[1]

While interest in the arts was common in prosperous Victorian households, the Hopkins family were again somewhat exceptional. Manley Hopkins, a marine adjuster, not only wrote practical shipping manuals but found time to sketch, compose songs, write a novel and numerous poems.[2] His wife had evidently studied Italian and knew some German.[3] She was also fond of music. Their eight children were encouraged to follow the arts; one of Gerard's sisters became a good musician, while among his brothers one became a noted Chinese linguist and two others had careers as artists and illustrators.

Manley Hopkins seems to have exerted considerable influence on his eldest son, who followed him in most of his artistic hobbies. Gerard also remained fascinated throughout his life by ships and the sea. It is striking that twice his periods of poetic aridity were broken by the news of shipwrecks, calling from him two of his longest poems, 'The Wreck of the Deutschland' and 'The Loss of the Eurydice'. The influence is occasionally apparent in smaller, unexpected ways, as for example the image in the fragment, 'Moonrise', of the crescent moon touching a mountain: 'a fluke yet fanged him', a phrase used of anchors.

At Highgate School, which Gerard attended from the age of ten, for

[1] Gerard's eldest sister, Milicent (1849–1946), became a Sister of All Saints' Home in 1878 (see *J*, p. 361). (For abbreviations, see p. xiii.)

[2] For a list of Manley Hopkins's books, see *J*, p. 331.

[3] The list of books owned by Mrs Manley Hopkins includes several in Italian, a French–Italian dictionary, and one each in French and German (see Madelaine House, 'Books belonging to Hopkins and his family' in *Hopkins Research Bulletin*, no. 5 (1974), pp. 34–5). H. H. mentions that she had spent some time living with a German family in Hamburg (*J*, p. 305).

part of the time as a boarder, he followed a curriculum which was devoted primarily to Latin and Greek but also contained some arithmetic, history, and French. Unlike his brother, Cyril, Gerard did not study German.[4] Schoolfellows later described him as exceptionally honest with a keen eye for fair play and an attractive sense of humour. He participated in sport but excelled academically, winning several prizes.

One of these was given for a poem on the Escorial in which he revealed a precocious knowledge and observation of art and architecture; he contemplated becoming a professional poet-artist like Dante Gabriel Rossetti. Letters written a few years later, when he was at Oxford, describe various art exhibitions in detail with clear and reasoned preferences for particular artists or pictures. By then he was familiar with a number of different schools, from Ruskin, whose painstaking sketches seem to have influenced his own, to Giotto and Fra Angelico, Raphael, Dutch and Venetian painters, Gainsborough, and Sir Joshua Reynolds.

Hopkins entered Balliol with a college exhibition in April 1863. His early years at Oxford were probably amongst the happiest of his life. He was popular and enjoyed the breakfast parties, after-dinner 'wines', and long walks that were a large part of the social life there. He belonged to an essay society called the Hexameron (see *J*, pp. 328–9) and mentioned preparing papers for it on 'poetical criticism' and 'realism in the arts'. Literae Humaniores, the degree for which he read, included classical literature, history, philosophy, the gospels, and logic, which he studied with such tutors as Benjamin Jowett, T. H. Green, and Walter Pater, all of whom exacted careful analysis and precise expression. One of his friends, Robert Bridges, later described the great detail with which Hopkins went through texts assigned for the final examinations. Bridges, despairing of reading all the books in time, soon abandoned the attempt to work with Hopkins, but the latter's care paid off and he graduated with a first.[5]

He supplemented his official reading lists with his own 'books to be read' and to some of Shakespeare's plays gave the same careful attention that he paid classical texts. His reading extended from Dickens and George Eliot, George Herbert, Wordsworth, and Browning to books on religious topics and architecture. He also regularly read a number of the

---

[4] See Michael Allsopp, 'Hopkins at Highgate: Biographical Fragments', *Hopkins Quarterly*, vol. vi no. 1 (Spring 1979), pp. 3–10, and a letter to the German teacher at Highgate published by Graham Storey in ibid. no. 4 (1973), pp. 4–6.

[5] *The Poems of Digby Mackworth Dolben*, edited with a memoir by Robert Bridges (London, 1915), p. ci.

journals of the day, becoming acquainted with the work of contemporary writers and artists.

The breadth of this reading is reflected in Hopkins's numerous poems written while he was at Oxford, many of them incomplete. Their subjects range from biblical topics such as 'A Soliloquy of One of the Spies left in the Wilderness' and more personal religious poems like 'The Half-way House' and 'Nondum' to descriptions of Oxford, ballads, and satirical sketches. In these the influence of such poets as George Herbert, Tennyson, Milton, and the Pre-Raphaelites is clear. There are too, original poems in Latin and modified translations.

It is perhaps surprising to find how much continuity there is in the moods and characteristic traits of Hopkins's verse. 'Or try with eyesight to divide' (variant of *Floris in Italy*, see p. 38) and 'It was a hard thing to undo' are early attempts to include analysis of perception in poems (cf. 'The Lantern out of Doors'). In 'The Alchemist' there is the incipient despair at lack of productivity found in such late poems as 'To R.B.' and 'To his Watch'; 'My prayers must meet a brazen heaven' reveals a sense of sinfulness and battle with God leading to the self-splintering found with greater intensity in 'Carrion Comfort'.

Although the heyday of the Tractarian movement was over by the time Hopkins entered Oxford, religious controversy was not. No doubt Balliol was humming with the fact that Edward Pusey, one of the leaders of the Oxford Movement, was attempting to prosecute Benjamin Jowett, Hopkins's tutor in Greek at Balliol, for his liberal religious views. The dispute was also one of the complications in the attempt to raise the salary of the Regius Chair of Greek, which Jowett held, from £38 to £400 a year in line with that of other chairs, such as Hebrew. Pusey, who was the latter's incumbent, believed that the salary of the Professor of Greek should be raised but he wanted the change made in such a way that it was not a personal endorsement of Jowett. There were those, however, who urged that the money should be awarded in clear recognition of Jowett's dedication to his students; the quarrel was not settled until February 1865.

Pusey, whose motivating fear was that the tide of rationalism that he had witnessed years previously in Germany would sweep through England, came to see the main fight as one against unbelief, and he directed his efforts towards uniting the Catholic and English Churches. In order to establish what differences in dogma would need to be discussed he wrote an eirenicon in which he took the essence of Catholic doctrine to be that of the Council of Trent and showed that many of the practices commonly associated with Catholicism were not central dogma but popular. The document brought attacks from all sides. One of these, Frederick Oakeley's essay on the fundamental differences as seen from the Catholic point of view, Hopkins noted in his Journal late in 1865.

Hopkins was invited but decided not to join the Brotherhood of the Holy Trinity, a group of High Anglicans whose code was to 'rise early, use prayer, public and private, be moderate in food and drink, and avoid speaking evil of others'.[6] In fact he became more strict with himself than this, gathering for his confession even such trivialities as talking too much or unkindly or staying up too late. He began to confess regularly in March 1865, normally to Canon Liddon, the popular leader of the High Anglicans at Oxford, but on occasion to Pusey instead.

Hopkins's note of his regime for Lent 1866 (see p. 189) reveals the very privileged life that he was leading and the deliberate, almost self-conscious, nature of his religious practices, also suggested in his undergraduate poems. His impatience with the unintellectual piety of his family's church increased (see L III, p. 18) and undoubtedly he was in part swept along towards conversion by the youthful comradeship and excitement of the controversies at Oxford; three of his friends became Catholics within a few months of each other. The liveliness of the debate meant that he heard or could read the distinct beliefs held by agnostics such as his tutors T. H. Green and Walter Pater, the questioning liberal, Jowett, and the High Anglicans whose sermons and evening discussions he attended.

Hopkins's poem, 'The Half-way House' (1864), reveals his doubts about the Church of England, which were aggravated by the schisms that he could not help but see in it. But his letter to Liddon shortly after his conversion (L III, pp. 32–3) and a plainer account to Urquhart (L III, p. 40) indicate that his main concern was whether in having broken away from Roman Catholicism the Church of England was heretical and those who belonged to it damned. It had been for this reason that Newman had seceded, and once similarly convinced, Hopkins wrote to him asking if he would receive him into the Roman Catholic communion, which he did on 21 October 1866.

Such faith, however, placed Hopkins in a difficult position. There was considerable fear and dislike of Catholicism in England. Anticipating his parents' distress, Hopkins delayed telling them of his conversion until very late, and once they had accepted its inevitability they welcomed him back into the family with the plea that he should not try to convert his brothers and sisters. Similarly, the Bridges family, with whom Hopkins stayed while waiting to see Newman about his conversion, had refused to

[6] See J, pp. 305–6.

allow a relative, Digby Dolben, to be tutored with their children when he showed Catholic leanings. Oxford was still a university of nominally one religion; it was not easy to attend Catholic services and missing Anglican ones brought fines (see Hopkins's letter to his father, pp. 223–6).

Among Hopkins's friends at Oxford was Robert Bridges. They were not particularly close when at the University; while Hopkins became increasingly concerned with his religious life, Bridges was developing an interest in science. Bridges's stepfather was an Anglican minister and Bridges retained a lifelong antipathy to Catholicism but, despite their irreconcilable religious beliefs, Bridges and Hopkins later became good friends, influencing each other's poetry. Alone among Hopkins's acquaintances, it was Bridges who valued his poems enough to make extensive collections and transcriptions of them.

In February 1865 Bridges's 'cousin', Digby Dolben, visited him at Oxford and met Hopkins. At that time Hopkins and Dolben had in common their considerable interest in the Roman Catholic Church and in the writing of poetry. Hopkins obviously admired Dolben and was keen to continue their friendship. They were at an age and in an institution in which male relationships were exceedingly close, but evidence is now appearing which suggests that in the past too much has been made of Hopkins's feelings for Dolben. For example, two poems allegedly written with him in mind have now convincingly been shown to have quite different origins and implications (see 'Where art thou friend' and 'Not kind! to freeze me' (pp. 63, 92)). Hopkins was aware of how easily sexual drive could be aroused (see a late poem, 'To what serves Mortal Beauty?') and no doubt disquiet at this unruly side of himself played a part in channelling his feelings into religious fervour. It is clear, however, that despite a certain suppressed sexuality and frustration of ambition, evident in the mature poems, he was able to develop great warmheartedness, wit, and exceptional honesty.

On leaving Oxford Hopkins took a post as a teacher at the Oratory, a school near Birmingham established by Newman. He found his days there long and busy, divided as they were between teaching the fifth form, two private pupils, and more general supervision of preparation and games. Newman played the violin and musical evenings were held at the school. Always ready to try anything artistic, Hopkins began to learn the instrument. On the whole, however, he missed the stimulating atmosphere and leisure of Oxford and was soon contemplating alternative forms of employment.

Shortly after a retreat at Easter 1868 he decided to join the Society of Jesus, but before entering the Jesuit Novitiate at Manresa House in

London he spent a month walking in Switzerland with Edward Bond, a friend from Oxford. At the time Switzerland did not allow Jesuits to enter the country and Hopkins seized what he realized was the one opportunity he would have of seeing it. His Journal notes of the trip show his detailed observations of nature. The descriptions are highly compressed, with some words used in personal, idiosyncratic ways as in his mature poetry. The terms 'inscape' and 'instress' appear quite frequently. The first examples we have of these words actually occur in Hopkins's notes on Parmenides written in February 1868, but the manner in which they are used there suggests that they were by then well established in his vocabulary. Unlike 'idiom', for example, apparently used for the first time in February 1870 (see p. 201), they are not defined, and at times are scarcely definable in context. Both terms appear to have several meanings, and since they were used in private notes they are not applied with philosophical rigour. 'Inscape' is often used of the characteristic shape of a thing or species. An artist's analysis, it is sometimes linked with comparisons that are superficial and only visual as if to give practical hints to help in drawing (see *J*, p. 176: 'Rushing streams may be described as inscaped ordinarily in pillows—and upturned troughs', for instance). More importantly on other occasions it is used of the crucial features that form or communicate the inner character, essence, or 'personality' of something; one portrait is preferable to another, for example, because it conveys not just a passing mood but the personality of which that mood is part (*J*, p. 245).

Unlike 'inscape', which is the result of mental analysis and perception, 'instress' is more nebulous, often, although not always, associated with feeling; it is the identifying impression a thing can communicate to a careful and receptive observer (see *J*, pp. 213–14). Hopkins also uses the term to mean 'the stress within', the force which binds something or a person into a unit (see p. 217).[7]

The first ten days of his novitiate would have been spent in learning about the nature and requirements of a Jesuit's life and in considering the decision to enter the Society. This would have been followed by a short retreat designed to facilitate further thought about aptitude. Those who elected to stay then began their training. The novices' days began early and were divided into periods of prayer, meditation, and learning about the Institute and carrying out the manual chores necessary to keep the house running.[8] A retreat of thirty days, which Hopkins began on 16 September 1868, would have introduced him to the Spiritual Exercises

[7] For more specialized meanings in Hopkins's spiritual writings, see *S*, pp. 283–4.
[8] *Catholic Encyclopedia* (1912), vol. xiv, p. 83.

of St Ignatius. These, as Father Rickaby states in his edition of them, 'direct a man, first of all, to choose his state of life in view of God and salvation, solely; secondly, when his state of life is fixed, to order the details of his daily conduct on the same principle. To carry this out will be found to involve much overcoming of oneself' (p. 2). The necessity in fulfilling this of examining one's behaviour increasingly closely was an invitation to Hopkins's scrupulosity, and his superiors found it difficult to guide him away from dwelling on his faults and towards the intended consolations. The Exercises played an important part in forming in the young Jesuit a way of viewing the world in religious terms that is clear throughout his mature poetry. The Spiritual Exercises also lay emphasis on not seeking personal fame, which probably persuaded Hopkins against trying to get his poems published once it was clear that the Jesuit journal, the *Month*, did not want them. His prayer near the end of his life leaving them in God's hands to be used as He saw fit follows exactly the prescription of St Ignatius (see p. 301).

Even the form of the Exercises may have influenced the poems. Each Exercise starts by urging the exercitant to picture the scene associated with the subject of the meditation (see, for example, pp. 292–5 for Hopkins's notes on a Meditation on Hell) and to realize the scene as fully as possible by contemplating the sensations that it would give each of the senses in turn. From a full evocation of place the exercitant then turns to contemplating the significance of the event and its implications for his own life. Many of Hopkins's poems similarly begin with vivid descriptions of scenes and then proceed to the Christian and sometimes personal significance of what has been described. Meditation may also have increased a tendency to the internal dialogue seen in such sonnets as 'I wake and feel' or 'To his Watch'.

By and large Hopkins seems to have enjoyed Manresa. His letters home were cheerful and a number of his Oxford friends visited him. He noted in his Journal dialect forms used by others in the community and such physical experiences as dreams and visual impressions, a type of observation seen in 'Or try with eyesight to divide' and later to appear in poems like 'The Lantern out of Doors' and 'The Candle Indoors'. He recorded how the group of novices he was with, unlike some others who never smiled, would roar with laughter when anything went wrong. But a note of an unexpected flood of tears during the Long Retreat at Christmas 1869 suggests the tension that at times he was under (see pp. 200–1). Such outbursts were not uncommon in the Novitiate.

On 8 September 1870 Hopkins took vows of poverty, chastity, and obedience, and the following day travelled to the seminary at Stonyhurst in Lancashire, where he was to spend three years studying philosophy. Here winter was more severe and he often had colds. The descriptions of nature in his Journal from this period are very detailed, as if a magnifying glass were being passed slowly over the object, many features provoking unusual comparisons, as in his later poems.

Records are scanty, but Father Thomas suggests that in his first year Hopkins would have studied the works of Aquinas, Catholic doctrine, and the answers to possible objections to it, as well as mathematics.[9] In the second and third years ethics and psychology, special metaphysics, and some scientific principles were taught (Hopkins recorded spending a miserable morning over formulae for the lever, *L* III, p. 238). The influence of scientific study can be seen extending Hopkins's own acute observations noted in his Journal. For example, on 13 August 1874 he wrote, 'The laps of running foam striking the sea-wall double on themselves and return in nearly the same order and shape in which they came. This is mechanical reflection and is the same as optical: indeed all nature is mechanical, but then it is not seen that mechanics contain that which is beyond mechanics.' The final comment suggests Hopkins's position in the contemporary controversy over the respective truth of science and religion.

Novices and philosophy students served as catechists in the surrounding communities, and Hopkins's reaction to the poverty he saw was expressed in a letter to Bridges: 'I am afraid some great revolution is not far off. Horrible to say, in a manner I am a Communist. Their ideal bating some things is nobler than that professed by any secular statesman I know of . . . Besides it is just.—I do not mean the means of getting to it are. But it is a dreadful thing for the greatest and most necessary part of a very rich nation to live a hard life without dignity, knowledge, comforts, delight, or hopes in the midst of plenty—which plenty they make . . .' (*L* I, pp. 27–8). In calling himself a Communist Hopkins overstated his views. His indignation at the sufferings of the poor remained strong but he did not wish to see the structure of English society dismantled, merely sufficient redistribution of wealth to ease the lot of the poor (see 'Tom's Garland'). Bridges, however, did not reply to the letter. While undoubtedly he would have objected to Hopkins's remarks, his reasons for dropping the correspondence probably included the fact that he was busy as a medical student and heartily disliked the idea of his letters being read by Hopkins's superiors, who were entitled to censor mail although they seldom did so. He probably felt as well that he had increasingly little in common with the Jesuit. Hopkins had virtually

[9] Alfred Thomas, *Hopkins the Jesuit*, pp. 93–8.

ceased to write poetry and it was only after the publication of Bridges's first volume of verse in 1873 that Hopkins realized that Bridges, whom he knew composed music, also had literary interests. After that the correspondence flourished.

In 1872 Hopkins found a copy of the *Oxford Commentary* of Duns Scotus on the *Sentences* of Peter Lombard. The discovery greatly excited him and he noted that whenever he took in 'any inscape of the sky or sea' he thought of Scotus (see p. 211). Scotus (*c.*1266–1308) was known as the 'subtle doctor', partly because of his arguments and partly because his thought was obscured by the misunderstanding of subsequent generations. He was not a popular thinker in Hopkins's time, although some of the beliefs he defended, such as that of Mary's Immaculate Conception, had been made part of Catholic doctrine in the nineteenth century.

The idea in the *Oxford Commentary* that provoked Hopkins's outburst may well have been the defence of the reality of what man knows through his senses and perhaps, especially, a passage which A. B. Wolter translates, 'By grasping just what things are of themselves, a person separates the essences from the many additional incidental features associated with them in the sense image . . . and sees what is true . . . as a more universal truth.'[10] Such an idea has much in common with inscape.

It is clear from Hopkins's *Devotional Writings* that by 1883 he knew of *haecceitas* ('thisness'), a principle of individuation not mentioned by name in the *Oxford Commentary*. The individuality to which Scotus refers is more fundamental and abstract than that covered by inscape. He suggests that this uniqueness is part of God's concept of a person even before he has given him life and is far more radical than the incidental features by which we recognize individuals. Whereas Hopkins applies inscape to species, *haecceitas* is that which differentiates the individual from the species; thus when Hopkins says that Henry Purcell represented in his music 'the very make and species of man' it was the inscape of man that he revealed, not *haecceitas*.

Hopkins liked Scotus' defence of the idea that Christ would have been incarnated even if man had not sinned, since his incarnation enabled him to offer God the worship that he could not give while divine. Hopkins also admired, as he mentioned in a sermon at Bedford Leigh (*S*, pp. 43–5), the doctrine of the Immaculate Conception, which stated that because of and in advance of Christ's sacrifice the Virgin Mary was made completely free of all sin.

[10] *Opus oxoniense*, I, dist. III, q. iv 'The Fourth Way', in A. B. Wolter (ed.), *Duns Scotus: Philosophical Writings* (London, 1963), p. 129.

Because Hopkins's health during the philosophate caused his superiors some concern—he was, for example, forbidden to fast during Lent 1873—he was sent to London to teach Latin, Greek, and English to the Juniors, a comparatively undemanding task for someone with his qualifications. During the year he visited several museums and art galleries, making notes of musical instruments, gems, and Japanese crafts as well as pictures in which he identified those with inscape or instress (see pp. 218–19). He also began to teach himself to play the piano. As before in 1868 when Hopkins was teaching, the number of letters that he wrote dwindled and, although the year in London gave him a variety of outings and a holiday in Devon, he noted at the end of July that 'altogether perhaps my heart has never been so burdened and cast down as this year. The tax on my strength has been greater than I have felt before . . .' (*J*, pp. 249–50).

It was decided that Hopkins should not continue to teach, and late in August 1874 he was sent to St Beuno's to begin his 'theology', a study, he told his father, of dogmatic and moral theology, 'canon law, church history, scripture, Hebrew and what not' (*L* III, p. 124). Although the Catholic community in the area was small and generally not fervent, Hopkins remarked to his mother that he warmed to the Welsh in part because he considered himself 'half Welsh' (*L* III, p. 127). His letters record that he began to learn the language but the Journals note that when warned that the exercise could be undertaken only if he felt the call to convert the Welsh, he decided initially that he could not claim this and reluctantly gave up his lessons. Unfortunately this decision was shortly followed by the failure of another hobby. St Beuno's possessed only a disintegrating harmonium for practice and this grunting instrument vanquished his attempts to teach himself music. He felt the double loss keenly. However, the beauty of Wales was irresistible to someone of Hopkins's temperament, and it was not long before he was enjoying outings, gathering examples of local beliefs and dialect, and, by February 1875, again learning Welsh (*J*, p. 263). Early in 1877 Hopkins wrote to Baillie, 'I have learnt Welsh as you say: I can read easy prose and can speak stumblingly, but at present I find the greatest difficulty, amounting mostly to total failure, in understanding it when spoken and the poetry, which is quite as hard as the choruses in a Gk. play—and consider what those would be with none but a small and bad dictionary at command—I can make little way with' (*L* III, p. 241).[11]

---

[11] W. H. G. notes that St Beuno's library contained the following Welsh dictionaries: Thos. Richards's *Thesaurus* (1753), W. Spurrell's *English–Welsh Dictionary* (1848), W. Owen Pughe's *Welsh–English Dictionary*, 2 vols. (3rd edn., 1866–73), and

Between entering the novitiate and the end of 1875 Hopkins wrote, he told Dixon, only 'two or three little presentation pieces which occasion called for' (*L* III, p. 14). He had thought hard about metre in order to prepare his lectures at Roehampton, as Graham Storey has pointed out (*J*, pp. xxvi–xxvii), and had gathered in his Journals and memory a storehouse of close observations of nature, some already formed into poetic images. He had, too, been developing both in his theological knowledge and religious experience and it is the fusing of precise natural description with deep, personal religious feelings that distinguishes the major poems.

The first of these, and the longest poem Hopkins wrote, was 'The Wreck of the Deutschland'. The *Deutschland*, whose passengers included five exiled German nuns, ran aground near the mouth of the Thames. Because of a severe storm its distress signals were not spotted and the ship remained lodged fast, pounded by waves until finally little but its rigging was above high tide. When help arrived the following day, a quarter of the passengers, including the nuns, had drowned (for more details, see p. 334 and *L* III, pp. 439–43). Hopkins told his mother that the wreck had made a deep impression on him, 'more than any other wreck or accident' he had ever read of (*L* III, p. 135). The poem, however, as he later explained to Bridges, was 'an ode and not primarily a narrative' (*L* I, p. 49).

The story of its reception is familiar; its near publication and then rejection for the *Month* by Father Coleridge, Hopkins's oldest friend in the Society, even after he had complied with the editorial request to delete the stress marks; Bridges's vehement refusal to read it again for any money, and Father Splaine's assertion that it was simply unreadable. It was strikingly new, but not just for its sprung rhythm, with which Hopkins had already experimented in 'For a Picture of Saint Dorothea'. Each stanza reads as a unit, the lines frequently run-on and the shape moulded by the use of numerous types of rhyme within the lines. In this Hopkins was probably influenced by his study of Welsh, and Greek poetry, which can be highly melodious with internal rhyme and alliteration; it commonly has very little punctuation and uses compact epithets often suggesting physical action.[12] Modern readers' problems with 'The Wreck' probably lie more in its allusions to theology and in the obscurity that comes in part from Hopkins's style, in part because he 'was not over-

D. Silvan Evans's *English–Welsh Dictionary*, 2 vols. (1852–8). It is not known to which dictionary Hopkins was referring; most Welsh dictionaries of the day were based on Pughe's (see *Poems*, 4th edn., pp. 326–7).

[12] See W. H. Gardner, *Study*, vol. 2, pp. 120–30.

desirous that the meaning of all should be quite clear, at least unmistake-able' (*L* I, p. 50).

The rhythm, as Elisabeth Schneider has pointed out, can generally be scanned as anapaestic along lines made familiar by Swinburne, although when Bridges evidently made a similar suggestion, Hopkins protested that Swinburne's dactyls and anapaests were 'halting' to his ear (*L* I, p. 44).[13] His own verse, he maintained, was far stricter. Hopkins sent explanations of sprung rhythm to Bridges and Dixon (see pp. 228–9 and 242–4) and later wrote a more detailed analysis to precede the poem in the album of his poems that Bridges began for him in 1883 (see p. 107). It will be noticed that while the description sent to Dixon explains sprung rhythm in terms of rising rhythm (i.e. with the strong stress following the weak one(s)), the later preface is written in terms of falling rhythm (i.e. the strong beat beginning each foot).[14] Since rising rhythms predomin-ate in English verse the earlier system makes scansion easier in the many poems that Hopkins wrote in mixed rhythms.

The strength and weakness of sprung rhythm come from its flexibility. Stresses are dictated by the meaning, but where, as does happen in Hopkins's verse, the meaning is unclear and relative emphasis on words consequently uncertain, the metre too must follow the reader's inter-pretation. Hence Hopkins's distress when Father Coleridge and, later, Bridges wanted to dispense with his metrical marks.

Despite the rejection of 'The Wreck of the Deutschland', Hopkins now felt that he could resume the writing of poetry without compromis-ing his vocation and in the next two years he wrote a great variety of poems. Their subjects ranged from 'Penmaen Pool', a light-hearted advertisement for the village in Merionethshire where the Jesuits holidayed, to occasional poems in English, Latin, and Welsh celebrating the twenty-fifth anniversary of the Bishop of Shrewsbury's episcopate ('Silver Jubilee', 'Ad Episcopum . . .', 'Cywydd'), to criticism of con-temporary society ('The Sea and the Skylark') and personal moods from ecstasy ('Hurrahing in Harvest') to loneliness ('The Lantern out of Doors'). In these the degree of joy is related directly to how closely in touch with God the poet was feeling. They show Hopkins making use of a number of standard rhythms, sometimes combining these with feet of sprung rhythm or 'counterpointing' them, that is, imposing on an established rhythm a second rhythm in such a way that the reader perceives both. The second rhythm could be introduced by placing two feet of it successively, especially if one of these was the second foot in the

---

[13] Elisabeth Schneider, *The Dragon in the Gate*, p. 50
[14] N. H. M., *Reader's Guide*, p. 239.

feet of it successively, especially if one of these was the second foot in the line, a place that Hopkins noted was especially sensitive in pentameter verse (see p. 107). He experimented too with outrides (see p. 108), additional syllables that were not to be counted in scanning the main rhythm. Hopkins's descriptions of the metres which he intended for each poem and his metrical marks in the final version can be found in the notes to the poems (pp. 307–99). Welsh patterns of rhyme and alliteration, cynghanedd, also appear. Hopkins's favourite system divides a line roughly into three parts; in two of these syllables rhyme, while one of the rhyme words is linked to the third part by alliterating with a word there.[15]

Hopkins had expected to spend a fourth year studying theology at St Beuno's. Instead he and three other members of his class were sent out to teach. Father Joseph Feeney suggests reasons for the decision.[16] There had been a number of complaints about low academic standards at St Beuno's and the new Rector, Father Gallwey, had taken the first steps towards raising these in 1877. Examinations were made more difficult and light reading was made less accessible. In addition, with the expansion of Jesuit institutions during the decades after the restoration of the Catholic hierarchy in 1850, there was a constant shortage of trained Jesuits to teach in the schools and colleges and minister in the parishes. As Hopkins explained, 'Much change is inevitable, for every year so many people must begin and so many more must have ended their studies and it is plain that these can seldom step into the shoes left by those, so there is an almost universal shift. Then besides there are offices of fixed term . . . Add deaths, sicknesses, leavings, foreign missions, and what not and you will see that ours can never be an abiding city . . .' (L III, p. 142). Hopkins's emphasis on Scotus in his third-year examinations led to his doing poorly and he was grouped among those to be employed elsewhere in the system. While preparing for his March examination in moral theology Hopkins wrote with a certain facetiousness to Baillie, who was a lawyer, 'You see moral theology covers the whole of life and to know it it is best to begin by knowing everything, as medicine, law, history, banking. But law is what I should most like to know: if you were to come to learn moral theology you would find your knowledge of law very advantageous' (L III, p. 241).

---

[15] See, for example, 'The Sea and the Skylark': l. 4 'there . . . wear . . . wend'; l. 5 'hand . . . land . . . lark'.

[16] Joseph Feeney, SJ, 'Grades, Academic Reform, and Manpower: Why Hopkins Never Completed His Course in Theology', Hopkins Quarterly, vol. ix no. 1 (Spring 1982), pp. 21–31 and vol. xiii nos. 3 and 4 (1986–7), pp. 99–114.

After being ordained to the priesthood at St Beuno's in September 1877 Hopkins went the following month to Mount St Mary's, near Chesterfield. He expected that the work would be 'nondescript—examining, teaching, probably with occasional mission work and preaching or giving retreats attached . . . The number of scholars', he reported, 'is about 150, the community moderately small and family-like, the country round not very interesting but at a little distance is fine country, Sheffield is the nearest great town' (L III, p. 148). The master of eleven-to thirteen-year-olds fell ill and Hopkins was asked to add their tuition to his duties. He grew fond of the lads but found the additional work burdensome. Among the consolations of the post was the tuition of a pupil particularly gifted in the arts. Hopkins was delighted when he won three of the intercollegiate first prizes and, praising his hard work and precise memory, commented, 'I shall not easily have so good a pupil again' (L III, p. 150).

Hopkins's 'muse turned utterly sullen in the Sheffield smoke-ridden air' (L I, p. 48) and he found life there 'dank as ditch-water'. Some years earlier the headmaster had temporarily explored the possibility of moving the school to a healthier spot further from the coal-mines, but nothing had been done, and the new college buildings that were now being constructed left the grounds a muddy building site.[17] However, on 24 March 1878 accounts of the sinking of the *Eurydice* so moved Hopkins that he started to write again. Lines 73–84, the description of the drowned sailor, which he quoted in a letter to Bridges (see p. 230), may well have been the first section of the poem to be composed. He was encouraged when Bridges's response to it was more positive than it had been to 'The Wreck of the Deutschland', although Bridges protested, with reason, at some of the rhymes.

A second poem grew out of Hopkins's association with his pupils. During the Easter plays he observed the incident which when he had moved to Oxford he turned into the poem, 'Brothers'. Father Keegan suggests that the brothers involved were Henry and James Broadbent. Henry was eleven and his brother slightly younger.[18] Hopkins explained to Bridges that the poem was 'something in Wordsworth's manner; which is, I know, inimitable and unapproachable, still I shall be glad to know if you think it a success, for pathos has a point as precise as jest has . . .' (L I, p. 86). But Hopkins was himself soon dissatisfied with the poem and rewrote it in a new metre. He sent successive versions of it to Bridges

[17] Francis Keegan, SJ, 'Gerard Manley Hopkins at Mount St Mary's College, Spinkhill, 1877–1878', *Hopkins Quarterly*, vol. vi no 1 (Spring 1979), p. 30.
[18] Ibid., pp. 22–30.

and, in 1881, a copy to Canon Dixon, who had taught him briefly at school. As Hopkins then reported to Bridges, the Canon 'objected to the first four lines' while Bridges's objections began 'after them' (L I, p. 118).

In April 1878 Hopkins was abruptly moved to Stonyhurst, where a classical scholar was urgently needed to coach students for external examinations for the University of London.¹⁹ By May he knew that he would be sent to Mount Street in London in July. It was while he was at Stonyhurst that Hopkins began to correspond with Canon Dixon. The friendship was carried out by letter—they met only on one occasion—and as with Bridges, it was literature and the correspondents' own creative work that provided most of the topics for discussion.

The move to London, where he was acting curate at Mount Street, enabled Hopkins to meet Bridges several times. However, in December he was sent to join St Aloysius' church in Oxford, perhaps, as Father Devlin suggests, because it was hoped that with his background he would be able not only to minister to the town's Catholic community but also to attend the Catholic undergraduates without offending the University authorities. In fact Hopkins seems to have had little to do with the University. He saw Walter Pater a number of times and the Paravicinis often, but seems to have been afraid that stories of his undergraduate life, however innocent, might mar his dignity as a priest. Hopkins was always unusually sensitive about his personal dignity outside the Society and especially afraid of provoking any unseemly criticism or mockery of the priesthood. This, as will be seen presently, influenced his attitude to the publication of his poetry.

Hopkins told Bridges that while he was fond of the congregation at St Aloysius 'they had not as a body the charming and cheering heartiness of those Lancashire Catholics, which is so deeply comforting' (L I, p. 97). To his mother he wrote that they had 'a stiff respectful stand-off air which we can scarcely make our way through nor explain, but we believe it to be a growth of a University, where Gown holds itself above and aloof from Town and Town is partly cowed by, partly stands on its dignity against Gown' (L III, p. 152). The congregation were also, he thought, too ready to criticize their Church.

Although Hopkins was kept busy as curate, especially after his superior, Father Parkinson, became ill, his letters to Bridges are lively, full of criticisms of Bridges's poems, of explanations of poems he had sent to Bridges, and, after a while, of new pieces he was writing. Two or three of these came directly from his parish work: 'The Bugler's First

¹⁹ Ibid., p. 11.

Communion', reflecting the fact that he was curate for the young soldiers at Cowley barracks, 'The Handsome Heart', based on an incident in Holy Week, and, perhaps, 'The Candle Indoors', showing the weary priest's missionary concern. Hopkins's only totally allegorical poem, 'Andromeda', belongs to this period, as do 'Binsey Poplars'—his reaction to the felling of trees by the Thames—and several others. There are as well a number of incomplete pieces, among them 'Cheery Beggar', 'The furl of fresh-leaved dogrose', and, probably, 'How all is one way wrought!' ('On a Piece of Music').

The creativity of this spell in Oxford was partly due to Hopkins's friendship with Bridges, which grew much warmer while he was there. Although Hopkins was undoubtedly heartened by Dixon's praise of his poems and their religious vocations gave them common ground, it is clear from the letters that Hopkins came to express himself more freely to Bridges. The friends exchanged compliments, and criticism that was forthright to the point of rudeness. The one recurrent strain on the relationship was religion. Bridges's agnosticism disturbed Hopkins. After urging Bridges, who was a busy doctor in London hospitals, to give alms in the hope that this would save his soul, Hopkins gradually restricted his efforts to commenting that 'the meaning was bad' in any of Bridges's poems that showed his lack of belief. Bridges, in turn, hurt Hopkins by on occasion belittling Catholicism. Nevertheless, it was Hopkins whom Bridges asked to be his best man when he married, and it was to Bridges that Hopkins described his ambitions and fears.

From Oxford Hopkins was appointed to Liverpool but sent first to Bedford Leigh to do supply work. The post lasted for three months, from October to December 1879. 'Leigh', as Hopkins told Bridges, 'is a town smaller and with less dignity than Rochdale [where Bridges's family lived] and in a flat; the houses red, mean, and two storied; there are a dozen mills or so, and coalpits also; the air is charged with smoke as well as damp; but the people are hearty' (*L* I, p. 90). It was in Leigh that Hopkins was to preach his best sermons. By October he had begun his most ambitious creative project, a tragedy on St Winefred. The work was never to be finished but Hopkins added little bits to it for some seven years.

In October Canon Dixon proposed to send 'The Loss of the Eurydice' to one of the Carlisle papers. It is not clear from Dixon's letter whether he intended to ask Hopkins's permission before posting the poem off for publication. Hopkins's first reaction was that if the poem had been published it was too late to make a fuss. A week later, however, having become alarmed, he sent a second letter saying 'Pray do not send the piece to the paper: I cannot consent to it, I forbid its publication' (*L* II, p. 30). Such publication could not bring fame, 'but what is not near enough for public fame may be more than enough for private notoriety, which is what I dread' (*L* II, p. 31). He was concerned too because he did not have the consent of the Society's censor for publication and the poem had been rejected by the *Month*; it was not published.

Hopkins stayed in Liverpool for twenty months. He was assistant at St Francis Xavier's, a thriving church with a reputation for sermons that attracted capacity crowds. Hopkins preached a number of times (see *S*, pp. 50–104). Some of these sermons were theologically too difficult for the congregation. Criticized for this, Hopkins wrote sermons that were simpler and homely, and was then taken to task for using the word 'sweetheart'; some of his best homilies were never even delivered because of changes in the church schedules. Though the church was successful, many of the areas in Hopkins's charge were extremely poor. The experience wrought from him the outburst: 'one is so fagged, so harried and gallied up and down. And the drunkards go on drinking, the filthy, as the scripture says, are filthy still: human nature is so inveterate. Would that I had seen the last of it' (*L* I, p. 110), and after he had started his tertianship he wrote, 'my Liverpool and Glasgow experience laid upon my mind a conviction, a truly crushing conviction, of the misery of town life to the poor ... of the misery of the poor in general, of the degradation even of our race, of the hollowness of this century's civilization: it made even life a burden to me to have daily thrust upon me the things I saw' (*L* II, p. 97). Liverpool and Glasgow were two of the poorest towns in the country, Glasgow with the worst statistics for crime and drunkenness in Britain and Liverpool with enormous housing problems. No doubt the hardship was further exacerbated by the exceptionally cold winter of 1880–1.

Although Hopkins wrote little while he was stationed at Liverpool— only 'Felix Randal', about the illness of one of his parishioners at Bedford Leigh, and 'Spring and Fall', composed while returning from saying mass at a country house—he devoted rather more of his energy to music, started to teach himself harmony using Stainer's *Primer*, and wrote a number of tunes for his own poems and those of Bridges and Dixon. In June Bridges contracted pneumonia. The attack was severe and complicated by empyema, and it was some eighteen months before he recovered. Hopkins's letters to Bridges during his illness reveal his gentleness and the depth of the friendship between the two men. Hopkins's own situation improved slightly when he was moved from Liverpool to Glasgow, a town where he preferred both buildings and people to those of Liverpool. From Glasgow he was able to spend two days visiting the Highlands and at Loch Lomond drafted in his tiny travelling diary the poem 'Inversnaid' incorporating ideas about the wilderness that he had mentioned to Bridges in 1879 (L I, pp. 73-4).

Within a month of beginning his tertianship, a year spent in similar ways to the novitiate and, like that, at Manresa House in London, Hopkins wrote to Bridges that 'it is a great rest to be here and I am in a very contented frame of mind'. Initially he read no newspapers but only spiritual books and he decided not to compose any poems. He commented occasionally on work sent to him by Bridges and Dixon but apart from an ode on Thomas Campion which has since been lost, he confined his own writing to notes towards a commentary on the Spiritual Exercises (see S, pp. 107 ff.).

Between September 1882 and February 1884 Hopkins was again at Stonyhurst College, teaching Latin, Greek, and English to secular students studying for external degrees of the University of London. The Provincial was eager that Hopkins should be able to write about the subjects he had in mind such as Greek dramatic metre and style. He did succeed in writing several letters to *Nature* describing in detail shadow effects at sunset and the vivid colours that followed the volcanic eruption of Krakatoa (see L II, appendix II and the *Hopkins Research Bulletin*, no. 2 (1971), pp. 5-7). However, with his usual conscientiousness he gave most of his energy to his teaching, reporting to Dixon that although his time was 'not so closely employed but that someone else in his place might not do a good deal', he could not (L II, p. 108). It was this feeling of insufficient productivity, increasingly to haunt him during the rest of his life, that marred his time in Lancashire. He liked the well-equipped college and admired the bleak, 'solemn and beautiful landscape'. His poems written there include 'Ribblesdale', 'The Blessed Virgin compared to the Air we Breathe', and the completion of 'The Leaden Echo and the Golden Echo'. In August 1883 he met Coventry Patmore, with whom he exchanged letters on literary topics till near the end of his life. They remained, however, on formal terms—Mr Hopkins and Mr Patmore.

Late in 1882 and throughout 1883 Bridges evidently asked Hopkins for explanations of some of his earlier sonnets. The questions may have arisen as Bridges transcribed many of the poems that Hopkins had sent him into a second album which he was prepared to trust to the postal service. Bridges, Hopkins, and Dixon all attempted to interest whoever they could in each other's work, and often had to trust that volumes would not be lost in the post. Bridges was aware that, since Hopkins was haphazard at keeping his poems, he possessed the only final version of a number of them. There is in the letters no reaction from Hopkins to a proposal that a transcribed volume be undertaken until the somewhat gruff response when Patmore asked to see some of them: 'I had not meant Mr. Patmore to know I wrote poetry, but since it has come naturally and unavoidably about there is no more to be said and you may therefore send me your book and I will point it and make a few corrections . . .' (L I, p. 189). However, in a more cheerful frame of mind in August 1884, he wrote 'that book could be the greatest boon to me, if you are so good as to offer it—a godsend and might lead to my doing more' (L I, p. 195).

By this time Hopkins was in Dublin as Professor of Greek and Latin Literature at University College. The college had been expanded from a small 'cramming' institution and moved in November 1883 into the crumbling buildings of the Catholic University founded by Newman in the 1850s.[20] It was one of several colleges forming the Royal University of Ireland and had not come into existence without opposition. The principal opponent to it, Dr W. Walsh, was keen that educational opportunities in Ireland should be provided for people of various backgrounds and that institutions should therefore be scattered throughout the country. Father Delaney, the moving power behind the formation of the new college, wanted to establish a Catholic rival to the Protestant Trinity College. His educational ideals were closer to those of the older English universities and he was consequently keen to obtain well-qualified staff, where necessary from other European countries. Despite vehement opposition Delaney managed to secure university funds for posts in classics and mathematics. An Irish Jesuit, Robert Curtis, was an uncontroversial candidate for the chair in mathematics but for the classicist Delaney decided that there were no Irish scholars appropriate and sought an English Jesuit. The post was offered to Hopkins. Letters exchanged by Father Delaney and Father Purbrick, the Provincial of the English province, show confidence in Hopkins's classical knowledge but suggest that he was not considered to be among the indispensable scholars of the province; there was in fact acknowledged difficulty in finding appropriate employment for him. The Irish saw Hopkins's nationality as the largest objection to appointing him. Nevertheless Delaney again succeeded and, despite an attempt to refuse, Hopkins moved to Dublin on 18 February 1884.

The buildings he found himself in were in serious need of repair and

[20] Norman White, 'Gerard Manley Hopkins and the Irish Row', *Hopkins Quarterly*, vol. ix no. 3 (Fall 1982), pp. 91–107.

modernization: they were much too cold and their plumbing was inadequate. Hopkins remarked that for purposes of study the college was 'very nearly naked' (*L* I, p. 190); he had more money to buy books than space to house them. The main problem, however, was that he was responsible for setting and marking by himself six examinations a year, each with hundreds of candidates. The difficulty was compounded by the scrupulous fairness with which he tried to mark, calling for endless fine decisions over the awarding of part marks out of hundreds. After his first year he was also required to lecture, a task he preferred to marking. Besides the work, which, with Hopkins's conscientiousness, placed too great a strain on him physically, he found himself in the midst of a political situation that left him miserable. His letters frequently allude to the Irish desire for Home Rule. Although it would be a blow for England, Hopkins was convinced that granting independence was the only solution to the problem. He wrote to Baillie, 'be assured of this, that the mass of the Irish people own no allegiance to any existing law or government. And yet they are not a worthless people; they have many true and winning virtues. But their virtues do not promote civil order and it has become impossible to govern them' (*L* III, p. 283). The national strife affected Hopkins's own situation directly. 'It is impossible to say', he reported to his mother, 'what a mess Ireland is and how everything enters into that mess. The Royal University is in the main, like the London University, an examining board. It does the work of examining well; but the work is not worth much. This is the first end I labour for and see little good in. Next my salary helps to support this college. [As a Jesuit Hopkins gave his annual salary of £400 to the college.] The college is . . . rather a failure than a success, and there is less prospect of success now than before. Here too, unless things are to change, I labour for what is worth little. And in doing this most fruitless work I use up all opportunity of doing any other ' (*L* III, p. 185).

He felt that his position required that he produce scholarly publications. His letters show that, even more than in his earlier life, he thought of topic after topic from critical essays on Sophocles (rejected by the *Classical Review*) and Pindar's verse to more scientific papers on statistics, light, and aether. Yet nothing was published except translations into Latin of some of the songs from Shakespeare's plays. Letters to Baillie are filled with speculations about the influence of Egypt on Greek language and culture. But among these many suggested projects creep increasing complaints of flagging energy, lack of inspiration, and an inability to complete started tasks. He struggled to improve his grasp of music, and did some drawing (although he considered an 1866 sketch of a valley at Shanklin much better). A fortnight's holiday in Wales in October 1886 enabled him to write more of his drama, *St. Winefred's Well*. He composed a number of poems, among them 'Harry Ploughman' and 'Tom's Garland', 'Spelt from Sibyl's Leaves', 'To his Watch', 'To what serves Mortal Beauty?', 'That Nature is a Heraclitean Fire . . .', 'St. Alphonsus Rodriguez', 'The shepherd's brow', 'The Soldier', 'Thou art indeed just', 'To R. B.', and the six sonnets of desolation. He started an elegy, 'On the Portrait of Two Beautiful Young People' and an epithalamion for his brother, Everard. Considering the high standard of most of the poetry it is not an inconsiderable accomplishment. But it did not satisfy Hopkins, in correspondence as he was with Dixon, Bridges, and Patmore, all of whom published greater quantities. That he experienced severe depression and anxiety is clear from all his writing: letters, poetry, and meditation notes. He had good Irish friends and as many holidays as the College could afford and the examining timetable allow, but it was not enough. Early in 1889 he contracted typhoid, to which he succumbed on 8 June.

ACKNOWLEDGEMENTS

Anyone working on Hopkins today has good reasons to be grateful for the scholarship of the editors of the four editions of his poems, his journals, sermons, and letters. Of these I am most indebted to the late W. H. Gardner and to N. H. MacKenzie, whose fourth edition of the *Poems* provided excellent solutions to many of the editorial problems I faced and whose notes I found invaluable in the drawing up of those in this edition.

I greatly appreciate the help given to me by Professor Peter Wiseman and Mr Jack Osborne in checking translations and have benefitted from discussions with Professor Frank Kermode, Dr H. H. Erskine-Hill, Mr Myrddin Jones, and Professor Joaquin Kuhn. Professor MacKenzie kindly pointed out some places in the text as established by the fourth edition where changes were needed. I am grateful to him too for his help with the chronology. The sequence of poems in the diaries was largely settled by Humphry House and Graham Storey. I provided approximate dates for later undated manuscripts from Hopkins's handwriting but, with few exceptions, all the rest of the work in forming the chronology used here was done by Professor MacKenzie.

I am grateful to the Society of Jesus for their permission to use copyright material and have much appreciated the hospitality and help of the community at Campion Hall and most especially of the Revd Paul Edwards, SJ, and the Revd Peter Hackett, SJ, successively Masters of the Hall. I am grateful also for the prompt assistance of Thomas Lord Bridges and for his permission to use MS *A* and Hopkins's letters to Bridges and Dixon.

The library staff of the Bodleian, Cambridge University, and the University of Exeter have all given me a great deal of help throughout my research. I have also appreciated the care which Hilary Feldman and Richard Jeffery of Oxford University Press have lavished on the book and have greatly benefitted from Richard Jeffery's knowledgeable advice on the notes. I would also like to express my thanks for the financial assistance of the British Academy, the facilities provided for me by the University of Exeter, and the constant help and support given to me by my husband.

* * *

Third impression (1990): I am pleased to have had an opportunity of making some corrections to the edition and am grateful to Professor N. H. MacKenzie and Dr Norman White for the trouble they have taken to suggest improvements. A list of the most important emendations can be found on p. 415.

CHRONOLOGY

1844   28 July, born at Stratford, Essex. Gerard was the first of eight children. His father, Manley Hopkins, was a marine adjuster and Ambassador for Hawaii in London.

1854–62   Gerard attends Highgate School. He does well academically, winning five prizes, among them the School Poetry Prize for 'The Escorial' (1860), the Governors' Gold Medal for Latin Verse, and a school Exhibition.

1862   Wins an exhibition to Balliol College, Oxford.

1863   April, enters Balliol.

1866   July, decides to join Catholic Church.

     21 October, received by Newman into the Catholic communion.

1867   June, graduates with first class degree.

     Sept.–April 1868, teaches at the Oratory, Birmingham.

1868   2 May, decides to become a priest although unsure whether to join the Benedictines or Jesuits.

     11 May, burns copies of his poems, indicating his new, vocational goal.

     3 July–1 Aug., walking holiday in Switzerland with Edward Bond.

     7 Sept., enters the Jesuit Novitiate at Manresa House, Roehampton (London).

1870   9 Sept., begins three years of philosophy at St Mary's Hall, Stonyhurst, Lancashire.

1872   Reads the *Oxford Commentary* of Duns Scotus on the *Sentences* of Peter Lombard.

1873   From September teaches rhetoric at Roehampton.

1874   Aug., begins three years of theology at St Beuno's, Wales.

1875   Dec., begins to write 'The Wreck of the Deutschland'.

1876   Writes 'Silver Jubilee', 'Ad Episcopum', 'Cywydd', 'Penmaen Pool'.

1877   Feb.–Sept., writes 'God's Grandeur', 'The Starlight Night', 'As kingfishers catch fire', 'Spring', 'The Sea and the Skylark', 'In the Valley of the Elwy', 'The Winhover', 'Pied Beauty', 'Hurrahing in Harvest', 'The Lantern out of Doors'.

     23 Sept., ordained.

     Oct., sent to Mount St Mary's College, Chesterfield where a classical scholar was required as teacher.

1878   April, moved to Stonyhurst to prepare students for the University of London examinations. 'The Loss of the Eurydice' and 'The May Magnificat' written here.

     July–Nov., acting curate at Mount Street, London.

     Dec., becomes curate at St Aloysius' church, Oxford.

1879   Feb.–Oct., writes nine complete poems ('Duns Scotus's Oxford', 'Binsey Poplars', 'Henry Purcell', 'The Candle Indoors', 'The Handsome Heart', 'The Bugler's First Communion', 'Andromeda',

'Morning, Midday, and Evening Sacrifice', and 'Peace') and a number of fragments, and begins to compose music.

Oct.–Dec., curate at St Joseph's, Bedford Leigh, where he writes 'At the Wedding March'.

30 Dec., becomes Select Preacher at St Francis Xavier's, Liverpool.

1880   Writes 'Felix Randal' and 'Spring and Fall'.

1881   Sept., becomes assistant in Glasgow. Visits Loch Lomond and there writes 'Inversnaid'.

Oct., starts tertianship at Roehampton; composes no extant poetry during the year but writes notes towards a commentary on the Spiritual Exercises.

1882   Sept., sent to Stonyhurst College to teach classics. There he completes 'The Leaden Echo and the Golden Echo' and writes 'Ribblesdale'.

1883   R. B. begins his second collection of Hopkins's poems (MS *B*).

Hopkins writes 'The Blessed Virgin compared to the Air we Breathe'.

Aug., meets Coventry Patmore.

1884   Feb., moves to Dublin as Fellow in Classics and Professor of Greek and Latin Literature at the newly formed University College. His duties at first were as examiner in Greek.

Oct.–April 1885, writes most of the extant passages of *St. Winefred's Well*.

1885   May well have written most of the poems called 'The Sonnets of Desolation' as well as 'To what serves Mortal Beauty?', 'The Soldier', 'To his Watch', and 'The times are nightfall'.

1886   May, meets R. B. while on holiday in England.

Completes 'Spelt from Sibyl's Leaves', writes 'On the Portrait of Two Beautiful Young People', translates 'Songs from Shakespeare'.

1887   Aug., holiday in England.

Writes 'Harry Ploughman', 'Tom's Garland', and, perhaps, 'Ashboughs'.

1888   Begins 'Epithalamion', writes 'That Nature is a Heraclitean Fire. . .', 'What shall I do for the land that bred me', and 'St. Alphonsus Rodriguez'.

Aug., holiday in Scotland.

1889   Jan., retreat at Tullabeg. Writes 'Thou art indeed just, Lord', 'The shepherd's brow', and 'To R. B.'.

8 June, dies of typhoid; buried at Glasnevin, Dublin.

NOTE ON THE TEXT

WHEN, shortly after Hopkins's death, Robert Bridges started to edit his friend's poems, it was the second time that he had contemplated an edition of them. His first attempt to print Hopkins's poetry was with that of four other poets and was begun in 1880. The attempt faltered because the chosen publisher, a friend of Bridges's later to publish a number of volumes of his poetry, was just starting his press and was not confident of being able to carry through the project. Jean-Georges Ritz speculates that Hopkins also raised objections himself. In 1889 the plan failed, in part because Bridges at that time considered a lengthy biographical introduction necessary and he was too upset by Hopkins's depression and his own feeling that his friend's life had been wasted to write it, and partly because he did not consider contemporary poetic taste ready to accept such rhythms and diction.

Between 1889 and 1918, when Bridges edited the first edition of Hopkins's verse, he published sixteen of Hopkins's poems or parts of them in various collections. In his edition he was less concerned to present an accurate version of the poems as Hopkins had left them than to win acceptance for the poetry. Consequently, he chose those versions or combinations of them that he thought most appealing, even when, as in the case of 'The Handsome Heart', Hopkins had himself cancelled the copy. The introduction was made editorial rather than biographical; in it Bridges was critical of sources of ambiguity, such as the omission of the subjective relative pronoun, 'that', and the use of homophones in grammatically uncertain positions. He also condemned what he saw as faults of taste in rhyme and metaphor which today few people would find objectionable.

The edition of 750 copies sold slowly and it was not until 1930 that Charles Williams, a poet himself and house-editor at Oxford University Press, was asked to produce an enlarged second edition. W. H. Gardner then carried this work further in 1948 and 1956, adding also a biographical introduction. Much of the editing of the early poems was done by Humphry House and published in *The Notebooks and Papers of Gerard Manley Hopkins* (1937). The fourth edition of the poems, edited jointly by W. H. Gardner and N. H. MacKenzie in 1967, and reprinted with corrections by N. H. Mackenzie a number of times since, caught numerous errors that had crept into the text and restored many of Hopkins's readings in place of those of Bridges.

The Oxford Authors edition discards the subsections in which the poems have previously been placed and presents them in chronological order. Absolute chronology cannot be achieved. The dates Hopkins assigned were normally those of the poem's inception, but he revised his work and left many of his poems without dates. It is hoped that, despite the imperfections, a rough chronology will give a fuller idea of the breadth of Hopkins's poetic interests at each stage. The order used here is one to which a number of other critics have also contributed, among them Humphry House and Graham Storey, W. H. Gardner, and, most of all, N. H. MacKenzie. There are, however, disagreements as to the most likely sequence.

This volume has been prepared from the manuscripts. Many of Hopkins's early poems are contained in two tiny diaries, *C* I and *C* II, where pencil drafts are fragmented by prose entries. These, and a number of poems written on loose leaves, are housed in Campion Hall, Oxford. Most of the manuscripts are in the Bodleian Library. The majority of these belong to the four collections described by Robert Bridges:

A is my own collection, a MS. book made up of autographs—by which word I denote poems in the author's handwriting—pasted into it as they were received from him, and also of contemporary copies of other poems. These autographs and copies date from '67 to '89, the year of his death . . .

B is a MS. book into which, in '83, I copied from *A* certain poems of which the author had kept no copy. He was remiss in making fair copies of his work, and his autograph of *The Deutschland* having been (seemingly) lost, I copied that poem and others from *A* at his request. After that date he entered more poems in this book as he completed them, and he also made both corrections of copy and emendations of the poems which had been copied into it by me. Thus, if a poem occur in both *A* and *B*, then *B* is [generally] the later, and, except for overlooked errors of copyist, the better authority . . .

D is a collection of the author's letters to Canon Dixon . . . they contain autographs of a few poems with late corrections.

H is the bundle of posthumous papers that came into my hands at the author's death. These were at the time examined, sorted, and indexed; and the more important pieces—of which copies were taken—were inserted into a scrap-book. That collection is the source of a series of his most mature sonnets, and of almost all the unfinished poems and fragments. Among these papers were also some early drafts.

Some of these poems have subsequently been bound in other volumes by the staff of the Bodleian; among these are MS *C*, an important and miscellaneous collection of Hopkins's poems, and *F*, a few pages taken from letters to his mother.

There are, in addition, a couple of manuscripts in the British Library and some in private hands.

In all cases the version which I believe to be that last written has been taken for text. This policy has been followed because Hopkins's poetic powers were far from spent when he died and although questions of the influence of Bridges and Dixon arise, it is clear that Hopkins did not simply follow their advice but considered it and, even when conceding their objections, normally found his own solutions to them.

Editorial intervention could have been still further reduced by showing all uncancelled variants in the text but it has been felt that while this is appropriate for scholars (and will be done by N. H. MacKenzie in the Oxford English Texts edition) it is not a suitable introduction to the poetry. Many of the variants can be found in the notes, and in poems where extensive changes have resulted from choosing the final version as text, as for example in 'The Handsome Heart', and 'St. Alphonsus Rodriguez', the earlier but better-known versions are also printed complete. Among the fragments I have in general followed a similar policy, presenting the lines in the sequence the manuscripts suggest was the one that Hopkins favoured at the time the poem was abandoned.

Throughout the edition layout of poems and marks of punctuation have been changed as a result of scrutiny of the manuscripts. In MS *B*, many pages of which contain writing by both Bridges and Hopkins, slight differences in ink-colour and magnification of pen-tracks suggest that some punctuation marks thought in the past to have been introduced by Bridges were in fact made by Hopkins. As in the fourth edition, layout has generally been taken from MS *A* since that used in MS *B* was chosen by Bridges.

The metrical marks that appear in this volume are all Hopkins's own and come for the most part from MS *B*, which was Hopkins's compromise between the more prolific markings of MS *A* and the absence of any guidance. Cost and editorial opinion at Oxford University Press have restricted metrical marks in the text to simple stresses. In some poems written out after 1881 Hopkins introduced a scheme differentiating stress. The full system is marked in the notes but in the text single stresses have been used to show the heaviest accents. The only exception is 'Tom's Garland', where the one double stress marked in MS *B* has been incorporated into the text.

A list of alterations made to the 1984 reprinting of the fourth edition of the poems can be found in Appendix C. There are a few small changes in the printed prose, the largest of which is the restoration of the rest of the final sentence in the first paragraph of 'The Last Sacraments' (*S*.,

p. 248). The passages from the prose have been chosen to elucidate the poetry and to give some idea of Hopkins's personality and the breadth of his interests.

The degree sign (°) indicates a note at the end of the book. More general notes and headnotes are not cued.

The Escorial

Βάτραχος δὲ ποτ' ἀκρίδας ὥς τις ἐρίσδω°

1

There is a massy pile above the waste
Amongst Castilian barrens mountain-bound;
A sombre length of grey; four towers placed
At corners flank the stretching compass round;
A pious work with threefold purpose crown'd—
A cloister'd convent first, the proudest home
Of those who strove God's gospel to confound
With barren rigour and a frigid gloom—
Hard by a royal palace, and a royal tomb.

2

They tell its story thus; amidst the heat   10
Of battle once upon St. Lawrence' day
Philip took oath, while glory or defeat
Hung in the swaying of the fierce melée,
'So I am victor now, I swear to pay
The richest gift St. Lawrence ever bore,
When chiefs and monarchs came their gifts to lay
Upon his altar, and with rarest store
To deck and make most lordly evermore.'

3

For that staunch saint still prais'd his Master's name
While his crack'd flesh lay hissing on the grate;   20
Then fail'd the tongue; the poor collapsing frame,
Hung like a wreck that flames not billows beat—
So, grown fantastic in his piety,
Philip, supposing that the gift most meet,
The sculptur'd image of such faith would be,
Uprais'd an emblem of that fiery constancy.

St. 2: At the battle of St Quentin, between the French and Spaniards, Philip II
vowed the Escorial to St Laurence, the patron saint of the day, if he gained the victory.
St. 3: St Laurence is said to have been roasted to death on a gridiron.

4

He rais'd the convent as a monstrous grate;
The cloisters cross'd with equal courts betwixt
Formed bars of stone; Beyond in stiffen'd state
The stretching palace lay as handle fix'd.   30
Then laver'd founts and postur'd stone he mix'd.
—Before the sepulchre there stood a gate,
A faithful guard of inner darkness fix'd—
But open'd twice, in life and death, to state,
To newborn prince, and royal corse inanimate.

5

While from the pulpit in a heretic land
Ranters scream'd rank rebellion, this should be
A fortress of true faith, and central stand
Whence with the scourge of ready piety
Legates might rush, zeal-rampant, fiery,   40
*Upon the stubborn Fleming; and the rod
Of forc'd persuasion issue o'er the free.—
For, where the martyr's bones were thickest trod,
They shrive themselves and cry, 'Good service to our God.'

6

No finish'd proof was this of Gothic grace
With flowing tracery engemming rays
Of colour in high casements face to face;
And foliag'd crownals (pointing how the ways
Of art best follow nature) in a maze
Of finish'd diapers, that fills the eye
And scarcely traces where one beauty strays
And melts amidst another; ciel'd on high
With blazoned groins, and crowned with hues of majesty.   50

St. 4: The Escorial was built in the form of a gridiron,—the rectangular convent was the grate, the cloisters the bars, the towers the legs inverted, the palace the handle.

The building contained the royal Mausoleum; and a gate which was opened only to the newborn heir apparent, and to the funeral of a monarch.

St. 5: * Philip endeavoured to establish the Inquisition in the Netherlands.

St. 6: Philip did not choose the splendid luxuriance of the Spanish Gothic as the style of architecture fitted for the Escorial,

7

This was no classic temple order'd round
With massy pillars of the Doric mood
Broad-fluted, nor with shafts acanthus-crown'd,
Pourtray'd along the frieze with Titan's brood
That battled Gods for heaven; brilliant-hued,*
With golden fillets and rich blazonry,
Wherein beneath the cornice, horsemen rode†   60
With form divine, a fiery chivalry—
Triumph of airy grace and perfect harmony.

8

*Fair relics too the changeful Moor had left
Splendid with phantasies aerial,
Of mazy shape and hue, but now bereft
By conqu'rors rude of honor; and not all
Unmindful of their grace, the Escorial
Arose in gloom, a solemn mockery
Of those gilt webs that languish'd in a fall.
This to remotest ages was to be   70
The pride of faith, and home of sternest piety.

9°

.   .   .   .

10

He rang'd long corridors and cornic'd halls,
And damasqu'd arms and foliag'd carving piled.—
With painting gleam'd the rich pilaster'd walls—.
*Here play'd the virgin mother with her Child
In some broad palmy mead, and saintly smiled,

St. 7: Nor the Classic.
* The Parthenon &c. were magnificently coloured and gilded.
† The horsemen in the Panathenaic processions.
St. 8: * The Alhambra &c.
St. 8: The Architect was Velasquez; the style Italian Classic, partly Ionic partly
Doric. The whole is sombre in appearance, but grand, and imposing.
St. 10: The interior was decorated with all the richest productions of art and
nature. Pictures, statues, marble, fountains, tapestry, &c. (*He* refers to Philip.)
* In one of Raphael's pictures the Madonna and St Joseph play with their Child in a
wide meadow; behind is a palm-tree.

And held a cross of flowers, in purple bloom;
†He, where the crownals droop'd, himself reviled
And bleeding saw.—Thus hung from room to room
The skill of dreamy Claude, and Titian's mellow gloom.   80

11

Here in some‡ darken'd landscape Paris fair
Stretches the envied fruit with fatal smile
To golden-girdled Cypris;—Ceres there
Raves through Sicilian pastures many a mile;
¶But, hapless youth, Antinous the while
Gazes aslant his shoulder, viewing nigh
Where Phoebus weeps for him whom Zephyr's guile
‖Chang'd to a flower; and there, with placid eye
§Apollo views the smitten Python writhe and die.

        .   .   .   .

12

Then through the afternoon the summer beam   90
Slop'd on the galleries; upon the wall
Rich Titians faded; in the straying gleam
The motes in ceaseless eddy shine and fall
Into the cooling gloom; till slowly all
Dimm'd in the long accumulated dust;
Pendant in formal line from cornice tall
*Blades of Milan in circles rang'd, grew rust
And silver damasqu'd plates obscur'd in age's crust.°

---

† Alluding to Raphael's 'Lo Spasimo', which is, I believe, in the Escorial.
St. 11: ‡ Alluding to the dark colouring of landscapes to be seen in Rubens, Titian &c.
¶ A beautiful youth drowned in the Nile; the statue has the position described.
‖ Hyacinthus.   § The Belvidere Apollo.
St. 12: The Escorial was adorned by succe[e]ding kings, until the Peninsular war, when the French as a piece of revenge for their defeats, sent a body of dragoons under La Houssaye, who entered the Escorial, ravaged and despoiled it of some of its greatest treasures. The monks then left the convent. Since that time it has been left desolate and uninhabited. The 12th stanza describes this.
* Alluding to the practise (*sic*) of arranging swords in circles, radiating from their hilts.

13

†But from the mountain glens in autumn late
  Adown the clattering gullies swept the rain;   100
  The driving storm at hour of vespers beat
  Upon the mould'ring terraces amain;
  The Altar-tapers flar'd in gusts; in vain
  Louder the monks dron'd out Gregorians slow;
  Afar in corridors with painèd strain
  Doors slamm'd to the blasts continuously; more low,°
  Then pass'd the wind, and sobb'd with mountain-echo'd woe.

14

Next morn a peasant from the mountain side
  Came midst the drizzle telling how last night
  Two mazèd shepherds perish'd in the tide;   110
  But further down the valley, left and right,
  Down-splinter'd rocks crush'd cottages.—Drear sight—
  An endless round of dead'ning solitude:
ˣTill, (fearing ravage worse than in his flight,
  What time the baffled Frank swept back pursu'd
  Fell on the palace, and the lust of rabble rude,)

15

*Since trampled Spain by royal discord torn
  Lay bleeding, to Madrid the last they bore,
  The choicest remnants thence;—such home forlorn
  The monks left long ago: Since which no more   120
†Eighth wonder of the earth, in size, in store
  And art and beauty: Title now too null—
  More wondrous to have borne such hope before
  It seems; for grandeur barren left and dull
  Than changeful pomp of courts is aye more wonderful.

St. 13: † The Escorial is often exposed to the attacks of the storms which sweep down from the mountains of Guadarrama.
St. 14: ˣ Some years ago, fearing that the Carlists would plunder the Escorial, they removed the choicest remaining treasures to Madrid.
St. 15: * The civil wars of late years in Spain.
† The Spaniards call it 8th wonder of the world.

Aeschylus: Promêtheus Desmotês

Lines 88–127

PROMÊTHEUS

Divinity of air, fleet-feather'd gales,
Ye river-heads, thou billowy deep that laugh'st
A countless laughter, Earth mother of all,
Thou sun, allseeing eyeball of the day,
Witness to me! Look you I am a god,
And these are from the gods my penalties.
   Look with what unseemliness
   I a thousand thousand years
   Must watch down with weariness
     Fallen from my peers.   10
   The young chief of the bless'd of heaven
   Hath devis'd new pains for me
     And hath given
   This indignity of chains.
   What is, and what is to be,
   All alike is grief to me;
   I look all ways but only see
The drear dull burthen of unending pains.

.   .   .   .

   Ah well a day!—
What was that echo caught anigh me,   20
That scent from breezes breathing by me,
Sped of gods, or mortal sign,
Or half-human, half-divine?
To the world's end, to the last hill
Comes one to gaze upon my ill;
Be this thy quest or other, see
A god enchain'd of destiny,
Foe of Zeus and hate of all
That wont to throng Zeus' banquet-hall,
Sith I lov'd and lov'd too well   30
The race of man; and hence I fell.

    Woe is me, what do I hear?
    Fledgèd things do rustle near;
    Whispers of the mid-air stirring
    With light pulse of pinions skirring,
And all that comes is fraught to me with fear.

Il Mystico

Hence sensual gross desires,
Right offspring of your grimy mother Earth!
    My Spirit hath a birth
Alien from yours as heaven from Nadir-fires:
    You rank and reeking things,
Scoop you from teeming filth some sickly hovel,
    And there for ever grovel
'Mid fever'd fumes and slime and cakèd clot:
    But foul and cumber not
The shaken plumage of my Spirit's wings.　　　　10
    But come, thou balm to aching soul,
    Of pointed wing and silver stole,
    With heavenly cithern from high choir,
    Tresses dipp'd in rainbow fire,
    An olive-branch whence richly reek
    Earthless dews on ancles sleek;
    Be discover'd to my sight
    From a haze of sapphire light,
    Let incense hang across the room
    And sober lustres take the gloom;　　　　20
    Come when night clings to what is hers
    Closer because faint morning stirs;
    When chill woods wake and think of morn,
    But sleep again ere day be born;
    When sick men turn, and lights are low,
    And death falls gently as the snow;
    When wholesome spirits rustle about,
    And the tide of ill is out;
    When waking hearts can pardon much
    And hard men feel a softening touch;　　　　30

When strangely loom all shapes that be,
And watches change upon the sea;
Silence holds breath upon her throne,
And the waked stars are all alone.
    Come because then most thinly lies
The veil that covers mysteries;
And soul is subtle and flesh weak
And pride is nerveless and hearts meek.

.   .   .   .

Touch me and purify, and shew
Some of the secrets I would know.   40

.   .   .   .

Grant that close-folded peace that clad
The seraph brows of Galahad,
Who knew the inner spirit that fills
Questioning winds around the hills;
Who made conjecture nearest far
To what the chords of angels are;
And to the mystery of those Things

.   .   .   .

Shewn to Ezekiel's open'd sight°
On Chebar's banks, and why they went
Unswerving through the firmament;   50
Whose ken through amber of dark eyes
Went forth to compass mysteries;
Who knowing all the sins and sores
That nest within close-barrèd doors,
And that grief masters joy on earth,
Yet found unstinted place for mirth;
Who could forgive without grudge after
Gross mind discharging foulèd laughter;
To whom the common earth and air
Were limn'd about with radiance rare   60
Most like those hues that in the prism
Melt as from a heavenly chrism;

Who could keep silence, tho' the smart
Yawn'd like long furrow in the heart;

    .   .   .   .

  Or, like a lark to glide aloof°
Under the cloud-festoonèd roof,
That with a turning of the wings
Light and darkness from him flings;
To drift in air, the circled earth
Spreading still its sunnèd girth;   70
To hear the sheep-bells dimly die
Till the lifted clouds were nigh;
In breezy belts of upper air
Melting into aether rare;
And when the silent height were won,°
And all in lone air stood the sun,
To sing scarce heard, and singing fill
The airy empire at his will;
To hear his strain descend less loud
On to ledges of grey cloud;   80
And fainter, finer, trickle far
To where the listening uplands are;
To pause—then from his gurgling bill
Let the warbled sweetness rill,
And down the welkin, gushing free,
Hark the molten melody;
In fits of music till sunset
Starting the silver rivulet;
Sweetly then and of free act
To quench the fine-drawn cataract;   90
And in the dews beside his nest
To cool his plumy throbbing breast.
  Or, if a sudden silver shower
Has drench'd the molten sunset hour,
And with weeping cloud is spread
All the welkin overhead,
Save where the unvexèd west
Lies divinely still, at rest,
Where liquid heaven sapphire-pale
Does into amber splendours fail,   100
And fretted clouds with burnish'd rim,

Phoebus' loosen'd tresses, swim;
While the sun streams forth amain
On the tumblings of the rain,
When his mellow smile he sees
Caught on the dark-ytressèd trees,
When the rainbow arching high
Looks from the zenith round the sky,
Lit with exquisite tints seven
Caught from angels' wings in heaven,   110
Double, and higher than his wont,
The wrought rim of heaven's font,—
Then may I upwards gaze and see
The deepening intensity
Of the air-blended diadem,
All a sevenfold-single gem,
Each hue so rarely wrought that where
It melts, new lights arise as fair,
Sapphire, jacinth, chrysolite,°
The rim with ruby fringes dight,   120
Ending in sweet uncertainty
'Twixt real hue and phantasy.
Then while the rain-born arc glows higher
Westward on his sinking sire;
While the upgazing country seems
Touch'd from heaven in sweet dreams;
While a subtle spirit and rare
Breathes in the mysterious air;
While sheeny tears and sunlit mirth
Mix o'er the not unmovèd earth,—   130
Then would I fling me up to sip
Sweetness from the hour, and dip
Deeply in the archèd lustres,
And look abroad on sunny clusters
Of wringing tree-tops, chalky lanes,
Wheatfields tumbled with the rains,
Streaks of shadow, thistled leas,
Whence spring the jewell'd harmonies
That meet in mid-air; and be so
Melted in the dizzy bow   140
That I may drink that ecstacy
Which to pure souls alone may be . . . .

A windy day in summer

The vex'd elm-heads are pale with the view
Of a mastering heaven utterly blue;
Swoll'n is the wind that in argent billows
Rolls across the labouring willows;
The chestnut-fans are loosely flirting,
And bared is the aspen's silky skirting;
The sapphire pools are smit with white
And silver-shot with gusty light;
While the breeze by rank and measure
Paves the clouds on the swept azure.   10

A fragment of anything you like

Fair, but of fairness as a vision dream'd;
Dry were her sad eyes that would fain have stream'd;
She stood before a light not hers, and seem'd

The lorn Moon, pale with piteous dismay,
Who rising late had miss'd her painful way
In wandering until broad light of day;

Then was discover'd in the pathless sky
White-faced, as one in sad assay to fly
Who asks not life but only place to die.

A Vision of the Mermaids

Rowing, I reach'd a rock—the sea was low—
Which the tides cover in their overflow,
Marking the spot, when they have gurgled o'er,
With a thin floating veil of water hoar.
A mile astern lay the blue shores away;
And it was at the setting of the day.
Plum-purple was the west; but spikes of light
Spear'd open lustrous gashes, crimson-white;
(Where the eye fix'd, fled the encrimsoning spot,
And gathering, floated where the gaze was not;)   10

And thro' their parting lids there came and went
Keen glimpses of the inner firmament:
Fair beds they seem'd of water-lily flakes
Clustering entrancingly in beryl lakes:
Anon, across their swimming splendour strook,
An intense line of throbbing blood-light shook
A quivering pennon; then, for eye too keen,
Ebb'd back beneath its snowy lids, unseen.
   Now all things rosy turn'd: the west had grown
To an orb'd rose, which, by hot pantings blown   20
Apart, betwixt ten thousand petall'd lips
By interchange gasp'd splendour and eclipse.
The zenith melted to a rose of air;
The waves were rosy-lipp'd; the crimson glare
Shower'd the cliffs and every fret and spire
With garnet wreaths and blooms of rosy-budded fire.
   Then, looking on the waters, I was ware
Of something drifting thro' delighted air,
—An isle of roses,—and another near;—
And more, on each hand, thicken, and appear   30
In shoals of bloom; as in unpeopled skies,
Save by two stars, more crowding lights arise,
And planets bud where'er we turn our mazèd eyes.
I gazed unhinder'd: Mermaids six or seven,
Ris'n from the deeps to gaze on sun and heaven,
Cluster'd in troops and halo'd by the light,
Those Cyclads made that thicken'd on my sight.
   This was their manner: one translucent crest
Of tremulous film, more subtle than the vest
Of dewy gorse blurr'd with the gossamer fine,   40
From crown to tail-fin floating, fringed the spine,
Droop'd o'er the brows like Hector's casque, and sway'd
In silken undulation, spurr'd and ray'd
With spikèd quills all of intensest hue;
And was as tho' some sapphire molten-blue
Were vein'd and streak'd with dusk-deep lazuli,
Or tender pinks with bloody Tyrian dye.
From their white waists a silver skirt was spread
To mantle-o'er the tail, such as is shed
Around the Water-Nymphs in fretted falls,   50
At red Pompeii on medallion'd walls.

A tinted fin on either shoulder hung;
Their pansy-dark or bronzen locks were strung
With coral, shells, thick-pearlèd cords, whate'er
The abysmal Ocean hoards of strange and rare.
Some trail'd the Nautilus; or on the swell
Tugg'd the boss'd, smooth-lipp'd, giant Strombus-shell.
Some carried the sea-fan; some round the head
With lace of rosy weed were chapleted;
One bound o'er dripping gold a turquoise-gemm'd   60
Circlet of astral flowerets—diadem'd
Like an Assyrian prince, with buds unsheath'd
From flesh-flowers of the rock; but more were wreath'd
With the dainty-delicate fretted fringe of fingers
Of that jacinthine thing, that, where it lingers,
Broiders the nets with fans of amethyst
And silver films, beneath with pearly mist,
The Glaucus cleped; others small braids encluster'd
Of glassy-clear Aeolis, metal-lustred
With growths of myriad feelers, crystalline   70
To shew the crimson streams that inward shine,
Which, lightening o'er the body rosy-pale,
Like shiver'd rubies dance or sheen of sapphire hail.
   Then saw I sudden from the waters break
Far off a Nereid company, and shake
From wings swan-fledged a wheel of watery light
Flickering with sunny spokes, and left and right
Plunge orb'd in rainbow arcs, and trample and tread
The satin-purfled smooth to foam, and spread
Slim-pointed sea-gull plumes, and droop behind   80
One scarlet feather trailing to the wind;
Then, like a flock of sea-fowl mounting higher,
Thro' crimson-golden floods pass swallow'd into fire.
   Soon—as when Summer of his sister Spring
Crushes and tears the rare enjewelling,
And boasting 'I have fairer things than these'
Plashes amidst the billowy apple-trees
His lusty hands, in gusts of scented wind
Swirling out bloom till all the air is blind
With rosy foam and pelting blossom and mists   90
Of driving vermeil-rain; and, as he lists,
The dainty onyx-coronals deflowers,

A glorious wanton;—all the wrecks in showers
Crowd down upon a stream, and, jostling thick
With bubbles bugle-eyed, struggle and stick
On tangled shoals that bar the brook—a crowd
Of filmy globes and rosy floating cloud:—
So those Mermaidens crowded to my rock,
And thicken'd, like that drifted bloom, the flock
Sun-flush'd, until it seem'd their father Sea   100
Had gotten him a wreath of sweet Spring-broidery.
    Careless of me they sported: some would plash
The languent smooth with dimpling drops, and flash
Their filmy tails adown whose length there show'd
An azure ridge; or clouds of violet glow'd
On prankèd scale; or threads of carmine, shot
Thro' silver, gloom'd to a blood-vivid clot.
Some, diving merrily, downward drove, and gleam'd
With arm and fin; the argent bubbles stream'd
Airwards, disturb'd; and the scarce troubled sea   110
Gurgled, where they had sunk, melodiously.
Others with fingers white would comb among
The drenchèd hair of slabby weeds that swung
Swimming, and languish'd green upon the deep
Down that dank rock o'er which their lush long tresses weep.
    But most in a half-circle watch'd the sun;
And a sweet sadness dwelt on everyone;
I knew not why,—but know that sadness dwells
On Mermaids—whether that they ring the knells
Of seamen whelm'd in chasms of the mid-main,   120
As poets sing; or that it is a pain
To know the dusk depths of the ponderous sea,
The miles profound of solid green, and be
With loath'd cold fishes, far from man—or what;—
I know the sadness but the cause know not.
Then they, thus ranged, 'gan make full plaintively
A piteous Siren sweetness on the sea,
Withouten instrument, or conch, or bell,
Or stretch'd chords tuneable on turtle's shell;
Only with utterance of sweet breath they sung   130
An antique chaunt and in an unknown tongue.
Now melting upward thro' the sloping scale
Swell'd the sweet strain to a melodious wail;

Now ringing clarion-clear to whence it rose
Slumber'd at last in one sweet, deep, heart-broken close.
   But when the sun had lapsed to ocean, lo
A stealthy wind crept round seeking to blow,
Linger'd, then raised the washing waves and drench'd
The floating blooms and with tide flowing quench'd
The rosy isles: so that I stole away   140
And gain'd thro' growing dusk the stirless bay;
White loom'd my rock, the water gurgling o'er,
Whence oft I watch but see those Mermaids now no more.
                        The End.

Winter with the Gulf Stream

The boughs, the boughs are bare enough
But earth has never felt the snow.
Frost-furred our ivies are and rough

With bills of rime the brambles shew.
The hoarse leaves crawl on hissing ground
Because the sighing wind is low.

But if the rain-blasts be unbound
And from dank feathers wring the drops
The clogged brook runs with choking sound

Kneading the mounded mire that stops   10
His channel under clammy coats
Of foliage fallen in the copse.

A simple passage of weak notes
Is all the winter bird dare try.
The bugle moon by daylight floats

So glassy white about the sky,
So like a berg of hyaline,
And pencilled blue so daintily,

I never saw her so divine.
But through black branches, rarely drest   20
In scarves of silky shot and shine,

The webbed and the watery west
Where yonder crimson fireball sets
Looks laid for feasting and for rest.

I see long reefs of violets
In beryl-covered fens so dim,
A gold-water Pactolus frets°

Its brindled wharves and yellow brim,
The waxen colours weep and run,
And slendering to his burning rim   30

Into the flat blue mist the sun
Drops out and all our day is done.

Spring and Death

I had a dream. A wondrous thing:
It seem'd an evening in the Spring;
—A little sickness in the air
From too much fragrance everywhere:—
As I walk'd a stilly wood,
Sudden, Death before me stood:
In a hollow lush and damp,
He seem'd a dismal mirky stamp
⎧ On the flowers that were seen
⎨ His charnelhouse-grate ribs between,   10
⎩ And with coffin-black he barr'd the green.
'Death,' said I, 'what do you here
At this Spring season of the year?'
'I mark the flowers ere the prime
Which I may tell at Autumn-time.'
Ere I had further question made
Death was vanish'd from the glade.
Then I saw that he had bound
Many trees and flowers round

With a subtle web of black,   20
And that such a sable track
Lay along the grasses green
From the spot where he had been.
   But the Spring-tide pass'd the same;
Summer was as full of flame;
Autumn-time no earlier came.
And the flowers that he had tied,
As I mark'd, not always died
Sooner than their mates; and yet
Their fall was fuller of regret: 30
It seem'd so hard and dismal thing,
Death, to mark them in the Spring.

(*Fragments*)

(i)

The wind, that passes by so fleet,
Runs his fingers through the wheat,
And leaves the blades, where'er he will veer,
Tingling between dusk and silver.

(ii)

Whose braggart 'scutcheon, whose complaisant crest
Catch sunlight and one strain of stupid praise.
.   .   .   .
The villain shepherds and misguided flock.

(iii)

The sparky air
Leaps up before my vision,—thou art gone.

(iv)

—and on their brittle green quils
Shake the balanced daffodils.

Fragments of *Pilate*

2

Unchill'd I handle stinging snow;°
The sun whose vast afflictive heat
Does lay men low with one blade's sudden blow
  Cleaves not my brain, burns not my feet,
When the fierce skies are blue to black, albeit
The shearing rays contract me with their blaze
  Most dead-alive upon those days.

3

Then I seek out the shadow and stones
And to those stones become akin
My several moans come distant in their tones   10
  As though they were not from within
And for that fearful hour life is more thin
And numbs and starves, as between icy wharves
  A freezing runnel sobs and dwarfs.

[4]

Sometimes I see the summit stake
High up the balanced stony air
In whose dead lake even a voice may make
  The hanging snows rush down and bare
Their rocky lodges. Then the weather rare
Allows the sound of bells in hamlets round   20
  To come to me from the underground.

5

Often when winds impenitent
Beat, heave and the strong mountain tire
I can stand pent in the monstrous element
  And feel no blast.—O fretful fire
Breathe o'er my bare nerve rather. I desire
They swathe and lace the shroud-plaits o'er my face,
  But to be ransom'd from this place.

6

Whatever time this vapourous roof,
    The screen of my captivity,   30
Folds off aloof, that signal is and proof
    Not of clear skies, but storm to be.
But then I make an eager shift to see
Houses that make abode beside the lake,
    And then my heart goes near to break.

7

Then clouds come, like ill-balanced crags,
    Shouldering. Down valleys smokes the gloom.
The thunder brags. In joints and sparkling jags
    The lightnings leap. The day of doom!
I cry 'O rocks and mountain make me room'   40
And yet I know it would be better so,
    Ay, sweet to taste beside this woe.

.   .   .   .

The pang of Tartarus, Christians hold,
    Is this, from Christ to be shut out.
This outer cold, my exile from of old
    From God and man, is hell no doubt.
Would I could hear the other Pilates shout.
But yet they say Christ comes at the last day.
    Then will he keep in this stay?

There is a day of all the year   50
    When life revisits me, nerve and vein.
They all come here and stand before me clear
    I try the Christus o'er again.
Sir! Christ! against this multitude I strain.—
Lord, but they cry so loud. And what am I?
    And all in one say 'Crucify!'

Before that rock, my seat, He stands;
    And then—I choke to tell this out—
I give commands for water for my hands;
    And some of those who stand about,—   60

Vespillo my centurion hacks out
Some ice that locks the glacier to the rocks
    And in a bason brings the blocks.

    I choose one; but when I desire
    To wash before the multitude
The vital fire does suddenly retire
    From hands now clammy with strange blood.
My frenzied working is not understood.
Now I grow numb. My tongue strikes on the gum
    And cleaves, I struggle and am dumb.   70

    I hear the multitude tramp by.
    O here is the most piteous part,
For He whom I send forth to crucify,
    Whispers 'If thou have warmth at heart
Take courage; this shall need no further art.'

       .   .   .   .

    I have a hope if so it be,
    A hope of an approved device;
I will break free from the Jews' company,
    And find a flint, a fang of ice,
Or fray a granite from the precipice: 80
When this is sought trees will be wanting not,
    And I shall shape one to my thought.

    Thus I shall make a cross, and in't
    Will add a footrest there to stand,
And with sharp flint will part my feet and dint
    The point fast in, and my left hand
Lock with my right; then knot a barken band
To hold me quite fix'd in the selfsame plight;
    And thus I will thrust in my right:—

    I'll take in hand the blady stone   90
    And to my palm the point apply,
And press it down, on either side a bone,
    With hope, with shut eyes, fixedly;
Thus crucified as I did crucify.°

       .   .   .   .

'She schools the flighty pupils'

She schools the flighty pupils of her eyes,
With levell'd lashes stilling their disquiet;
And puts in leash her pair'd lips lest surprise
Bare the condition of a realm at riot.
If he suspect that she has ought to sigh at
His injury she'll avenge with raging shame.
She kept her love-thoughts on most lenten diet,
And learnt her not to startle at his name.

Richard

He was a shepherd of the Arcadian mood,
That not Arcadia knew nor Haemony.
Affinèd to the earnest solitude,
The winds and listening downs he seem'd to be.°

He went with listless strides, disorderedly.
And answer'd the dry tinkles of his sheep
With piping unexpected melody.
With absent looks inspired as drinking deep°
True nectar filter'd thro' the thymy leaves of sleep.°

He rested on the forehead of the down°   10
Shaping his outlines on a field of cloud.
His sheep seem'd to step from it, past the crown
Of the hill grazing:

A Soliloquy of One of the Spies left in the Wilderness

He feeds me with His manna every day:
My soul does loathe it and my spirit fails.
A press of wingèd things comes down this way:
    The gross flock call them quails.
Into my hand he gives a host for prey,
    Come up, Arise and slay.

Sicken'd and thicken'd by the glare and sand,
Who would drink water from a stony rock?
Are all the manna-bushes in the land
    A shelter for this flock?
Behold at Elim wells on every hand!
    And seventy palms there stand.

Egypt, the valley of our pleasance, there!
Most wide ye are who call this gust Simoom.
Your parchèd nostrils snuff Egyptian air.
    The comfortable gloom
After the sandfield and the unreinèd glare!
    Goshen is green and fair.

Not Goshen. Wasteful wide huge-girthèd Nile
Unbakes my pores, and streams, and makes all fresh.
I gather points of lote-flower from an isle
    Of leaves of greenest flesh.
Ye sandblind! Slabs of water many a mile
    Blaze for him all this while.

In beds, in gardens, in thick plots I stand.
Handle the fig, suck the full-sapp'd vine-shoot.
From easy runnels the rich-piecèd land
    I water with my foot.
Must you be gorged with proof? Did ever sand
    So trickle from your hand?

Strike timbrels, sing, eat, drink, be full of mirth.
Forget the waking trumpet, the long law.
Spread o'er the swart face of this prodigal earth.
    Bring in the glistery straw.
Here are sweet messes without price or worth,
    And never thirst or dearth.

Give us the tale of bricks as heretofore;
To plash with cool feet the clay juicy soil.
Who tread the grapes are splay'd with stripes of gore.
    And they who crush the oil
Are spatter'd. We desire the yoke we bore,
    The easy burden of yore.

Who is this Moses? Who made him, we say,
To be a judge and ruler over us?
He slew the Egyptian yesterday. To-day
  In hot sands perilous
He hides our corpses dropping by the way
  Wherein he makes us stray.

Your hands have borne the tent-poles: on you plod:
The trumpet waxes loud: tired are your feet.   50
Come by the flesh-pots: you shall sit unshod
  And have your fill of meat;
Bring wheat-ears from the loamy stintless sod,
  To a more grateful god.

    . . . .

Go then: I am contented here to lie.
Take Canaan with your sword and with your bow.
Rise: match your strength with monstrous Talmai
  At Kirjath-Arba: go.—
Sure, this is Nile: I sicken, I know not why,
  And faint as though to die.   60

The Lover's Stars

The destined lover, whom his stars
 More golden than the world of lights,
O'er passes bleak, o'er perilous bars
 Of rivers, lead, thro' storms and nights,

Or if he leave the West behind,°
 Or father'd by the sunder'd South,
Shall, when his star is zenith'd, find
 Acceptance round his mistress' mouth:

Altho' unchallenged, where she sits,   10
 Three rivals throng her garden chair,
And tho' the silver seed that flits
 Above them, down the draught of air,

And keeps the breeze and clears the seas
   And tangles on a down of France,
Yet leaves him in ungirdled ease
   8000 furlongs in advance.

But in the other's horoscope
   Bad Saturn with a swart aspect
Fronts Venus.—His ill-launchèd hope
   In unimperill'd haven is wreck'd.   20

He meets her, stintless of her smile;
   Her choice in roses knows by heart;
Has danced with her: and all the while
   They are Antipodes apart.

His sick stars falter. More he may
   Not win, if this be not enough.
He meets upon Midsummer day
   The stabbing coldness of rebuff.

'During the eastering'

During the eastering of untainted morns,
In the ascendancy of rainbow's horns,
In the first signals of the several drops
That lick the shelly leaves which floor the copse,
In the quick fragrance of tall rolling pines,
Under the cloister-light of greenhouse vines,

.   .   .   .

'—Hill / Heaven'

                         —Hill,
Heaven and every field, are still
  { As a self-embraced sweet thought:
  { — – —-caress'd —— ——:
  { And the thin stars tremble not.
  { —— — lessen'd stars ray ——.

'Distance / Dappled'

Distance
Dappled with diminish'd trees
Spann'd with shadow every one.

The peacock's eye

Mark you how the peacock's eye°
Winks away its ring of green,
Barter'd for an azure dye,
And the piece that's like a bean,
The pupil, plays its liquid jet°
To win a look of violet.

Love preparing to fly

He play'd his wings as though for flight;
They webb'd the sky with glassy light.°
His body sway'd upon tiptoes,
Like a wind-perplexèd rose;
In eddies of the wind he went
At last up the blue element.

Barnfloor and Winepress

And he said, If the Lord do not help thee, whence shall I help thee?
Out of the barnfloor, or out of the winepress?

2 KINGS vi. 27.

Thou that on sin's wages starvest,
Behold we have the joy in harvest:
For us was gather'd the first-fruits,
For us was lifted from the roots,
Sheaved in cruel bands, bruised sore,
Scourged upon the threshing-floor;
Where the upper mill-stone roof'd His head,
At morn we found the heavenly Bread,

And, on a thousand altars laid,
Christ our Sacrifice is made!   10

Thou whose dry plot for moisture gapes,
We shout with them that tread the grapes:
For us the Vine was fenced with thorn,
Five ways the precious branches torn;
Terrible fruit was on the tree
In the acre of Gethsemane;°
For us by Calvary's distress
The wine was rackèd from the press;
Now in our altar-vessels stored
Is the sweet Vintage of our Lord.   20

In Joseph's garden they threw by°
The riv'n Vine, leafless, lifeless, dry:
On Easter morn the Tree was forth,
In forty days reach'd Heaven from earth;
Soon the whole world is overspread;
Ye weary, come into the shade.

The field where He has planted us
Shall shake her fruit as Libanus,°
When He has sheaved us in His sheaf,
When He has made us bear His leaf.—   30
We scarcely call that banquet food,
But even our Saviour's and our blood,
We are so grafted on His wood.

New Readings

Although the letter said
On thistles that men look not grapes to gather,°
    I read the story rather
How soldiers platting thorns around CHRIST'S Head
    Grapes grew and drops of wine were shed.

Though when the sower sowed°
The wingèd fowls took part, part fell in thorn
    And never turned to corn,

Part found no root upon the flinty road,—
    CHRIST at all hazards fruit hath shewed.   10

    From wastes of rock He brings°
Food for five thousand: on the thorns He shed
    Grains from His drooping Head;
And would not have that legion of winged things°
    Bear Him to heaven on easeful wings.

'He hath abolished the old drouth'

He hath abolished the old drouth,°
And rivers run where all was dry,
The field is sopp'd with merciful dew.
He hath put a new song in my mouth,°
The words are old, the purport new,
And taught my lips to quote this word°
That I shall live, I shall not die,
But I shall when the shocks are stored°
See the salvation of the Lord.
We meet together, you and I,   10
Meet in one acre of one land,
And I will turn my looks to you,
And you shall meet me with reply,
We shall be sheavèd with one band°
In harvest and in garnering,
When heavenly vales so thick shall stand°
With corn that they shall laugh and sing.

Heaven-Haven
(a nun takes the veil)

    I have desired to go
        Where springs not fail,
To fields where flies no sharp and sided hail
        And a few lilies blow.

    And I have asked to be
        Where no storms come,
Where the green swell is in the havens dumb,
        And out of the swing of the sea.

'I must hunt down the prize'

I must hunt down the prize
   Where my heart lists.
Must see the eagle's bulk, render'd in mists,
   Hang of a treble size.

   Must see the waters roll
     Where the seas set
Towards wastes where round the ice-blocks tilt and fret
     Not so far from the pole.

or

   Must see the green seas roll
     Where waters set
Towards those wastes where the ice-blocks tilt and fret,
     Not so far from the pole.

'Why should their foolish bands'

Why should their foolish bands, their hopeless hearses
Blot the perpetual festival of day?
Ravens, for prosperously-boded curses
Returning thanks, might offer such array.
Heaven comfort sends, but harry it away.
Gather the sooty plumage from Death's wings
And the poor corse impale with it and fray,
If so it be, an angel's hoverings.°
And count the rosy cross with bann'd disastrous things.

'Why if it be so'

Why if it be so, for the dismal morn
Into his hollow'd palm should moan the blast;
And in grey bands the sun should lie still born;
And straight showers parallel should follow fast;
And, swarter still, the rolling pines should cast
Their heads together in a stormy blot.

'Or else their cooings'

Or else their cooings came from bays of trees,
Like a contented wind, or gentle shocks
Of falling water. This and all of these
We tunèd to one key and made their harmonies.

'It was a hard thing'

It was a hard thing to undo this knot.
The rainbow shines, but only in the thought
Of him that looks. Yet not in that alone,
For who makes rainbows by invention?
And many standing round a waterfall
See one bow each, yet not the same to all,
But each a hand's breadth further than the next.
The sun on falling waters writes the text
Which yet is in the eye or in the thought.
It was a hard thing to undo this knot.   10

'Glimmer'd along the square-cut steep'

Glimmer'd along the square-cut steep.
They chew'd the cud in hollows deep;
Their cheeks moved and the bones therein.
The lawless honey eaten of old
Has lost its savour and is roll'd
Into the bitterness of sin.

What would befal the godless flock
Appear'd not for the present, till
A thread of light betray'd the hill
Which with its lined and creased flank   10
The outgoings of the vale does block.
Death's bones fell in with sudden clank
As wrecks of minèd embers will.

'Late I fell in the ecstacy'

Late I fell in the ecstacy
And saw the men before the flood°
Which once were disobedient

'Think of an opening page'

Think of an opening page illuminèd
With the ready azure and high carmine:—think
Her face was such, as being diaperèd
With loops of veins; not of an even pink,

'Miss Story's character'

Miss Story's character! too much you ask,
When 'tis the confidante that sets the task.
How dare I paint Miss Story to Miss May?
And what if she my confidence betray!
What if my Subject, seeing this, resent
What were worth nothing if all compliment!
No: shewn to her it cannot but offend;
But candour never hurt the dearest *friend*.
    Miss Story has a moderate power of will,
But, having that, believes it greater still: 10
And, hide it though she does, one may divine
She inly nourishes a wish to shine;
Is very capable of strong affection
Tho' apt to throw it in a strange direction;
Is fond of flattery, as any she,
But has not learnt to take it gracefully;
Things that she likes seems often to despise,
And loves—a fatal fault—to patronize;
Has wit enough, if she would make it known°
And charms—but they should be more freely shewn.   20
She's framed to triumph in adversity;
Prudence she has, but wise she'll never be;
About herself she is most sensitive,

Talks of self-sacrifice, yet can't forgive;
Her character she does not realize,
And cannot see at all with others' eyes;
(And, well supplied with virtues on the whole,
Is slightly selfish in her inmost soul)
Believes herself religious, and is not;
And, thinking that she thinks, has never thought;   30
Married, will make a sweet and matchless wife,
But single, lead a misdirected life.

'Her prime of life'

Her prime of life—cut down too soon
By death—as th'morning flower at noon:
Her loving husband lives t'deplore:
Yet hopes she'll flourish evermore.
                    Jane Green.
Wife of Jonathan Green, of this Parish, Baker.
                    20th 1848.
                    Aged 52 years.

'Did Helen steal'

Did Helen steal my love from me?
    She never had the wit.
Or was it Jane? But she's too plain,
    And could not compass it.

A bad verse in the middle, then—

It might be Helen, Jane, or Kate,
    It might be none of the three:
But I'm alone, for my love's gone
    That should have been true to me.

(*Woods in Spring*)

(*a*)

—the shallow folds of the wood
We found were dabbled with a colouring growth,
In lakes of bluebells, pieced with primroses.

(*b*)

In the green spots of that wood
Were eyes of central primrose: bluebells ran
In skeins about the brakes.

'Like shuttles fleet the clouds'

Like shuttles fleet the clouds, and after
A drop of shade rolls over field and flock;
The wind comes breaking here and there with laughter:
The violet moves and copses rock.

.　　.　　.　　.　　.　　.　　.　　.

When the wind drops you hear the skylarks sing;
From Oxford comes the throng and hum of bells
Breaking the .　　.　　.　　. air of spring.

(*Seven Epigrams*)

(i)

Of virtues I most warmly bless,
Most rarely see, Unselfishness.
And to put graver sins aside
I own a preference for Pride.

(ii)

You ask why can't Clarissa hold her tongue.
Because she fears her fingers will be stung.

(iii)
On a dunce who had not a word to say for himself

He's all that's bad, I know; a knave, a flat,
But his effrontery's not come to that.

(iv)
Modern Poets

Our swans are now of such remorseless quill,
Themselves live singing and their hearers kill.

(v)
By one of the old school who was bid to follow Mr. Browning's flights

To rise you bid me with the lark
With me 'tis rising in the dark.

(vi)
On a Poetess

Miss M.'s a nightingale. 'Tis well
    Your simile I keep.
It is the way with Philomel
    To sing while others sleep.

(vii)
On one who borrowed his sermons

Herclots's preachings I'll no longer hear:
They're out of date—lent sermons all the year.

By Mrs. Hopley

He's wedded to his theory, they say.°
If that were true, it could not live a day.
And did the children of his brains enjoy
But half the pains he spends upon his boy,
You may depend that ere a week was fled,
There would not be a whole place in his head.

On seeing her children say Goodnight to their father.

Bid your Papa Goodnight. Sweet exhibition!
They kiss the Rod with filial submission.

(34)

Io

Forward she leans, with hollowing back, stock-still,
Her white weed-bathèd knees are shut together,
Her silky coat is sheeny, like a hill
Gem-fleeced at morn, so brilliant is the weather.
Her nostril glistens; and her wet black eye
Her lids half-meshing shelter from the sky.°

Her finger-long new horns are capp'd with black;
In hollows of her form the shadow clings;
Her milk-white throat and folded dew-lap slack
Are still; her neck is creased in close-ply rings;
Her hue's a various brown with creamy lakes,°
Like a cupp'd chestnut damask'd with dark breaks.

Backward are laid her pretty black-fleeced ears;
The feathery knot of locks upon her head°
Plays to the breeze; where now are fled her fears,
Her jailor with his vigil-organ dead?°
Morn does not now new-basilisk his stare,°
Nor night is blown with flame-rings everywhere.

(*Fragments*)

(i)

{ Thick-fleeced / Out-fleeced } bushes like a { heifer's / spaniel's } ear.

(ii)

There is an island, wester'd in the main,°
Around it balances the level sea.

(iii)

The time was late and the wet yellow woods
Told off their leaves along the piercing gale,

(iv)

... in her cheeks that dwell
Centred like meteors, bright like pimpernel.

(v)

—now the rain,
A brittle sheen, runs upward like a cliff,
Flying a bow.

The rainbow

See on one hand
He drops his bright roots in the water'd sward,°
And rosing part, on part dispenses green;
But with his other foot three miles beyond
He rises from the flocks of villages
That bead the plain; did ever Havering church-tower
Breathe in such ether? or the Quickly elms
{ With such a violet slight their distanced green?
{ Slight with such violet their bright-mask'd green?

or

Mask'd with such violet disallow their green?

'—Yes for a time'

A. —Yes for a time they held as well
Together, as the criss-cross'd shelly cup
Sucks close the acorn; as the hand and glove;
As water moulded to the duct it runs in;
As keel locks close to kelson—

B.   Let me now
Jolt and unset your morticed metaphors.
The hand draws off the glove; the acorn-cup
Drops the fruit out; the duct runs dry or breaks;
The stranded keel and kelson warp apart;   10
And your two etc.

Fragments of *Floris in Italy*

(i)

It does amaze me, when the clicking hour
Clings on the stroke of death, that I can smile.
Yet when my unset tresses hung loose-traced
Round this unsexing doublet,—while I set
This downy counterfeit upon my lip,

    .   .   .   .   .   .

                    —Lately I fear'd
My signalling tears might ring up Floris; now
Methinks there is more peril from my laughter.°
Well, I know not. But all things seem to-night
Double as sharp, meaning and forcible,   10
With twice as fine a sense to apprehend them,
As ever I remember in my life.
Laughing or tears. I think I could do either—
So strangely elemented is my mind's weather,
That tears and laughter are hung close together.
(*Comes to the bed.*)
Sleep Floris while I rob you. Tighten, O sleep,
Thy impalpable oppression. Pin him down,
Ply fold on fold across his dangerous eyes,
Lodge his eyes fast; but as easy and light
As the laid gossamers of Michaelmas   20
Whose silver skins lie level and thick in field.
Hold him.—
I must not turn the lantern on his face.—
No I'll not hazard it. Only his hand,
(*Turns the lantern on Floris' hand.*)

    .   .   .   .   .   .

(*Trying on the ring.*)
It is too large for me. What does that mean?
No time to think. I'll knot it on this ribbon,
And wear it thus, a pectoral, by my heart.
—

             Did I say but lately
That I was so near laughter? Alas now
I find I am as ready with my tears   30
As the fine morsels of a dwindling cloud
That piece themselves into a race of drops
To spill o'er fields of lilies. So could I
So waste in tears over this bed of sweetness,
This flower, this Floris, this dear majesty,
This royal manhood.—'Tis in me rebellion
To speak so, yet I'll speak it for this once;
Deep shame it were to be discover'd so,
Worse than when Floris found me in the garden
Weeping,—Even now I curse myself remembering;—   40
No, let that go; I have said Goodnight to shame.
Now let me see you, you large princely hand,
Since on the face it is unsafe to look;
Yet this could be no other's hand than his,
'Tis so conceived in his true lineament.°
I have wrong'd it of its coronet, and now
I outrage it with treasonable kissing.
Ah Floris, Floris, let me speak this little

            —

What I do now is but the least least thing.
But since I have no scope for benefits   50
Though ill-contented, precious precious Floris,
Most ill-content, this least least thing I do.
Now one word more and then I am gone indeed,
Warn'd by the bright procession of the stars.
My cousin will not love you as I love,
Floris; she will not hit thy sum of worth,
Thou jacinth; nor have skill of all thy virtues,
Floris, thou late-found All-heal;

            —

             With what bold grace
This sweeter Deserter lists herself anew   60
Enroll'd and sexèd with our ruder files°
And marching to false colours! those few strokes
That forge her title of inheritance

To Manhood, on the upper lip,—they look'd
Most like the silver plighted tuft about°
The mouthèd centre of a violet.

(ii)

—O Guinevere
I read that the recital of thy sin,
Like knocking thunder all round Britain's welkin,
Jarr'd down the balanced storm; the bleeding heavens
Left not a rood with curses unimpregnate;
There was no crease or gather in the clouds
But dropp'd its coil of woes: and Arthur's Britain
The mint of current courtesies, the forge
Where all the virtues were illustrated
In blazon, gilt and mail'd shapes of bronze,°   10
Abandon'd by her saints, turn'd black and blasted,
Like scalded banks topp'd once with principal flowers:
Such heathenish misadventure dogg'd one sin.°

(iii)

Floris in Italy. Floris, having found by chance that Giulia loves him, reasons with himself (or perhaps with Henry) in defence of his not returning her love. Her beauty is urged.

Say beauty lies but in the meet of lines,°
In careful-spacèd sequences of sound.
These rather are the arc where beauty shines,
The temper'd soil where only her flower is found.
Allow at least it has one term and part
Beyond, and one within the looker's eye;
And I must have the centre in my heart
To turn the compass on the all-starr'd sky:°
For only try by gazing to divide°
One star by daylight from the strong blue air,   10
And find it will not therefore be descried
Because its place is known and charted there.
No, love prescriptive, love with place assign'd,°
Love by monition, heritage, or lot,
Love by prenatal serfdom still confined
Even to the tillage of the sweetest spot,—

It is a regimen on the imperfect wind,°
Piecing the elements out by plan and plot.
Though self-made bands at last may true love bind,
New love is free love, or true love 'tis not.   [*exit* 20

Henry. Such spider's web he ties across his sight,°
And gives for tropes his judgment all away,
Gilds with some sparky fancies blinding night,
And stumbling swears he walks by light of day.
A learned fool indeed and well-bred churl
That swinishly refuses such a pearl!

(*Fragments*)

(*a*) Stars

(i)

How looks the night? There does not miss a star.
The million sorts of unaccounted motes
Now quicken, sheathed in the yellow galaxy.
There is no parting or bare interstice
Where the stint compass of a skylark's wings
Would not put out some tiny golden centre.

(ii)

Stars waving their indivisible rays.
Sky fleeced with the milky way.

(iii)

Night's lantern
Pointed with piercèd lights, and breaks of rays
Discover'd everywhere.

(iv)

The sky minted into golden sequins.
Stars like gold tufts.
Stars like golden bees.
Stars like golden rowels.
Sky peak'd with tiny flames.
Stars like tiny-spoked wheels of fire.

(v)

His gilded rowels
Now stars of blood.

(*b*)

(i)

They came
Next to meadows abundant, pierced with flowers,
With sulphur-colour'd lilies, brittle in stalk,
And seals of red carnation which had each
Two tongues like butterflies.

(ii)

Dewy fields in the morning under the sun
Stand shock and silver-coated.

'I am like a slip of comet'

—I am like a slip of comet,
Scarce worth discovery, in some corner seen
Bridging the slender difference of two stars,
Come out of space, or suddenly engender'd
By heady elements, for no man knows:
But when she sights the sun she grows and sizes
And spins her skirts out, while her central star
Shakes its cocooning mists; and so she comes
To fields of light; millions of travelling rays
Pierce her; she hangs upon the flame-cased sun,   10
And sucks the light as full as Gideon's fleece:°
But then her tether calls her; she falls off,
And as she dwindles shreds her smock of gold
Amidst the sistering planets, till she comes
To single Saturn, last and solitary;°
And then goes out into the cavernous dark.
So I go out: my little sweet is done:
I have drawn heat from this contagious sun:
To not ungentle death now forth I run.

'No, they are come'

No, they are come; their horn is lifted up;
They stand, they shine in the sun; Fame has foregone
All quests save the recital of their greatness;
Their clarions from all corners of the field
With potent lips call down cemented towers;
Their harness beams like scythes in morning grass;
Like flame they gather on our cliffs at evening,
At morn they come upon our lands like rains;
They plough our vales; you see the unsteady flare°
Flush thro' their heaving columns; when they halt,°   10
They seem to fold the hills with golden capes;
They draw all coverts, cut the fields, and suck
The treasure from all cities. etc.

'Now I am minded'

Now I am minded to take pipe in hand
And yield a song to the decaying year;
Now while the full-leaved hursts unalter'd stand,°
    And scarcely does appear
The Autumn yellow feather in the boughs;
    While there is neither sun nor rain;
And a grey heaven does the hush'd earth house,
And bluer grey the flocks of trees look in the plain.

So late the hoar green chestnut breaks a bud,
And feeds new leaves upon the winds of Fall;   10
So late there is no force in sap or blood;
    The fruit against the wall
Loose on the stem has done its summering;
These should have starv'd with the green broods of spring,
    Or never been at all;
Too late or else much, much too soon,
Who first knew moonlight by the hunters' moon.°

A Voice from the World

Fragments of 'An answer to Miss Rossetti's *Convent Threshold*'

At last I hear the voice well known;
Doubtless the voice: now fall'n, now spent,
Now coming from the alien eaves,—
You would not house beneath my own;
To alien eaves you fled and went,—
Now like the bird that shapes alone
A turn of seven notes or five,
When skies are hard as any stone,
The fall is o'er, told off the leaves,
'Tis marvel she is yet alive.   10
Once it was scarce perceivèd Lent
For orience of the daffodil;
Once, jostling thick, the bluebell sheaves°
The peacock'd copse were known to fill;
Through other bars it used to thrill,°
And carried me with ravishment,
Your signal, when apart we stood,
Tho' far or sick or heavy or still
Or thorn-engaged, impalèd and pent
With just such sweet-potential skill,   20
Late in the green weeks of April
Cuckoo calls cuckoo up the wood.
Five notes or seven, late and few.
From parts unlook'd-for, alter'd, spent,°
At last I hear the voice I knew.

·   ·   ·   ·

I plead: familiarness endears
My evil words thorny with pain:
I plead: and you will give your tears:
I plead: and ah! how much in vain!

·   ·   ·   ·

I know I mar my cause with words: 30
So be it; I must maim and mar.
Your comfort is as sharp as swords;
And I cry out for wounded love.

And you are gone so heavenly far
You hear nor care of love and pain.
My tears are but a cloud of rain
My passion like a foolish wind
Lifts them a little way above.
But you, so spherèd, see no more—
You see but with a holier mind—   40
You hear and, alter'd, do not hear
Being a stoled apparel'd star.
You should have been with me as near
As halves of sweet-pea-blossom are;
But now are fled, and hard to find
As the last Pleiad, yea behind
Exilèd most remote El Khor.

                    .   .   .   .   .

The love of women is not so strong,—
'Tis falsely given—as love in men;
A thing that weeps, enduring long: 50
But mine is dreadful leaping pain,
Phrenzy, but edged and clear of brain
Ruinous heart-beat, wandering, death.
I walk towards eve our walks again;
When lily-yellow is the west.
Say, o'er it hangs a water-cloud
And ravell'd into strings of rain.
At once I struggle with my breath.
'The light was so, the wind so loud
No louder, when I was with you.   60
Always the time remembereth
His very looks in other years,
Only with us is old and new.'
I fall, I tear and shower the weed,
I bite my hands, my looks I shroud;
My cry is like a bleat; a few
Intolerable tears I bleed.
Then is my misery full indeed:
I die, I die, I do not live.—
Alas! I rave, where calm is due;   70
I would remember. Love, forgive.
I cannot calm, I cannot heed.

I storm and shock you. So I fail.
And like a self-outwitted blast
Fling to the convent wicket fast.
Who would not shelter from the hail?
But is there a place for tenderness,
There was a charm would countervail
The spell of woe if any could.
Once in a drawer of Indian wood   80
You folded (did you not?) your dress.
The essence ne'er forgot the fold;
And I esteem'd the sandal good
And now I get some precious slips.

     .   .   .   .

. . . upon you dreamed:°
I [dream'd] my counterpart. It seem'd
[A bell] at midnight woke the town
[And all] into the Duomo ran:
You met me, I had hasten'd down:
That night the judgment day began: 90
'Twas said of none but all men knew:
Nocturns I thought were hurried through.
Some knelt, some stood: I seem'd to feel
Who knelt were for the Lord's right hand;°
They are the goats who stand, said I.
I stood; but does she stand or kneel?
I strove to look; I lost the trick
Of nerve; the clammy ball was dry.
Then came the benediction.
His lips moved fast in sense too thick!   100
The others heard; I could not hear
Him .   .   .   .   .   .   °
Save me: and you were standing near.

An angel came: 'The judgment done,
Mercy is left enough for one:
Choose, one for hell and one for heaven!'
You cried 'But I have served thee well,
O Lord; but I have wrought and striven;
Duly, dear Lord, my prize is won.
I did repent; I am forgiven.   110

Give him the gift.' I cannot tell
But all the while it seem'd to me
I reason'd the futility.
—

Or this, or else I do not love,
I inly said; but could not move
My fast-lodged tongue. '[To her the gift]°
I yield' I would have cried. At last,
Something I said; I swooned and fell.
The angel lifted us above.
The bitterness of death was past,   120
My love; and all was sweet and well.

.   .   .   .   .

.   .   .   .

Who say that angels, in your ear
Are heard, that cry 'She does repent',
Let charity thus begin at home,—
Teach me the paces that you went
I can send up an Esau's cry;°
Tune it to words of good intent.
This ice, this lead, this steel, this stone,
This heart is warm to you alone;
Make it to God. I am not spent   130
So far but I have yet within
The penetrative element
That shall unglue the crust of sin.
Steel may be melted and rock rent.
Penance shall clothe me to the bone.
Teach me the way: I will repent.

.   .   .   .   .

But grant my penitence begun:
I need not, love, I need not break
Remember'd sweetness. For my thought°
No house of Rimmon may I take,   140
To bow but little, and worship not?
Is not some little Bela set
Before the mountain?—No, not one,
The heaven-enforcèd answer comes,
Yea, to myself I answer make:
Who can but barter slender sums

By slender losses are undone;
They breathe not who are late to run.—
O hideous vice to haggle yet
For more with Him who gives thee all,   150
Freely forgives the monstrous debt!
Having the infinitely great
Therewith to hanker for the small!

Knowledge is strong but love is sweet.—
I found the ways were sown with salt
Where you and I were wont to tread;
Not further'd far my travell'd feet
For all the miles that they were sped;
No flowers to find, no place to halt,
No colour in the overhead,   160
No running in the river-bed;
And passages where we used to meet,—
Fruit-cloistering hyacinth-warding woods,
I call'd them and I thought them then—
When you were learner and I read,
Are waste, and had no wholesome foods,°
Unpalateable fruits to eat.
What have I more than other men,
For learning stored and garnerèd?
And barely to escape the curse,   170
I who was wise would be untaught,
And fain would follow I who led.
How shall I search, who never sought?
How turn my passion-pastured thought
To gentle manna and simple bread?

'The cold whip-adder'

The cold whip-adder unespied
With wavèd passes there shall glide
Too near thee, and thou must abide
The ringèd blindworm hard beside.

(*Fragments*)

(i)

The ends of the crisp buds she chips
And the flower strips,
The breaking leaves of gold are curl'd upon her lips.

(ii)

A pure gold lily, but by the pure gold lily
We will charge our flocks that they not feed.
Leave it with its grove hard by
'Some are pretty enough, and some are poor indeed.'
Give us our green lots in another mead
Fit for flowers, water-pierced and rilly.
Lead shepherd, now we follow, shepherd lead.

(iii)

We live to see
How Shakspere's England weds with Dante's Italy.

(iv)

Although she be more white,
More white,
Than a skeinèd, than a skeinèd waterfall,
And better veinèd than pea blossoms all
And though she be so light
As thin-spun whirling bats' wings in the air etc.

(v)

A star most spiritual, principal, preeminent
Of all the golden press.

(vi)

Or ever the early stirrings of skylark
Might cover the neighbour downs with a span of singing,
While Phosphor, risen upon the shallowing dark,
In the ruddied county of the day's upbringing
Stood capital, eminent, . . . gonfalon bearer
To all the starry press.—

For a Picture of Saint Dorothea

I bear a basket lined with grass;
I am so light, I am so fair,
That men must wonder as I pass
And at the basket that I bear,
Where in a newly-drawn green litter
Sweet flowers I carry,—sweets for bitter.

Lilies I shew you, lilies none,
None in Caesar's gardens blow,—
And a quince in hand,—not one
Is set upon your boughs below;   10
Not set, because their buds not spring;
Spring not, 'cause world is wintering.

But these were found in the East and South
Where Winter is the clime forgot.—
The dewdrop on the larkspur's mouth
O should it then be quenchèd not?
In starry water-meads they drew
These drops: which be they? stars or dew?

Had she a quince in hand? Yet gaze:
Rather it is the sizing moon.   20
Lo, linkèd heavens with milky ways!
That was her larkspur row.—So soon?
Sphered so fast, sweet soul?—We see
Nor fruit, nor flowers, nor Dorothy.

'Proved Etherege'

Proved Etherege prudish, selfish, hypocrite, heartless,
No scholar, a would be critic, a *dillentante*,
Cream-laid, a surface, who could quote, to startle us,
The Anatomy, Politian, a little Dante,—
And so forth. Then for his looks—like pinkish paper:
Features? A watermark; other claims as scanty.
    In such wise did the gentle . . . vapour.

Fragments of *Richard*

(i)

As void as clouds that house and harbour none,
Whose gaps and hollows are not browzed upon,
As void as those the gentle downs appear
On such a season of the day and year.
There was no bleat of ewe, no chime of wether,
Only the bellèd foxgloves lisp'd together.
Yet there came one who sent his flock before him,
Alone upon the hill-top, heaven o'er him,
And where the brow in first descending bow'd
He sat and wrought his outline on a cloud.   10
His sheep seem'd to come from it as they stept,
One and then one, along their walks, and kept
Their changing feet in flicker all the time
And to their feet the narrow bells gave rhyme.
Affinèd well to that sweet solitude,   15
He was a shepherd of the Arcadian mood
That not Arcadia knew nor Haemony.
His tale and telling has been given to me.

   .   .   .   .

(ii)

But what drew shepherd Richard from his downs,
And bred acquaintance of unusèd towns?
What put taught graces on his country lip,
And brought the sense of gentle fellowship,
That many centres found in many hearts?   5
What taught the humanities and the round of arts?°
And for the tinklings on the falls and swells
Gave the much music of our Oxford bells?

   .   .   .   .

(iii)

'Sylvester, come, Sylvester; you may trust
Your footing now to the much-dreaded dust,°
Crisp'd up and starchy from a short half-hour
Of standing to the blossom-hitting shower

That still makes counter-roundels in the pond.
A rainbow also shapes itself beyond
The shining slates and houses. Come and see.
You may quote Wordsworth, if you like, to me.'
Sylvester came: they went by Cumnor hill,°
Met a new shower, and saw the rainbow fill   10
From one frail horn that crumbled to the plain
His steady wheel quite to the full again.
They watched the brush of the swift stringy drops,
Help'd by the darkness of a block of copse
Close-rooted in the downward-hollowing fields;
Then sought such leafy shelter as it yields,
And each drew bluebells up, and for relief
Took primroses, their pull'd and plotted leaf
Being not forgotten, for primroses note
The blue with brighter places not remote.   20

(iv)

There was a meadow level almost: you traced
The river wound about it as a waist.
Beyond, the banks were steep; a brush of trees
Rounded it, thinning skywards by degrees,
With parallel shafts,—as upward-parted ashes,—
Their highest sprays were drawn as fine as lashes,
With centres duly touch'd and nestlike spots,—
And oaks,—but these were leaved in sharper knots.
Great butter-burr-leaves floor'd the slope corpse ground
Beyond the river, all the meadow's round,   10
And each a dinted circle. The grass was red
And long, the trees were colour'd, but the o'er-head,
Milky and dark, with an attuning stress
Controll'd them to a grey-green temperateness,
Making the shadow sweeter. A spiritual grace,
Which Wordsworth would have dwelt on, about the place
Led Richard with a sweet undoing pain
To trace some traceless loss of thought again.
Here at the very furthest reach away
(The furthest reach this side, on that the bay   20
Most dented) lay Sylvester, reading Keats'
Epistles, while the running pastoral bleats

Of sheep from the high fields and other wild
Sounds reach'd him. Richard came. Sylvester smiled
And said 'I like this: it is almost isled,
The river spans it with so deep a hip.
I hope that all the places on our trip
Will please us so.'

.   .   .   .

'All as that moth'

All as that moth call'd Underwing, alighted,
Pacing and turning, so by slips discloses
Her sober simple coverlid underplighted°
To colour as smooth and fresh as cheeks of roses,
  { Her showy leaves staid watchet counterfoiling
  { Her showy leaves with gentle watchet foiling
Even so my thought the rose and grey disposes

The Queen's Crowning

1.   They were wedded at midnight
         By shine of candles three,
      And they were bedded till daylight
         Before he went to sea.

2.   'When are you home, my love,' she said,
         'When are you home from sea?'
      'You may look for me home, my love,' he said,
         'In two years or in three.'

3.   'Heaven make the time be short,' she said,
         'Although it were years three.   10
      Heaven make it sweet to you,' she said,
         'And make it short to me.

4.   And what is your true name?' she said,
         'Your name and your degree?
      How shall I call my love,' she said,
         'When he is over the sea?'

5. 'O I am the king's son,' he said,
       'Lord William they call me.
    I give you my love and I give you my land,
       When I come home from sea.'   20

6. He yearn'd, he yearn'd to have his love,
       For two years and for three.
    Then he set sail in a golden ship
       With a golden company.

7. Or ever he set his foot to the land
       He saw his brothers three.
    'O have you here a foreign lady
       Come with you from over the sea?'

8. 'O I have here no foreign lady
       Come with me from over the sea.'   30
    'Then will you wed with an English lady,
       As wedded you must be?'

9. Says 'Get you, get you a lady to wed
       That has both gold and fee.
    Ere you set sail the king was dead.
       The crown has come to thee.'

10. 'And if I chose a love to wed
       That was of low degree?
    The crown should be unto her head
       And what were that to thee?'   40

11. One has gone to the king's steward,
       Shewn him both gold and fee:
    Said 'Who then is this lowly woman,
       And truly tell to me.'

12. The king's friend told the thing that was hid
       Because of gold and fee.
    Said, it was not meet the king should wed
       With one of low degree.

13. They have held his eyes with blindfold bands
        Because he should not see.   50
     They have bound his feet, they have bound his hands:
        It was but one to three.°

14. They have taken out their long brands,°
        They made him kneel on knee.
     'It is for the shame of the lowly woman
        That this has come to thee.'

15. They have happ'd him with the sand and stone°
        That was beside the sea.
     In his heart said everyone
        The crown shall be for me.   60

16. Lowly Alice sat in her bower
        With a two years child at her knee.
     'I think it is seven days,' she said,
        'Thy father thou shalt see.'

17. Lowly Alice look'd abroad
        Over field and tree,
     And she was ware of a servingman
        Came running over the lea.

18. 'O what will you now, good servingman,
        O what will you now with me?'   70
     Says 'Are you not Lord William's love
        That is of low degree?'

19. 'I am Lord William's love,' she said,
        'And Alice they call me.'
     'Lord William comes hunting tomorrow morning,
        And he will come to thee.

20. But how will you Lord William know
        Beside his brothers three?'
     'Because he is my love,' she said,
        'And is so fair to see.'   80

21.   'Yet how will you Lord William know
          Beside his brothers three?
       His three brothers are each as tall
          And each as fair as he.

22.   If it be a white rose in his hand,
          A lily if it should be,
       In this wise you may know your lord
          Beside his brothers three:

23.   If he wear the crown upon his head
          Among his brothers three,   90
       If he wear a crown upon his head
          And bring a crown for thee.'

24.   She heard the hunt the morrow morning
          And she came out to see.
       And there she never saw the king,
          But saw his brothers three.

25.   She stood before them in the glen,
          She kneeled upon her knee.
       'O where is Lord William, my lords,' she said,
          'I pray you tell to me.'   100

26.   Two made answer in one breath
          And each said 'I am he.'
       'Fie, you are not Lord William,' she said;
          'O fie that this should be.'

27.   Then up and spake the third brother,
          Said 'Listen now to me.
       Lord William is king of all this land
          And thou of low degree.'

28.   'Fie,' she said unto them all,
          'No truth between you three.   110
       If he were king of all this land
          He would have come for me.'

29. As she lay weeping at the night
       She heard but knockings three.
    'It is as cold as death without:
       Open the door to me.'

30. Said 'Who is this that stands without?'
       Said 'Open, open to me.'
    When she had made the door wide
       Her true-love she might see.   120

31. 'O why art thou so wan,' she said,
       'And why so short with me?
    And art thou come from English land,
       Or come from over the sea?'

32. 'I am not come from English land,
       Nor yet from over the sea.
    If I were come from Paradise,
       It were more like to be.'

33. 'Is it a lily in your hand,
       Is it a rose I see?   130
    Did you pull it in the king's garden
       When you came forth for me?'

34. 'I did not pull it in king's garden
       When I came forth for thee.
    If it were a flower of Paradise,
       It were more like to be.'

35. 'Is that the King's crown on your head,
       And have you a crown for me?'
    'If it were a crown of Paradise,
       It were more like to be.'   140

36. The more she ask'd, the more he spoke,
       The fairer waxèd he.
    The more he told, the less she spoke,
       The wanner wanèd she.

37.   'Wilt thou follow me, my true love,
         If I give thee kisses three?
       Wilt thou follow me, my true love?
         I have a crown for thee.'

38.   'O I will follow thee, my true love.
         Give me thy kisses three.   150
       Sweeter thy kisses, my own love,
         Than all the crowns to me.'

39.   He gave her kisses cold as ice;
         Down upon ground fell she.
       She has gone with him to Paradise.
         There shall her crowning be.

'Tomorrow meet you?'

Tomorrow meet you? O not tomorrow.
  I would not make the trial.
  Fear hindrance and espial
ˣAnd after that sad sorrow.
But with a sweet persistency
He dallies yet and yet with me
  And will not take denial.

ˣ [*or* Then severance and sorrow.]

Fragment of *Stephen and Barberie*

            —She by a sycamore,
Whose all-belated leaves yield up themselves
To the often takings of desirous winds,
Sits without consolation, marking not
The time save when her tears which still [descend]°
Her barrèd fingers clasp'd upon her eyes,
Shape on the under side and size and drop.°
Meanwhile a litter of the jaggèd leaves
Lies in her lap, which she anon sweeps off.
'This weary Martinmas, would it were summer'   10
I heard her say, poor poor afflictèd soul,—

'Would it were summer-time.' Anon she sang
The country song of *Willow*. 'The poor soul—°
(Like me)—*sat sighing by a sycamore-tree*.'
Perhaps it was for this she chose the place.

'Boughs being pruned'

'Boughs being pruned, birds preenèd, show more fair;
  To grace them spires are shaped with corner squinches;°
Enrichèd posts are chamfer'd; everywhere°
  He heightens worth who guardedly diminishes;
Diamonds are better cut; who pare, repair;
  Is statuary rated by its inches?
Thus we shall profit, while gold coinage still
Is worth and current with a lessen'd mill.'°

'A silver scarce-call-silver'

A silver scarce-call-silver gloss
{ Lighted the watery-plated leaves.
{ The watery-plated plane-leaves lit.

'I hear a noise of waters'

I hear a noise of waters drawn away,
And, headed always downwards, with less sounding
Work through a cover'd copse whose hollow rounding
Rather to ear than eye shews where they stray,
Making them double-musical. And they
Low-covered pass, and brace the woodland clods
With shining-hilted curves, that they may stay
The bluebells up whose crystal-ending rods
  .   .   .   .   . in their natural sods.

                  a standing fell   10
Of hyacinths   .   .   .   .
And pledgèd purply in a half-lit dell.

(*Dawn*)

(i)

In more precision now of light and dark
The heightening dawn with milky orience
Rounds its still-purpling centreings of cloud.

(ii)

Now more precisely touched in light and gloom,
The place of the east with earliest milky morn
Rounds its still-purpling centre-darks of cloud.

(iii)

Dawn that the pebbly low-down East
Covers with shallow silver, that unsets
The lock of clouds betimes and hangs the day.

Dawn that the low-down pebbly East
Covers with shallow silver, the lock of clouds
That early 'sperses, and high hangs the day.°

'*When eyes that cast*'

When eyes that cast about the heights of heaven
To canvass the retirement of the lark
(Because the music from his bill forth-driven
So takes the sister sense) can find no mark,
But many a silver visionary spark
Springs in the floating air and the skies swim,—
Then often the ears in a new fashion hark,
Beside them, about the hedges, hearing him:
At last the bird is found a flickering shape and slim.

At once the senses give the music back,   10
{ The proper sweet re-attributing above.
{ That sweetness re-attributing above.—

The Summer Malison

Maidens shall weep at merry morn,
And hedges break, and lose the kine,°
And field-flowers make the fields forlorn,
And noonday have a shallow shine,
And barley turn to weed and wild,
And seven ears crown the lodgèd corn,°
And mother have no milk for child,
    And father be overworn.

And John shall lie, where winds are dead,
And hate the ill-visaged cursing tars,   10
And James shall hate his faded red,
Grown wicked in the wicked wars.
No rains shall fresh the flats of sea,
Nor close the clayfield's sharded sores,
And every heart think loathingly
    Its dearest changed to bores.

St. Thecla

That his fast-flowing hours with sandy silt
Should choke sweet virtue's glory is Time's great guilt.
Who thinks of Thecla? Yet her name was known,
Time was, next whitest after Mary's own.
To that first golden age of Gospel times
And bright Iconium eastwards reach my rhymes.
Near by is Paul's free Tarsus, fabled where
Spent Pegasus down the stark-precipitous air°
Flung rider and wings away; though these were none,°
And Paul is Tarsus' true Bellerophon.°   10
They are neighbours; but (what nearness could not do)
Christ's only charity charmed and chained these two.
    She, high at the housetop sitting, as they say,
Young Thecla, scanned the dazzling streets one day;
Twice lovely, tinted eastern, turnèd Greek—
Crisp lips, straight nose, and tender-slanted cheek.

Her weeds all mark her maiden, though to wed,
And bridegroom waits and ready are bower and bed.
Withal her mien is modest, ways are wise,
And grave past girlhood earnest in her eyes.   20
    Firm accents strike her fine and scrollèd ear,
A man's voice and a new voice speaking near.
The words came from a court across the way.
She looked, she listened: Paul taught long that day.
He spoke of God the Father and His Son,
Of world made, marred, and mended, lost and won;
Of virtue and vice; but most (it seemed his sense)
He praised the lovely lot of continence:
All over, some such words as these, though dark,
The world was saved by virgins, made the mark.   30
    He taught another time there and a third.
The earnest-hearted maiden sat and heard,
And called to come at mealtime she would not:
They rose at last and forced her from the spot.

Easter Communion

Pure fasted faces draw unto this feast:
God comes all sweetness to your Lenten lips.
You striped in secret with breath-taking whips,°
Those crookèd rough-scored chequers may be pieced
To crosses meant for Jesu's; you whom the East°
With draught of thin and pursuant cold so nips
Breathe Easter now; you sergèd fellowships,
You vigil-keepers with low flames decreased,
God shall o'er-brim the measures you have spent
With oil of gladness, for sackcloth and frieze   10
And the ever-fretting shirt of punishment
Give myrrhy-threaded golden folds of ease.°
Your scarce-sheathed bones are weary of being bent:
Lo, God shall strengthen all the feeble knees.

'From any hedgerow'

From any hedgerow, any copse,
Bring me palm with pearlèd knops,
And primrose bring, and make a sheaf
With his pull'd and plotted leaf.

'O Death, Death'

O Death, Death, He is come.
O grounds of Hell make room.
Who came from further than the stars
   Now comes as low beneath.
   Thy ribbèd ports, O Death
Make wide; and Thou, O Lord of Sin,
   Lay open thine estates.
   Lift up your heads, O Gates;°
Be ye lift up, ye everlasting doors
   The King of Glory will come in.   10

'A basket broad'

A. A basket broad of woven white rods
     I have fill'd, that hard to fill is,
  With the multitude of the lily-buds
    Of the brakes of lilies.

B. And I come laden from such floods
    Of flowers that counting closes,
  With the warm'd and the water'd buds
    Of the press of roses.

(Fragments)

(i)

The sun just risen
Flares his wet brilliance in the dintless heaven.

(ii)

When cuckoo calls and I may hear,
And thrice and four times and again.

'Love me as I love thee'

Love me as I love thee. O double sweet!
But if thou hate me who love thee, albeit
    Even thus I have the better of thee:
Thou canst not hate so much as I do love thee.

To Oxford

New-dated from the terms that reappear,°
More sweet-familiar grows my love to thee,
And still thou bind'st me to fresh fealty
With long-superfluous ties, for nothing here
Nor elsewhere can thy sweetness unendear.
This is my park, my pleasaunce; this to me°
As public is my greater privacy,
All mine, yet common to my every peer.

Those charms accepted of my inmost thought,
The towers musical, quiet-wallèd grove,°   10
The window-circles, these may all be sought°
By other eyes, and other suitors move,
And all like me may boast, impeachèd not,
Their special-general title to thy love.

(continued)

Thus, I come underneath this chapel-side,
So that the mason's levels, courses, all
The vigorous horizontals, each way fall
In bows above my head, as falsified
By visual compulsion, till I hide°
The steep-up roof at last behind the small
Eclipsing parapet; yet above the wall
The sumptuous ridge-crest leave to poise and ride.

None besides me this bye-ways beauty try.
Or if they try it, I am happier then: 10
The shapen flags and drillèd holes of sky,
Just seen, may be [to] many unknown men°
The one peculiar of their pleasured eye,
And I have only set the same to pen.

'Where art thou friend'

Where art thou friend, whom I shall never see,
Conceiving whom I must conceive amiss?
Or sunder'd from my sight in the age that is°
Or far-off promise of a time to be;
Thou who canst best accept the certainty
That thou hadst borne proportion in my bliss,
That likest in me either that or this,—
Oh! even for the weakness of the plea
That I have taken to plead with,—if the sound
Of God's dear pleadings have as yet not moved thee,—   10
And for those virtues I in thee have found,
Who say that had I known I had approved thee,—
For these, make all the virtues to abound,—
No, but for Christ who hath foreknown and loved thee.°

'Bellisle'

Bellisle! that is a fabling name, but we
Have here a true one, echoing the sound;°
And one to each of us is holy ground;
But let me sing that which is known to me.

'Confirmed beauty'

Confirmed beauty will not bear a stress;—
Bright hues long look'd at thin, dissolve and fly:
Who lies on grass and pores upon the sky
Shall see the azure turn expressionless
And Tantalean slaty ashiness°

Like Pharaoh's ears of windy harvest dry,°
Dry up the blue and be not slaked thereby.
Ah! surely all who have written will profess
The sweetest sonnet five or six times read
Is tasteless nothing: and in my degree   10
I prove it; what then when these lines are dead
And coldly do belie the thought of thee?
I'll lay them by, and freshly turn instead
To thy not-staled uncharted memory.

The Beginning of the End

(i)

My love is lessened and must soon be past.
I never promised such persistency
In its condition. No, the tropic tree
Has not a charter that its sap shall last
Into all seasons, though no winter cast
The happy leafing. It is so with me:
My love is less, my love is less for thee.
I cease the mourning and the abject fast
And rise and go about my works again
And, save by darting accidents, forget.   10
But ah! if you could understand how then
That *less* is heavens higher even yet
Than treble-fervent *more* of other men,
Even your unpassion'd eyelids might be wet.

(ii)

I must feed Fancy. Show me any one
That reads or holds the astrologic lore,
And I'll pretend the credit given of yore;
And let him prove my passion was begun
In the worst hour that's measured by the sun,
With such malign conjunctions as before
No influential heaven ever wore;
That no recorded devilish thing was done

With such a seconding, nor Saturn took
Such opposition to the Lady-star   10
In the most murderous passage of his book;
And I'll love my distinction: Near or far
He says his science helps him not to look
At hopes so evil-heaven'd as mine are.

(iii)

You see that I have come to passion's end;
This means you need not fear the storms, the cries,
That gave you vantage when you would despise:
My bankrupt heart has no more tears to spend.
Else I am well assured I should offend
With fiercer weepings of these desperate eyes
For poor love's failure than his hopeless rise.
But now I am so tired I soon shall send

Barely a sigh to thought of hopes forgone.
Is this made plain? What have I come across   10
That here will serve me for comparison?
The sceptic disappointment and the loss
A boy feels when the poet he pores upon
Grows less and less sweet to him, and knows no cause.

The Alchemist in the City

My window shews the travelling clouds,
Leaves spent, new seasons, alter'd sky,
The making and the melting crowds:
The whole world passes; I stand by.

They do not waste their meted hours,
But men and masters plan and build:
I see the crowning of their towers,
And happy promises fulfill'd.

And I—perhaps if my intent
Could count on prediluvian age,   10
The labours I should then have spent
Might so attain their heritage,

But now before the pot can glow
With not to be discover'd gold,
At length the bellows shall not blow,
The furnace shall at last be cold.

Yet it is now too late to heal
The incapable and cumbrous shame
Which makes me when with men I deal
More powerless than the blind or lame.   20

No, I should love the city less
Even than this my thankless lore;
But I desire the wilderness
Or weeded landslips of the shore.

I walk my breezy belvedere
To watch the low or levant sun,
I see the city pigeons veer,
I mark the tower swallows run

Between the tower-top and the ground
Below me in the bearing air;   30
Then find in the horizon-round
One spot and hunger to be there.

And then I hate the most that lore
That holds no promise of success;
Then sweetest seems the houseless shore,
Then free and kind the wilderness,

Or ancient mounds that cover bones,
Or rocks where rockdoves do repair
And trees of terebinth and stones
And silence and a gulf of air.   40

There on a long and squarèd height
After the sunset I would lie,
And pierce the yellow waxen light
With free long looking, ere I die.

'But what indeed'

But what indeed is ask'd of me?
Not this. Some spirits, it is told,
Have will'd to be disparadised
For love and greater glory of Christ.
But I was ignorantly bold
To dream I dared so much for thee.
This was not ask'd, but what instead?
Waking I thought; and it sufficed:
My hopes and my unworthiness,
At once perceivèd, with excess   10
Of burden came and bow'd my head.

    .   .   .   .

Yea, crush'd my heart, and made me dumb.
I thought: Before I gather strength. . . .

'Myself unholy'

Myself unholy, from myself unholy
To the sweet living of my friends I look—
Eye-greeting doves bright-counter to the rook,
Fresh brooks to salt sand-teasing waters shoaly:—°
And they are purer, but alas! not solely
The unquestion'd readings of a blotless book.
And so my trust confusèd, struck, and shook°
Yields to the sultry siege of melancholy.
He has a sin of mine, he its near brother,°
Knowing them well I can but see the fall.   10
This fault in one I found, that in another:
And so, though each have one while I have all,
No *better* serves me now, save *best*; no other,
Save Christ: to Christ I look, on Christ I call.

To Oxford

As Devonshire lette.s, earlier in the year
Than we in the East dare look for buds, disclose
Smells that are sweeter-memoried than the rose,
And pressèd violets in the folds appear,
So is it with my friends, I note, to hear
News from Belleisle, even such a sweetness blows°
(I know it, knowing not) across from those
Meadows to them inexplicably dear.

'As when a soul laments, which hath been blest'—°
I'll cite no further what the initiate know.
I never saw those fields whereon their best
And undivulgèd love does overflow.

'See how Spring opens'

See how Spring opens with disabling cold,
And hunting winds and the long-lying snow.
Is it a wonder if the buds are slow?
Or where is strength to make the leaf unfold?
Chilling remembrance of my days of old
Afflicts no less, what yet I hope may blow,
That seed which the good sower once did sow,°
So loading with obstruction that threshold

Which should ere now have led my feet to the field.
It is the waste done in unreticent youth
Which makes so small the promise of that yield
That I may win with late-learnt skill uncouth
From furrows of the poor and stinting weald.
Therefore how bitter, and learnt how late, the truth!°

Continuation of R. Garnett's *Nix*

She mark'd where I and Fabian met;
She loves his face, she knows the spot;
And there she waits with locks unwet
For Fabian that suspects her not.

I see her riving fingers tear
A branch of walnut-leaves, and that
More sweetly shades her stolen hair
Than fan or hood or strawy plait.

He sees her, O but he must miss   10
A something in her face of guile,
×And relish not her loveless kiss
And wonder at her shallow smile.

× [*or* And half mislike her loveless kiss.]

Ah no! and she who sits beside
Bids him this way his gazes fix.
×Then she seems sweet who seems his bride,
She sour who seems the slighted Nix.

× [*or* Then sweetest seems the seeming bride
When maddest looks the slighted Nix.]

I know of the bored and bitten rocks
Not so far outward in the sea:
There lives the witch shall win my locks
And my blue eyes again for me.   20

Alas! but I am all at fault,
Nor locks nor eyes shall win again.
I dare not taste the thickening salt,
I cannot meet the swallowing main.

Or if I go, she stays meanwhile,
Who means to wed or means to kill,
And speeds uncheck'd her murderous guile
Or wholly winds him to her will.

'A noise of falls'

A noise of falls I am possessèd by
{ Of streams; and clouds like mesh'd and parted moss
{ Of water. Clouds like parted moss
Attain the windy levels of the sky
Which between ash-tops suffers loss
Of its concavity.

'O what a silence'

O what a silence is this wilderness!
Might we not think the sweet(?) and daring rises
Of the flown skylark, and his traverse flight°
At highest when he seems to brush the clouds,
†Had been more fertile and had sown with notes

The unenduring fallows of the heaven?
Or take it thus—that the concording stars
Had let such music down, without impediment
Falling along the breakless pool of air,
*As struck with rings of sound the close-shut palms   10
Of the wood-sorrel and all things sensitive?

————

† [*or* Had been effectual to have sown with notes]

* [As might have struck and shook the close-shut palms]   12

A. As the wood-sorrel and all things sensitive
   That thrive in the loamy greenness of this place?
B. What spirit is that makes stillness obsolete
   With ear-caressing speech? Where is the tongue
   Which drives this stony air to utterance?—
   Who is it? how come to this forgotten land?

'Mothers are doubtless'

Mothers are doubtless happier for their babes
And risen sons: yet are the childless free
From tears shed over children's graves.
So those who [hold ?] Thee°
Take their peculiar thorns and natural pain
Among the lilies and thy good domain.

Daphne

Who loves me here and has my love,
I think he will not tire of me,
But sing contented as the dove
That comes again to the woodland tree.

He shall have summer sweets and dress
His pleasure to the changing clime,
And I can teach him happiness
That shall not fail in winter-time.

[or He shall have summer goods and trim
His pleasure to the changing clime,
And I shall know of sweets for him
That are not less in winter-time.]

His cap shall be shining fur,
And stain'd, and knots of golden thread,   10
He shall be warm with miniver
Lined all with silk of juicy red.

In spring our river-banks are topt
With yellow flags will suit his brow,
In summer are our orchards knopt°
With green-white apples on the bough.

But if I cannot tempt his thought
With wealth that mocks his high degree,
The shepherds, whom I value not,
Have told me I am fair to see.   20

Fragments of *Castara Victrix*

(i) *Scene: a bare hollow between hills. Enter Castara and her Esquire.*

C.   What was it we should strike the road again?
E.   There was a wood of dwarf and sourèd oaks
    Crept all along a hill upon our left,
    A wonder in the country, and a landmark
    They said we could not miss. A pushing brook
    Ran through it, following which we should have sight
    Of mile-long reaches of our road below us.
    My thought was, there to rest against the trees
    And watch until our horses and the men
    Circled the safe flanks of the bulky hills.   10
C.   And how long was the way?
E.   This shorter way?
    Two miles indeed.
C.   We have come four, do you think?
    Somewhere we slipt astray, you cannot doubt.
E.   True, madam. I am sorry now to see
    I better'd all our path with sanguine eyes.

          .   .   .   .

(ii) *At the picnic or whatever we call it. Daphnis, Castara.*

D.   —Can I do any harm?
C.   If you are silent, that I know of, none.
D.   Ill meant, yet true. I best should flatter then,
    In copying well what you have well begun.
C.   In copying? how?
D.   Must I give tongue again?
    In copying your sweet silence.
C.   Am I so
    Guilty of silence?
D.   Quite, as ladies go.
    Yet what you are, the world would say, remain:
    It never yet so sweetly was put on
    By any lauded statue, nor again   10
    By speech so sweetly broken up and gone.
C.   What if I hated flattery?

D.   Say you do:
The hatred comes with a good grace from you:
Flattery's all out of place where praise is true.

.   .   .   .

(iii) *Valerian, Daphnis.*

V.   Come, Daphnis.
D.   Good Valerian, I will come. (*exit V.*
Why should I go because Castara goes?
I do not, but to please Valerian.
But why then should Castara weigh with me?
Why, there's an interest and sweet soul in beauty
Which makes us eye-attentive to the eye
That has it; and she is fairer than Colomb,
Selvaggia, Orinda, and Adela, and the rest.
Fairer? These are the flaring shows unlovely
That make my eyes sore and cross-colour things   10
With fickle spots of sadness; accessories
⎧ Familiar and so hated by the sick;
⎩ Hated and too familiar to — —;
These are my very text of discontent;
These names, these faces? They are customary
And kindred to my lamentable days,
Of which I say there is no joy in them.
To these Castara is rain or breeze or spring,
— —— —— – dew, is dawn, is day,
Sheet lightning to the stifling lid of night
Bright-lifting with a little-lasting smile
And breath on it. That is, her face is this.   20
And if it is why there is cause enough
To say I go because Castara goes.
Yet I'd not say it is her face alone
That this is true of: 'tis Castara's self;
But this distemper'd court will change it all:—
Which says at least then go while all is fresh,—
Much cause to go because Castara goes.

'My prayers must meet a brazen heaven'

My prayers must meet a brazen heaven
And fail and scatter all away.
Unclean and seeming unforgiven
My prayers I scarcely call to pray.
I cannot buoy my heart above;
Above I cannot entrance win.
I reckon precedents of love,
But feel the long success of sin.

My heaven is brass and iron my earth:
Yea, iron is mingled with my clay,   10
So harden'd is it in this dearth
Which praying fails to do away.
Nor tears, nor tears this clay uncouth
Could mould, if any tears there were.
A warfare of my lips in truth,
Battling with God, is now my prayer.

Shakspere

In the lodges of the perishable souls
He has his portion. God, who stretch'd apart
Doomsday and death—whose dateless thought must chart
All time at once and span the distanced goals,
Sees what his place is; but for us the rolls
Are shut against the canvassing of art.
Something we guess or know: some spirits start
Upwards at once and win their aureoles.

. . . .

'Trees by their yield'

Trees by their yield°
Are known; but I—
My sap is sealed,
My root is dry.

If life within
I none can shew
(Except for sin),
Nor fruit above,—
It must be so—   10
I do not love.

Will no one show
I argued ill?
Because, although
Self-sentenced, still
I keep my trust.
If He would prove
And search me through
Would He not find
(What yet there must
Be hid behind   20

.   .   .   .

'Let me be to Thee'

Let me be to Thee as the circling bird,
Or bat with tender and air-crisping wings
That shapes in half-light his departing rings,
From both of whom a changeless note is heard.
I have found my music in a common word,
Trying each pleasurable throat that sings
And every praisèd sequence of sweet strings,
And know infallibly which I preferred.

The authentic cadence was discovered late
Which ends those only strains that I approve,   10
And other science all gone out of date
And minor sweetness scarce made mention of:
I have found the dominant of my range and state—
Love, O my God, to call Thee Love and Love.

(76)

The Half-way House

Love I was shewn upon the mountain-side
And bid to catch Him ere the drop of day.
See, Love, I creep and Thou on wings dost ride;
Love, it is evening now and Thou away;
Love, it grows darker here and Thou art above;
Love, come down to me if Thy name be Love.

My national old Egyptian reed gave way;°
I took of vine a cross-barred rod or rood.°
Then next I hungered: Love when here, they say,
Or once or never took Love's proper food;
But I must yield the chase, or rest and eat.—
Peace and food cheered me where four rough ways meet.

Hear yet my paradox: Love, when all is given,
To see Thee I must [see] Thee, to love, love;
I must o'ertake Thee at once and under heaven
If I shall overtake Thee at last above.
You have your wish; enter these walls, one said:°
He is with you in the breaking of the bread.

A Complaint

I thought that you would have written: my birthday came and went,
And with the last post over, I knew no letter was sent.
And now if at last you write it never can be the same:
What *would* be a birthday letter that after the birthday came?

I know what you will tell me, Neglectful you were not:
But is not that my grievance—you promised and you forgot?
It's the day that makes the charm; no after-words can succeed
Though they took till the seventeenth of next October to read.

Think this, my birthday falls in a saddening time of year;
Only the dahlias blow, and all is Autumn here.
Hampstead was never bright, and whatever Miss Cully's charms,
It's hardly a proper treat for a birthday to rest in her arms.

Our sex should be born in April perhaps or the lily time,
But the lily is past, as I say, and the rose is not in its prime:
What I *did* ask then was a circle of rose-red sealing-wax
And a few leaves not lily-white but charactered over with blacks.

But late is better than never. You see you have managed so,
You have made me quote almost the dismalest proverb I know;
For a letter comes at last (shall I say before Christmas is come?)
And I must take your amends, cry 'Pardon', and then be dumb.   20

'Moonless darkness'

Moonless darkness stands between.
Past, the Past, no more be seen!
But the Bethlehem-star may lead me
To the sight of Him Who freed me
From the self that I have been.
Make me pure, Lord: Thou art holy;
Make me meek, Lord: Thou wert lowly;
Now beginning, and alway:
Now begin, on Christmas day.

'The earth and heaven'

The earth and heaven, so little known,
Are measured outwards from my breast.
I am the midst of every zone
And justify the East and West;

The unchanging register of change
My all-accepting fixèd eye,
While all things else may stir and range,
All else may whirl or dive or fly.

The swallow, favourite of the gale,°
Will on the moulding strike and cling,   10
Unvalve or shut his vanèd tail
And sheathe at once his leger wing.

He drops upon the wind again;
His little pennon is unfurled.
In motion is no weight or pain,
Nor permanence in the solid world.

There is a vapour stands in the wind;
It shapes itself in taper skeins:
You look again and cannot find,
Save in the body of the rains.   20

And these are spent and ended quite;
The sky is blue, and the winds pull
Their clouds with breathing edges white
Beyond the world; the streams are full

And millbrook-slips with pretty pace
Gallop along the meadow grass.—
O lovely ease in change of place!
I have desired, desired to pass

'As it fell upon a day'

As it fell upon a day
There was a lady very gay,
She was dressed in silk attire
For all to see and to admire.

.   .   .   .

But the boatman on the green
Told of the wonders he had seen.

'In the staring darkness'

In the staring darkness
I can hear the harshness
Of the cold wind blowing.
I am warmly clad,
And I'm very glad
That I've got a home.

'The stars were packed so close'

The stars were packed so close that night
  They seemed to press and stare
And gather in like hurdles bright
  The liberties of air.

The Nightingale

'From nine o'clock till morning light
The copse was never more than grey.
The darkness did not close that night
  But day passed into day.
And soon I saw it shewing new
Beyond the hurst with such a hue
As silky garden-poppies do.

A crimson East, that bids for rain.
So from the dawn was ill begun
The day that brought my lasting pain   10
  And put away my sun.
But watching while the colour grew
I only feared the wet for you
Bound for the Harbour and your crew.

I did not mean to sleep, but found
I had slept a little and was chill.
And I could hear the tiniest sound,
  The morning was so still—
The bats' wings lisping as they flew
And water draining through and through   20
The wood: but not a dove would coo.

You know you said the nightingale
In all our western shires was rare,
That more he shuns our special dale
  Or never lodges there:
And I had thought so hitherto—
Up till that morning's fall of dew,
And now I wish that it were true.

For he began at once and shook
My head to hear. He might have strung   30
A row of ripples in the brook,
   So forcibly he sung,
The mist upon the leaves have strewed,
And danced the balls of dew that stood
In acres all above the wood.

I thought the air must cut and strain
The windpipe when he sucked his breath
And when he turned it back again
   The music must be death.
With not a thing to make me fear,   40
A singing bird in morning clear
To me was terrible to hear.

Yet as he changed his mighty stops
Betweens I heard the water still
All down the stair-way of the copse
   And churning in the mill.
But that sweet sound which I preferred,
Your passing steps, I never heard
For warbling of the warbling bird.'

Thus Frances sighed at home, while Luke   50
Made headway in the frothy deep.
She listened how the sea-gust shook
   And then lay back to sleep.
While he was washing from on deck
She pillowing low her lily neck
Timed her sad visions with his wreck.

The Habit of Perfection

   Elected Silence, sing to me
   And beat upon my whorlèd ear,
   Pipe me to pastures still and be
   The music that I care to hear.

Shape nothing, lips; be lovely-dumb:
It is the shut, the curfew sent
From there where all surrenders come
Which only makes you eloquent.

Be shellèd, eyes, with double dark
And find the uncreated light: 10
This ruck and reel which you remark
Coils, keeps, and teases simple sight.

Palate, the hutch of tasty lust,
Desire not to be rinsed with wine:
The can must be so sweet, the crust
So fresh that come in fasts divine!

Nostrils, your careless breath that spend
Upon the stir and keep of pride,
What relish shall the censers send
Along the sanctuary side!   20

O feel-of-primrose hands, O feet
That want the yield of plushy sward,
But you shall walk the golden street
And you unhouse and house the Lord.

And, Poverty, be thou the bride
And now the marriage feast begun,
And lily-coloured clothes provide
Your spouse not laboured-at nor spun.

Nondum

'*Verily Thou art a God that hidest Thyself.*'
ISAIAH xlv. 15.

God, though to Thee our psalm we raise°
No answering voice comes from the skies;
To Thee the trembling sinner prays
But no forgiving voice replies;
Our prayer seems lost in desert ways,
Our hymn in the vast silence dies.

We see the glories of the earth°
But not the hand that wrought them all:
Night to a myriad worlds gives birth,
Yet like a lighted empty hall　　　　　　　　　　10
Where stands no host at door or hearth
Vacant creation's lamps appal.

We guess; we clothe Thee, unseen King,
With attributes we deem are meet;
Each in his own imagining
Sets up a shadow in Thy seat;
Yet know not how our gifts to bring,
Where seek Thee with unsandalled feet.

And still th'unbroken silence broods°
While ages and while aeons run,　　　　　　　　　　20
As erst upon chaotic floods
The Spirit hovered ere the sun
Had called the seasons' changeful moods
And life's first germs from death had won.

And still th'abysses infinite
Surround the peak from which we gaze.
Deep calls to deep, and blackest night°
Giddies the soul with blinding daze
That dares to cast its searching sight　　　　　　　　　　30
On being's dread and vacant maze.

And Thou art silent, whilst Thy world
Contends about its many creeds
And hosts confront with flags unfurled
And zeal is flushed and pity bleeds
And truth is heard, with tears impearled,
A moaning voice among the reeds.

My hand upon my lips I lay;
The breast's desponding sob I quell;
I move along life's tomb-decked way
And listen to the passing bell　　　　　　　　　　40
Summoning men from speechless day
To death's more silent, darker spell.

Oh! till Thou givest that sense beyond,
To shew Thee that Thou art, and near,
Let patience with her chastening wand
Dispel the doubt and dry the tear;
And lead me child-like by the hand
If still in darkness not in fear.

Speak! whisper to my watching heart
One word—as when a mother speaks   50
Soft, when she sees her infant start,
Till dimpled joy steals o'er its cheeks.
Then, to behold Thee as Thou art,
I'll wait till morn eternal breaks.

Easter

Break the box and shed the nard;°
Stop not now to count the cost;
Hither bring pearl, opal, sard;
Reck not what the poor have lost;
Upon Christ throw all away:
Know ye, this is Easter Day.

Build His church and deck His shrine,
Empty though it be on earth;
Ye have kept your choicest wine—
Let it flow for heavenly mirth;   10
Pluck the harp and breathe the horn:
Know ye not 'tis Easter morn?

Gather gladness from the skies;
Take a lesson from the ground;
Flowers do ope their heavenward eyes
And a Spring-time joy have found;
Earth throws Winter's robes away,
Decks herself for Easter Day.

Beauty now for ashes wear,
Perfumes for the garb of woe,   20
Chaplets for dishevelled hair,
Dances for sad footsteps slow;

Open wide your hearts that they
Let in joy this Easter Day.

Seek God's house in happy throng;
Crowded let His table be;
Mingle praises, prayer, and song,
Singing to the Trinity.
Henceforth let your souls alway
Make each morn an Easter Day.   30

Lines for a Picture of St. Dorothea

Dorothea and Theophilus

I bear a basket lined with grass.
I' am so' light' and fair'
Men are amazed to watch me pass
With' the básket I bear',
Which in newly drawn green litter
Carries treats of sweet for bitter.

See my lilies: lilies none,°
None in Caesar's garden blow.
Quínces, look', when' not one'
Is set in any orchard; no,°   10
Not set because their buds not spring;°
Spring not for world is wintering.

But' they came' from' the South',
Where winter-while is all forgot.—
The dew-bell in the mallow's mouth
Is' it quénchèd or not'?
In starry, starry shire it grew:
Which' is it', star' or dew'?—

That a quince I pore upon?
O no it is the sizing moon.°   20
Now her mallow-row is gone°
In tufts of evening sky.—So soon?
Sphered so fast, sweet soul?—We see°
Fruit nor flower nor Dorothy.

How to name it, blessed it!
Suiting its grace with *him* or *her*?
Dorothea—or was your writ
Sérvèd bý méssenger′?
Your parley was not done and there!
You went into the partless air.   30

It waned into the world of light,
Yet made its market here as well:
My eyes hold yet the rinds and bright
Remainder of a miracle.
O this is bringing! Tears may swarm°
Indeed while such a wonder's warm.

Ah dip in blood the palmtree pen
And wordy warrants are flawed through.
More will wear this wand and then°
The warpèd world we shall undo.   40
Proconsul!—Is Sapricius near?—
I find another Christian here.

Summa

The best ideal is the true
    And other truth is none.
All glory be ascribèd to
    The holy Three in One.

Man is most low, God is most high.
    As sure as heaven it is°
There must be something to supply
    All insufficiencies.
For souls that might have blessed the time°
    And breathed delightful breath   10
In sordidness of care and crime
    The city tires to death.
And faces fit for leisure gaze
    And daylight and sweet air,
Missing prosperity and praise,
    Are never known for fair.

Jesu Dulcis Memoria

Jesus to cast one thought upon
Makes gladness after He is gone,
But more than honey and honeycomb
Is to come near and take Him home.

No music so can touch the ear,°
No news is heard of such sweet cheer,
Thought half so dear there is not one
As Jesus God the Father's Son.

Jesu, their hope who go astray,
So kind to those who ask the way,   10
So good to those who look for Thee,
To those who find what must Thou be?

Jesu, a springing well Thou art,°
Daylight to head and treat to heart,
And matched with Thee there's nothing glad
That men have wished for or have had.

To speak of that no tongue will do
Nor letters suit to spell it true:
But they can guess who have tasted of
What Jesus is and what is love.   20

Wish us Good morning when we wake
And light us, Lord, with Thy day-break.
Beat from our brains the thicky night
And fill the world up with delight.

Who taste of Thee will hunger more,
Who drink be thirsty as before:
What else to ask they never know
But Jesus' self, they love Him so.

— — — — — —

And a sweet singing in the ear   30
And in the mouth a honey zest
And drinks of heaven in the breast.

Thou art the hope, Jesu my sweet,
The soul has in its sighing-fit;
The loving tears on Thee are spent,
The inner cry for Thee is meant.

Be our delight, O Jesu, now
As by and by our prize art Thou,
And grant our glorying may be
World without end alone in Thee.   40

Inundatio Oxoniana

Verna diu saevas senserunt pascua nubes
Imbribus assiduis, et aquosi copia caeli
Ingruit et spretae direpto limine ripae
Fit mare per patulos ventisque ferentibus agros.
Interrupta locis candenti gramina surgunt
Laetius in pelago, pars lenibus edita dorsis
Quae viret: at vacuus jam caetera condidit humor.
Vix indiscretas proprio deducitur alveo
Isis aquas; liquidos exercent libera tractus
Flabra, vadisque novis Austro juvat ire secundo;   10
Invia velificant nemorum et penetratur opacas
In salices; inter discussae culmina silvae
Populus insolitis dat currere mersa carinis.
    At quinto tandem si sol equitaverit orbi
Per purum, toties si riserit igneus aether,
Deficient reduces undae. Tum saepe marinus
Fertur odor campo et madidas levis occupat auras,
Urbem qui subeat mediam lustretque domorum
Intima; tristem adeo non usquam averteris algam.
Hinc quota vis morbi, quoties adiisse querentur,   20
Tecta petis nostri vicinam obnoxia febrem.
At vicibus vertisse solum est, aegrosque calores
Jam fugere: his non perpetui versamur in umbris.
Pars ascripta solo sedes servabit avitas
Tutior, indigenae plebes assueta periclo:
Hinc almo certe submotae numine pestes,
Namque licet tepidos in nostra Favonius imbres
Arva iteret pernox, resupina impune fatigat

Ipse loca et campis obducitur aequor inerme.
Vix rubeant immo siccis sua lilia pratis,   30
Quot capita ad notos agitari videris amnes,
Debita ni paullum fecundo luserit unda
Diluvio interea, dubii se pandere fluctus
Ni poterint prius et limos posuisse sequaces.
Dulcia sic fluviis praetendit fortior arbos
Vimina; sic crescunt salices; eques avia quaerit
Aequora sic, tumidasque libentius itur in herbas.

Ecquis binas

O for a pair like turtles wear,
O wings my spirit could put on!
And where I see the sweet cross-tree
In instant time I would be gone.

Elegiacs: *Tristi tu, memini*

Tristi tu, memini, virgo cum sorte fuisti,
    Illo nec steterat tempore primus amor.
Jamque abeo: rursus tu sola relinqueris: ergo
    Tristior haec aetas; tristis et illa fuit.
Adsum gratus ego necopini apparitor ignis,
    Inter ego vacuas stella serena nives.

'Alget honos'

Alget honos frondum silvis dependitus, alget
    Quae fuit in solo forma relicta toro

'Quo rubeant'

Quo rubeant dulcesve rosae vel pomifer aestas?
    Est rubor in teneris virginis ille genis.

Elegiacs: after *The Convent Threshold*

(Paragraphs 1 and 9)

Fraterno nobis interluit unda cruore
    Et novus exstincti stat patris, Aule, cruor.
O mihi tu summe et semper suavissime rerum,
    Divisam longe jam cruor ille tenet.
It via per stellas sublimis et aureus ordo
    Excipiens noctem nocte dieque diem:
Hanc ingressa poli seras elabar ad arces
    Sub vitreasque domos ad vitreumque mare.
Candida quos perhibes praecellere lilia forma
    Purpurei infecta sunt male labe pedes.   10
Purpurea sunt labe pedes et tristibus exsto
    Indicio guttis criminis ipsa mei,
Gaudia quae fuerint et qui post gaudia fletus
    Et qui conciderit nec recidivus amor.
At neque habent illi tantum nec sanguis inhaeret
    Scilicet admotis ille abolendus aquis:
Si penitus caecum possim recludere pectus
    Haec penitus caeco pectore culpa latet.
Sed mare quod mixta rutilat flammaque vitroque—°
    Illud molle vitrum, limpidus ignis erat—   20
Afferat ah captis oro medicamina plantis
    Infectaeque notae suppositique doli.
Quumque sit exstructo monstratum limite caelum
    O adeas mecum quae subit astra viam.

.   .   .   .

Hesternae referam quot vidi insomnia noctis
    Ambiguaeque umbrae noxve diesve foret.
Plurima tum nobis gelido coma rore madebat:
    Creverat ex gelida ros liquefactus humo.
Huc ades atque tua num tangar imagine quaeris,
    An memor et requies hactenus illa tui.
Ista quod quondam saliebat imagine pectus   30
    Urgenti dictis stat tibi pulvis iners.
Percipio quaesita tamen, nec reddere vocem
    Non erat, et tardo pauca sopore dabam:

Sunt tristes thalami, funesta toralia nobis;
    Impositoque rigent frigida saxa toro.
Tu dulces thalamos, tu quaere novos hymenaeos,
    Adde, licet, grato mollia membra toro.
Est tibi quae melius te foverit altera conjunx,
    Suavior exstat amor qui sit amore meo.   40
Perculeras valide trepidas ad talia palmas
    Visaque sunt subitis membra labare modis.
Extrema haec sensi; crassae simul intima terrae
    Volvor et in vacuos praecipitata locos.
Ah neque te festis plausu dare signa choreis
    Nec rata sum nimio membra labare mero.

St. Dorothea

(lines for a picture)

The Angel

I bear a basket lined with grass.
I am só light and fair
Men must start to see me pass
And the básket I bear,
Which in newly-drawn green litter
Carries treats of sweet for bitter.

See my lilies: lilies none,
None in Caesar's gardens blow—
Quinces, loók, whén not one
Is set in any orchard, no;   10
Not set because their buds not spring;
Spring not, 'cause world is wintering.

The Protonotary Theophilus

Bút they cáme fróm the south,
Where winter's out and all forgot.

The Angel

The bell-drops in my mallow's mouth
Hów are théy quenchèd not?—
These drops in starry shire they drew:
Whích are théy? stars or dew?

A Catechumen

That a quince we pore upon?
O no, it is the sizing moon.   20
Now her mallow-row is gone
In floats of evening sky.—So soon?
Sphered so fast, sweet soul?—We see
Nor fruit nor flowers nor Dorothy.

Theophilus

How to name it, blessed it,
Suiting its grace by *him* and *her*?
Dorothea—or was your writ
Servèd by sweet seconder?—
Your parley was not done and there!
You fell into the partless air.   30

You waned into the world of light,
Yet made your market here as well:
My eyes hold yet the rinds and bright
Remainder of a miracle.
O this is bringing! Tears may swarm
While such a wonder's wet and warm!

Ah myrtle-bend never sit,
Sit no more these bookish brows!
I want, I want, if I were fit,
Whát the cóld mónth allows—   40
Nothing green or growing but
A pale and perished palmtree-cut.

Dip in blood the palmtree-pen
And wordy warrants are flawed through;
And more shall wear this wand and then
The warpèd world it will undo.—
Próconsul,—cáll him near—
I find another Christian here.

'Not kind! to freeze me'

Not kind! to freeze me with forecast,
Dear grace and girder of mine and me.
You to be gone and I lag last—
Nor I nor heaven would have it be.

Horace: *Persicos odi, puer, apparatus*

(ODES I. xxxviii)

Ah child, no Persian-perfect art!°
Crowns composite and braided bast°
They tease me. Never know the part
      Where roses linger last.
Bring natural myrtle, and have done:°
Myrtle will suit your place and mine:°
And set the glasses from the sun
      Beneath the tackled vine.°

Horace: *Odi profanum volgus et arceo*

(ODES III. i)

Tread back—and back, the lewd and lay!°—
Grace love your lips!—what never ear°
Heard yet, the Muses' man, today
I make the boys and maidens hear.°

Kings herd it on their subject droves
But Jove's the herd that keeps the kings—
Jove of the Giants: simple Jove's
Mere eyebrow rocks this round of things.

Say man than man may rank his rows°
Wider, more wholesale; one with claim
Of blood to our green hustings goes;
One with more conscience, cleaner fame;

10

One better backed comes crowding by:—
That level power whose word is Must
Dances the balls for low or high:
Her urn takes all, her deal is just.

Sinner who saw the blade that hung°
Vertical home, could Sicily fare°
Be managed tasty to that tongue?
Or bird with pipe, viol with air   20

Bring sleep round then? sleep not afraid°
Of country bidder's calls or low
Entries or banks all over shade
Or Tempe with the west to blow.°

Who stops his asking mood at par
The burly sea may quite forget
Nor fear the violent calendar
At Haedus-rise, Arcturus-set,°

For hail upon the vine nor break
His heart at farming, what between   30
The dog-star with the fields abake°
And spiting snows to choke the green.

Fish feel their waters drawing to°
With our abutments: there we see
The lades discharged and laded new,°
And Italy flies from Italy.

But fears, fore-motions of the mind,
Climb quits: one boards the master there
On brazèd barge and hard behind
Sits to the beast that seats him—Care.   40

O if there's that which Phrygian stone°
And crimson wear of starry shot°
Not sleek away; Falernian-grown°
And oils of Shushan comfort not,°

Why . . . . . .
    . . . .
Why should I change a Sabine dale
For wealth as wide as weariness?

The Elopement

All slumbered whom our rud red tiles°
Do cover from the starry spread,
When I with never-needed wiles
    Crept trembling out of bed.
Then at the door what work there was, good lack,
To keep the loaded bolt from plunging back.

When this was done and I could look
I saw the stars like flash of fire.
My heart irregularly shook,
    I cried with my desire.   10
I put the door to with the bolts unpinned,
Upon my forehead hit the burly wind.

No tumbler woke and shook the cot,
The rookery never stirred a wing,
At roost and rest they shifted not,
    Blessed be everything.
And all within the house were sound as posts,
Or listening thought of linen-winded ghosts.

The stars are packed so thick to-night°
They seem to press and droop and stare,°   20
And gather in like hurdles bright°
    The liberties of air.°
I spy the nearest daisies through the dark,
The air smells strong of sweetbriar in the park.

I knew the brook that parts in two
The cart road with a shallowy bed
Of small and sugar flints, I knew
    The footway, Stephen said,
And where cold daffodils in April are
Think you want daffodils and follow as far°   30

As where the little hurling sound
To the point of silence in the air
Dies off in hyacinthed ground,
   And I should find him there.
O heart, have done, you beat you beat so high,
You spoil the plot I find my true love by.

Oratio Patris Condren: O Jesu vivens in Maria

Jesu that dost in Mary dwell,
Be in thy servants' hearts as well,
In the spirit of thy holiness,
In the fulness of thy force and stress,°
In the very ways that thy life goes;°
And virtues that thy pattern shows,
In the sharing of thy mysteries;
And every power in us that is
Against thy power put under feet
In the Holy Ghost the Paraclete   10
    To the glory of the Father. Amen.

A.M.D.G.

In Festo Nativitatis

Ad Matrem Virginem

Hymnus Eucharisticus

Mater Jesu mei,
Mater magni Dei,
Doce me de Eo,
De parvo dulci Deo.
   Quantum amavisti
Quem tu concepisti,
Non concipiendum,
Dominum tremendum,
Sed in te contractum,
Verbum carnem factum?   10
Et contemnit idem
Ne cor meum quidem:

Meum cor indignum
Quod capiat tantum signum,
Indignum O quod gerat
Qui mane mecum erat,
Subit, O Maria,
In eücharistia.
Ipse vult intrare:
Nolo me negare.   20
Candens exemplare,
Doce me amare.
   Dic, ut plus ametur,
Qualis videretur
Vulva dum lateret,
Necdum appareret,
Cum tua fecit laetam
Vox Elisabetham,
Laetam matre matrem,°
Laetum fratre fratrem.   30
Doce me gaudere,
Rosa, tuo vere,
Virga, tuo flore,
Vellus, tuo rore,
Arca, tua lege,
Thronus, tuo rege,
Acies, tuo duce,
Luna, tua luce,
Stella, tuo sole,
Parens, tua prole.   40
Nam tumeo et abundo
Immundo adhuc mundo;
Sum contristatus Sanctum
Spiritum et planctum
Custodi feci meo
Cum exhiberem Deo
Laesum atque caesum
In mea carne Jesum.
   Demum quid sensisti
Ipsum cum vidisti   50
Tandem visu pleno
Parvulum in foeno,
Ecce tremebundum

Qui fixum firmat mundum
Et involutum pannos
Qui aeternos annos
Nondum natus de te
Volvebat in quiete?
Quae tu tum dicebas
Et quae audiebas?   60
Etsi fuit mutus
Tamen est locutus.
Da complecti Illum,
Mihi da pauxillum
Tuo ex amore
Et oscula ab ore.
Qui pro me vult dari,
Infans mihi fari,
Mecum conversari,
Tu da contemplari,   70
Mater magni Dei,
Mater Jesu mei.
        L. D. S.

'Haec te jubent'

Haec te jubent salvere, quod possunt, loca
        Diluta nimiis imbribus,
Multum, Pater, salvere deserens jubet
        Infecta prata foenisex.
Sed candidatus quem vides nostrum chorus
        Ipso colore prospera
Videtur augurari et ore optat meo
        Et gratias et gaudia.
Intonsus ergo hic cum suis pastoribus
        Bene vertat oro grex tuus°   10
Et quae tuae novella cura dexterae
        Remittitur provincia.

(*May Lines*)

Ab initio et ante saecula creata sum et usque ad futurum saeculum non desinam

O praedestinata bis
    Quae fuisti
A saeculorum saeculis
    Mater Christi,
Post praevisa merita
    Innocentis,
Iterum post scelera°
    Nostrae gentis,
Quamvis illa purior
    Sit corona   10
Magis haec commendat cor-
    di Dei dona.
Utique deiparam
    Te mirarer,
At non partu tuo tam°
    Delectarer;
Confiterer virginem
    Matrem factam
At non inter omnes sem-
    per te intactam.   20
Sed bifronti gloriae
    Tibi erunt
Haec quae stant et illa quae
    Conciderunt—
Et redempta scelera
    Nostrae gentis
Et praevisa merita
    Innocentis.

Ad Mariam

When a sister, born for each strong month-brother,
  Spring's one daughter, the sweet child May,
Lies in the breast of the young year-mother
  With light on her face like the waves at play,

Man from the lips of him speaketh and saith,
At the touch of her wandering wondering breath
Warm on his brow: lo! where is another
 Fairer than this one to brighten our day?

We have suffered the sons of Winter in sorrow
 And been in their ruinous reigns oppressed,   10
And fain in the springtime surcease would borrow
 From all the pain of the past's unrest;
And May has come, hair-bound in flowers,
With eyes that smile thro' the tears of the hours,
With joy for to-day and hope for to-morrow
 And the promise of Summer within her breast!

And we that joy in this month joy-laden,
 The gladdest thing that our eyes have seen,
Oh thou, proud mother and much proud maiden—
 Maid yet mother as May hath been—   20
To thee we tender the beauties all
Of the month by men called virginal
And, where thou dwellest in deep-groved Aidenn,°
 Salute thee, mother, the maid-month's Queen!

For thou, as she, wert the one fair daughter
 That came when a line of kings did cease,
Princes strong for the sword and slaughter,
 That, warring, wasted the land's increase,
And like the storm-months smote the earth
Till a maid in David's house had birth,   30
That was unto Judah as May, and brought her
 A son for King, whose name was peace.

Wherefore we love thee, wherefore we sing to thee,
 We, all we, thro' the length of our days,
The praise of the lips and the hearts of us bring to thee,
 Thee, oh maiden, most worthy of praise;
For lips and hearts they belong to thee
Who to us are as dew unto grass and tree,
For the fallen rise and the stricken spring to thee,
 Thee, May-hope of our darkened ways!   40

O Deus, ego amo te

O God, I love thee, I love thee—
Not out of hope of heaven for me
Nor fearing not to love and be
    In the everlasting burning.
Thou, thou, my Jesus, after me
    Didst reach thine arms out dying,
For my sake sufferedst nails and lance,°
Mocked and marrèd countenance,
    Sorrows passing number,
    Sweat and care and cumber,   10
Yea and death, and this for me,
    And thou couldst see me sinning:
Then I, why should not I love thee,
Jesu so much in love with me?
Not for heaven's sake; not to be
Out of hell by loving thee;
Not for any gains I see;
But just the way that thou didst me
I do love and I will love thee:
What must I love thee, Lord, for then?—   20
For being my king and God. Amen.

Rosa Mystica

1.

The rose in a mystery, where is it found?
Is it anything true? Does it grow upon ground?—
It was made of earth's mould but it went from men's eyes
And its place is a secret and shut in the skies.
Refrain—
In the gardens of God, in the daylight divine
Find me a place by thee, mother of mine.

2.

But where was it formerly? which is the spot
That was blest in it once, though now it is not?—

It is Galilee's growth: it grew at God's will
And broke into bloom upon Nazareth hill.   10
In the gardens of God, in the daylight divine
I shall look on thy loveliness, mother of mine.

3.

What was its season then? how long ago?
When was the summer that saw the bud blow?—
Two thousands of years are near upon past
Since its birth and its bloom and its breathing its last.
In the gardens of God, in the daylight divine
I shall keep time with thee, mother of mine.

4.

Tell me the name now, tell me its name.
The heart guesses easily: is it the same?—   20
Mary the Virgin, well the heart knows,
She is the mystery, she is that rose.
In the gardens of God, in the daylight divine
I shall come home to thee, mother of mine.

5.

Is Mary the rose then? Mary the tree?
But the blossom, the blossom there, who can it be?—
Who can her rose be? It could be but one:
Christ Jesus our Lord, her God and her son.
In the gardens of God, in the daylight divine
Shew me thy son, mother, mother of mine.   30

6.

What was the colour of that blossom bright?—
White to begin with, immaculate white.
But what a wild flush on the flakes of it stood
When the rose ran in crimsonings down the cross-wood!
In the gardens of God, in the daylight divine
I shall worship His wounds with thee, mother of mine.

7.

How many leaves had it?—Five they were then,
Five like the senses and members of men;

Five is their number by nature, but now
They multiply, multiply who can tell how?°   40
In the gardens of God, in the daylight divine
Make me a leaf in thee, mother of mine.

8.

Does it smell sweet too in that holy place?—
Sweet unto God, and the sweetness is grace:
O Breath of it bathes great heaven above
In grace that is charity, grace that is love.
To thy breast, to thy rest, to thy glory divine
Draw me by charity, mother of mine.

'Quique haec membra'

Quique haec membra malis vis esse obnoxia multis
   Ne nimium esse velis,
Non ego namque mea haec haerentia sorte repugno
   Aut memorem esse piget,
Intersit mediae tantum indulgentia poenae
   Quamque subire jubes
Sit tua crux: tecum, quod sum torquendus, et oro
   Torquear arte tua.
Sed miserere tuis tam multis millibus Indis,
   Iam miserere tuis,
Quamque rogare alium properant peccantque salutem°   10
   Da Deus interea.

On St. Winefred

besides her miraculous cures
filling a bath and turning a mill

As wishing all about us sweet,
She brims her bath in cold or heat;
She lends, in aid of work and will,
Her hand from heaven to turn a mill—
Sweet soul! not scorning honest sweat
And favouring virgin freshness yet.

In S. Winefridam

praeter miraculorum gratiam
operam dantem
et balneis et molae

Temperat aestiva fessis sua balnea membris,
    Hiberna rigidis temperat alma manu;
Quin etiam nostros dextra dignata labores
    Utilis assiduae, nec pudet esse, molaest;
Scilicet alta polo sordes non temnit honestas
    Virgineum quamvis suadeat ipsa decus.

Fragments on St. Winefred

(i)
'Iam si rite'

Iam si rite sequor prisci vestigia facti
    Haec sunt egregie numine plena sacro.°

(ii)
'Quin etiam'

Quin etiam nostros non aspernata labores
    Utilis assiduaest, nec pudet esse, molae,
Scilicet et sordes ut quae patiatur honestas
    Et quae virgineum suadeat ipsa decus.

(iii)
'Atque tribus'

Atque tribus primum quod flumen fontibus exit
    Haec est tergemini credita forma Dei,
Qui quod sincero juncti simul aequore crescunt
    En tibi simplicitas quam colis, alma fides.°
Quid quod ab occulta submissus origine sese
    Inque hominum adspectus fons agit inque diem?

. . . .

Quod puteal cernis distinctum cardine quino
    Qua   inclusas   fronte coronet aquas
Hoc est quod species, quae fiunt nuntia rerum,°
    Quinque subit mentem, qua datur ire, viis.°

10

'Miror surgentem'

Miror surgentem per puram Oriona noctem,
    Candida luna licet
Adstet et exiguis incumbat durior astris
    Nec simul esse sinat.
Verum hic Orion miror quam crescat in altum et
    Quam micet igne suo,
Non suus aetherium quem purpurat impetus, itque
    Molle reditque decus:
Quin versare aliquos septena cacumina ventos
    Turbine posse putas.   10
Miror item suaves adeo spirarier auras
    Egelidumque Notum
Atque hiemem tantum primasque tepere Kalendas
    Quas novus annus agit,
Namque ab eo qui jam pulcerrimus occidit anni
    Dicimus ire dies.
O Jesu qui nos homines caelestis et alta haec
    Contrahis astra manu,
Omnia sunt a te: precor a te currat et annus:
    Is bonus annus erit.   20
Omnia sunt in te: nostrum vivat genus in te,
    Quod tua membra sumus,
Omnes concessas inquam quot carpimus auras
    Suspicimusque polum.
Gratia deest sed enim multis: ut gratia desit,
    Omnibus alma tamen,
Alma etiam natura subest, cui tenditur ista
    Provida cunque manus.

S. Thomae Aquinatis
Rhythmus ad SS. Sacramentum

'Adoro te supplex, latens deitas'

Godhead, I adore thee fast in hiding; thou°
God in these bare shapes, poor shadows, darkling now:
See, Lord, at thy service low lies here a heart
Lost, all lost in wonder at the God thou art.

Seeing, touching, tasting are in thee deceived;
How says trusty hearing? that shall be believed:
What God's Son has told me, take for truth I do;
Truth himself speaks truly or there's nothing true.

On the cross thy godhead made no sign to men;
Here thy very manhood steals from human ken: 10
Both are my confession, both are my belief,
And I pray the prayer of the dying thief.

I am not like Thomas, wounds I cannot see,
But can plainly call thee Lord and God as he:
This faith each day deeper be my holding of,
Daily make me harder hope and dearer love.

O thou our reminder of Christ crucified,
Living Bread the life of us for whom he died,
Lend this life to me then: feed and feast my mind,
There be thou the sweetness man was meant to find.   20

Like what tender tales tell of the Pelican;°
Bathe me, Jesu Lord, in what thy bosom ran—
Blood that but one drop of has the worth to win
All the world forgiveness of its world of sin.

Jesu whom I look at veilèd here below,°
I beseech thee send me what I thirst for so,
Some day to gaze on thee face to face in light
And be blest for ever with thy glory's sight.

Author's Preface

The poems in this book° are written some in Running Rhythm, the common rhythm in English use, some in Sprung Rhythm, and some in a mixture of the two. And those in the common rhythm are some counterpointed, some not.

Common English rhythm, called Running Rhythm° above, is measured by feet of either two or three syllables and (putting aside the imperfect feet at the beginning and end of lines and also some unusual measures in which feet seem to be paired together and double or composite feet to arise) never more nor less.

Every foot has one principal stress or accent, and this or the syllable it falls on may be called the Stress of the foot and the other part, the one or two unaccented syllables, the Slack.° Feet (and the rhythms made out of them) in which the Stress comes first are called Falling Feet and Falling Rhythms, feet and rhythm in which the Slack comes first are called Rising Feet and Rhythms, and if the Stress is between two Slacks there will be Rocking Feet and Rhythms. These distinctions are real and true to nature; but for purposes of scanning it is a great convenience to follow the example of music and take the stress always first, as the accent or the chief accent always comes first in a musical bar. If this is done there will be in common English verse only two possible feet—the so-called accentual Trochee and Dactyl, and correspondingly only two possible uniform rhythms, the so-called Trochaic and Dactylic. But they may be mixed and then what the Greeks called a Logaoedic Rhythm arises.° These are the facts and according to these the scanning of ordinary regularly-written English verse is very simple indeed and to bring in other principles is here unnecessary.

But because verse written strictly in these feet and by these principles will become same and tame the poets have brought in licences and departures from rule to give variety, and especially when the natural rhythm is rising, as in the common ten-syllable or five-foot verse, rhymed or blank. These irregularities are chiefly Reversed Feet and Reversed or Counterpoint Rhythm, which two things are two steps or degrees of licence in the same kind. By a reversed foot I mean the putting the stress where, to judge by the rest of the measure, the slack should be and the slack where the stress, and this is done freely at the beginning of a line and, in the course of a line, after a pause; only scarcely ever in the second foot or place and never in the last, unless when the poet designs some extraordinary effect; for these places are characteristic and sensitive and cannot well be touched. But the reversal of the first foot and of some middle foot after a strong pause is a thing so natural that our poets have generally done it, from Chaucer down, without remark and it commonly passes unnoticed and cannot be said to amount to a formal change of rhythm, but rather is that irregularity which all natural growth and motion shews. If however the reversal is repeated in two feet running, especially so as to include the sensitive second foot, it must be due either to great want of ear or else is a calculated effect, the superinducing or *mounting* of a new rhythm upon the old; and since the new or mounted rhythm is actually heard and at the same time the mind naturally supplies the natural or standard foregoing rhythm, for we do not forget what the rhythm is that by rights we should be hearing, two rhythms are in some manner running at once and we have something answerable to counterpoint in music, which is two or more strains of tune going on together, and this is Counterpoint Rhythm.° Of this kind of verse Milton is the great master° and the choruses of *Samson Agonistes* are written throughout in it—but with the disadvantage that he does not let the reader clearly know what the ground-rhythm is meant to be and so they have struck most readers as merely irregular. And in fact if you counterpoint throughout, since one only of the counter rhythms is actually heard, the other is really destroyed or cannot come to exist and what is written is one rhythm only and probably Sprung Rhythm, of which I now speak.

Sprung Rhythm, as used in this book, is measured by feet of from one to four syllables, regularly, and for particular effects any number of weak or slack syllables may be used. It has one stress, which falls on the only syllable, if there is only one, or, if there are more, then scanning as above, on the first, and so gives rise to four sorts of feet, a monosyllable and the so-called accentual Trochee, Dactyl, and the First Paeon.° And there will be four corresponding natural rhythms; but nominally the feet are mixed and any one may follow any other. And hence Sprung Rhythm differs from Running Rhythm in having or being only one nominal rhythm, a mixed or 'logaoedic' one, instead of three, but on the other hand in having twice the flexibility of foot, so that any two stresses may either follow one another running or be divided by one, two, or three slack syllables. But strict Sprung Rhythm cannot be counterpointed. In Sprung Rhythm, as in logaoedic rhythm generally, the feet are assumed to be equally long or strong and their seeming inequality is made up by pause or stressing.

Remark also that it is natural in Sprung Rhythm for the lines to be *rove over*,° that is for the scanning of each line immediately to take up that of the one before, so that if the first has one or more syllables at its end the other must have so many the less at its beginning; and in fact the scanning runs on without break from the beginning, say, of a stanza to the end and all the stanza is one long strain, though written in lines asunder.

Two licences are natural to Sprung Rhythm. The one is rests, as in music; but of this an example is scarcely to be found in this book, unless in the *Echos*, second line.° The other is *hangers* or *outrides*, that is one, two, or three slack syllables added to a foot and not counting in the nominal scanning. They are so called because they seem to hang below the line or ride forward or backward from it in another dimension than the line itself, according to a principle needless to explain here.° These outriding half feet or hangers are marked by a loop underneath them, and plenty of them will be found.

The other marks are easily understood, namely accents, where the reader might be in doubt which syllable should have the stress; slurs, that is loops *over* syllables, to tie them together into the time of one; little loops at the end of a line to shew that the rhyme goes on to the first letter of the next line; what in music are called pauses ⌒, to shew that the syllable should be dwelt on; and twirls ⌣, to mark reversed or counterpointed rhythm.°

Note on the nature and history of Sprung Rhythm—Sprung Rhythm is the most natural of things. For (1) it is the rhythm of common speech and of written prose, when rhythm is perceived in them. (2) It is the rhythm of all but the most monotonously regular music, so that in the words of choruses and refrains and in songs written closely to music it arises. (3) It is found in nursery rhymes,° weather saws, and so on; because, however these may have been once made in running rhythm, the terminations having dropped off by the change of language, the stresses come together and so the rhythm is sprung. (4) It arises in common verse when reversed or counterpointed, for the same reason.

But nevertheless in spite of all this and though Greek and Latin lyric verse, which is well known, and the old English verse seen in *Pierce Ploughman* are in sprung rhythm, it has in fact ceased to be used since the Elizabethan age, Greene being the last writer who can be said to have recognized it. For perhaps there was not, down to our days, a single, even short, poem in English in which sprung rhythm is employed—not for single effects or in fixed places—but as the governing principle of the scansion. I say this because the contrary has been asserted: if it is otherwise the poem should be cited.

Some of the sonnets in this book° are in five-foot, some in six-foot or Alexandrine lines.

Nos. 1 and 25° are Curtal-Sonnets, that is they are constructed in proportions resembling those of the sonnet proper, namely 6+4 instead of 8+6, with however a half line tailpiece (so that the equation is rather $\frac{12}{2} + \frac{9}{2} = \frac{21}{2} = 10\frac{1}{2}$).

The Wreck of the Deutschland

Dec. 6, 7 1875

to the happy memory of five Franciscan nuns,
exiles by the Falck Laws, drowned between
midnight and morning of December 7.

Part the first

Thou mastering me°
God! giver of breath and bread;
World's strand, sway of the sea;
Lord of living and dead;
Thou hast bound bones and veins in me, fastened me flesh,°
And after it álmost únmade, what with dread,
　　Thy doing: and dost thou touch me afresh?
Over again I feel thy finger and find theé.°

2

I did say yes°
O at lightning and lashed rod;　　　　　　　　　　10
Thou heardst me truer than tongue confess
　　Thy terror, O Christ, O God;
Thou knowest the walls, altar and hour and night:
The swoon of a heart that the sweep and the hurl of thee trod
　　Hard down with a horror of height:
And the midriff astrain with leaning of, laced with fire of stress.

3

The frown of his face
Before me, the hurtle of hell
Behind, where, where was a, where was a place?
　　I whirled out wings that spell°　　　　　　　　20
And fled with a fling of the heart to the heart of the Host.°
My heart, but you were dovewinged, I can tell,
　　Carrier-witted, I am bold to boast,
To flash from the flame to the flame then, tower from the grace to the
　　　　grace.'°

4

I am sóft síft°
In an hourglass—at the wall°
Fast, but mined with a motion, a drift,
And it crowds and it combs to the fall;
I steady as a water in a well, to a poise, to a pane,°
But roped with, always, all the way down from the tall   30
Fells or flanks of the voel, a vein
Of the gospel proffer, a pressure, a principle, Christ's gift.

5

I kiss my hand°
To the stars, lovely-asunder
Starlight, wafting him out of it; and°
Glow, glory in thunder;
Kiss my hand to the dappled-with-damson west:
Since, though he is under the world's splendour and wonder,°
His mystery must be instressed, stressed;
For I greet him the days I meet him, and bless when I understand.   40

6

Not out of his bliss°
Springs the stress felt
Nor first from heaven (and few know this)
Swings the stroke dealt—
Stroke and a stress that stars and storms deliver,
That guilt is hushed by, hearts are flushed by and melt—
But it rides time like riding a river
(And here the faithful waver, the faithless fable and miss.)

7

It dates from day°
Of his going in Galilee;   50
Warm-laid grave of a womb-life grey;
Manger, maiden's knee;
The dense and the driven Passion, and frightful sweat;
Thence the discharge of it, there its swelling to be,
Though felt before, though in high flood yet—
What none would have known of it, only the heart, being hard at bay,°

8

Is out with it! Oh,
We lash with the best or worst°
Word last! How a lush-kept plush-capped sloe°
Will, mouthed to flesh-burst,   60
Gush!—flush the man, the being with it, sour or sweet,
Brim, in a flash, full!—Hither then, last or first,
To hero of Calvary, Christ,'s feet—
Never ask if meaning it, wanting it, warned of it—men go.

9

Be adored among men,°
God, three-numberèd form;
Wring thy rebel, dogged in den,
Man's malice, with wrecking and storm.
Beyond saying sweet, past telling of tongue,
Thou art lightning and love, I found it, a winter and warm;°   70
Father and fondler of heart thou hast wrung;
Hast thy dark descending and most art merciful then.

10

With an anvil-ding
And with fire in him forge thy will
Or rather, rather then, stealing as Spring
Through him, melt him but master him still:
Whether át ónce, as once at a crash Paul,°
Or as Austin, a lingering-out sweet skill,°
Make mercy in all of us, out of us all
Mastery, but be adored, but be adored King.   80

Part the second

11

'Some find me a sword; some
The flange and the rail; flame,°
Fang, or flood' goes Death on drum,
And storms bugle his fame.
But wé dréam we are rooted in earth—Dust!°
Flesh falls within sight of us, we, though our flower the same,
Wave with the meadow, forget that there must
The sour scythe cringe, and the blear share come.°

12

On Saturday sailed from Bremen,
    American-outward-bound,   90
Take settler and seamen, tell men with women,
    Two hundred souls in the round—
O Father, not under thy feathers nor ever as guessing°
The goal was a shoal, of a fourth the doom to be drowned;
    Yet díd the dark side of the bay of thy blessing°
Not vault them, the million of rounds of thy mercy not reeve even them
        in?°

13

Into the snows she sweeps,
    Hurling the haven behind,
The Deutschland, on Sunday; and so the sky keeps,
    For the infinite air is unkind,   100
And the sea flint-flake, black-backed in the regular blow,
Sitting Eastnortheast, in cursed quarter, the wind;
    Wiry and white-fiery and whírlwind-swivellèd snow
Spins to the widow-making unchilding unfathering deeps.

14

She drove in the dark to leeward,
    She struck—not a reef or a rock
But the combs of a smother of sand: night drew her°
    Dead to the Kentish Knock;°
And she beat the bank down with her bows and the ride of her keel;
The breakers rolled on her beam with ruinous shock;   110
    And canvass and compass, the whorl and the wheel°
Idle for ever to waft her or wind her with, these she endured.

15

Hope had grown grey hairs,
    Hope had mourning on,
Trenched with tears, carved with cares,
    Hope was twelve hours gone;
And frightful a nightfall folded rueful a day
Nor rescue, only rocket and lightship, shone,
    And lives at last were washing away:
To the shrouds they took,—they shook in the hurling and horrible airs.

        120

16

One stirred from the rigging to save
The wild woman-kind below,
With a rope's end round the man, handy and brave—
He was pitched to his death at a blow,
For all his dreadnought breast and braids of thew:
They could tell him for hours, dandled the to and fro
Through the cobbled foam-fleece. What could he do
With the burl of the fountains of air, buck and the flood of the wave?°

17

They fought with God's cold—
And they could not and fell to the deck   130
(Crushed them) or water (and drowned them) or rolled
With the sea-romp over the wreck.
Night roared, with the heart-break hearing a heart-broke rabble,
The woman's wailing, the crying of child without check—
Till a lioness arose breasting the babble,
A prophetess towered in the tumult, a virginal tongue told.

18

Ah, touched in your bower of bone
Are you! turned for an exquisite smart,
Have you! make words break from me here all alone,
Do you!—mother of being in me, heart.   140
O unteachably after evil, but uttering truth,
Why, tears! is it? tears; such a melting, a madrigal start!
Never-eldering revel and river of youth,°
What can it be, this glee? the good you have there of your own?

19

Sister, a sister calling
A master, her master and mine!—
And the inboard seas run swirling and hawling;°
The rash smart sloggering brine°
Blinds her; but shé that weather sees óne thing, one;
Has óne fetch ín her: she rears herself to divine°   150
Ears, and the call of the tall nun°
To the men in the tops and the tackle rode over the storm's brawling.

20

She was first of a five and came°
Of a coifèd sisterhood.
(O Deutschland, double a desperate name!
O world wide of its good!°
But Gertrude, lily, and Luther, are two of a town,°
Christ's lily and beast of the waste wood:°
From life's dawn it is drawn down,
Abel is Cain's brother and breasts they have sucked the same.)°   160

21

Loathed for a love men knew in them,
Banned by the land of their birth,
Rhine refused them, Thames would ruin them;
Surf, snow, river and earth
Gnashed: but thou art above, thou Orion of light;°
Thy unchancelling poising palms were weighing the worth,°
Thou martyr-master: in thý sight
Storm flakes were scroll-leaved flowers, lily showers—sweet heaven was
          astrew in them.

22

Five! the finding and sake°
And cipher of suffering Christ.   170
Mark, the mark is of man's make
And the word of it Sacrificed.
But he scores it in scarlet himself on his own bespoken,
Before-time-taken, dearest prizèd and priced—°
Stigma, signal, cinquefoil token
For lettering of the lamb's fleece, ruddying of the rose-flake.°

23

Joy fall to thee, father Francis,°
Drawn to the Life that died;°
With the gnarls of the nails in thee, niche of the lance, his
          Lovescape crucified   180
And seal of his seraph-arrival! and these thy daughters
And five-livèd and leavèd favour and pride,
Are sisterly sealed in wild waters,
To bathe in his fall-gold mercies, to breathe in his all-fire glances.°

24

Away in the loveable west,
    On a pastoral forehead of Wales,°
I was under a roof here, I was at rest,
    And they the prey of the gales;
She to the black-about air, to the breaker, the thickly
Falling flakes, to the throng that catches and quails   190
    Was calling 'O Christ, Christ, come quickly':
The cross to her she calls Christ to her, christens her wild-worst Best.°

25

The majesty! what did she mean?
    Breathe, arch and original Breath.°
Is it lóve in her of the béing as her lóver had béen?°
    Breathe, body of lovely Death.°
They were else-minded then, altogether, the men
Wóke thee with a *We are périshing* in the wéather of Gennésaréth.°
    Or ís it that she cried for the crown then,
The keener to come at the comfort for feeling the combating keen? 200

26

For how to the heart's cheering°
    The down-dugged ground-hugged grey
Hovers off, the jay-blue heavens appearing
    Of pied and peeled May!
Blue-beating and hoary-glow height; or night, still higher,
With belled fire and the moth-soft Milky Way,
    What by your measure is the heaven of desire,
The treasure never eyesight got, nor was ever guessed what for the
        hearing?°

27

Nó, but it was nót these.°
    The jading and jar of the cart,   210
Time's tásking, it is fathers that asking for ease
    Of the sodden-with-its-sorrowing heart,
Not danger, electrical horror; then further it finds
The appealing of the Passion is tenderer in prayer apart:
    Other, I gather, in measure her mind's
Burden, in wind's burly and beat of endragonèd seas.

28

But how shall I . . . make me room there:°
Reach me a . . . Fancy, come faster—
Strike you the sight of it? look at it loom there,
Thing that she . . . There then! the Master,   220
Ipse, the only one, Christ, King, Head:
He was to cure the extremity where he had cast her;
Do, deal, lord it with living and dead;
Let him ride, her pride, in his triumph, despatch and have done with
his doom there.

29

Ah! there was a heart right!
There was single eye!°
Read the unshapeable shock night
And knew the who and the why;
Wording it how but by him that present and past,
Heaven and earth are word of, worded by?—°   230
The Simon Peter of a soul! to the blast°
Tárpéïan-fast, but a blown beacon of light.°

30

Jesu, heart's light,°
Jesu, maid's son,
What was the feast followed the night
Thou hadst glory of this nun?—
Féast of the óne wóman withóut stáin.
For so conceivèd, so to conceive thee is done;°
But here was heart-throe, birth of a brain,°
Word, that heard and kept thee and uttered thee óutríght.   240

31

Well, shé has thée for the pain, for the°
Patience; but pity of the rest of them!
Heart, go and bleed at a bitterer vein for the
Comfortless unconfessed of them—°
No not uncomforted: lovely-felicitous Providence
Fínger of a ténder of, O of a féathery délicacy, the bréast of the°
Maiden could obey so, be a bell to, ring óf it, and
Startle the poor sheep back! is the shipwrack then a harvest, does
tempest carry the grain for thee?

32

I admire thee, master of the tides,
Of the Yore-flood, of the year's fall;°   250
The recurb and the recovery of the gulf's sides,°
The girth of it and the wharf of it and the wall;
Stanching, quenching ocean of a motionable mind;°
Ground of being, and granite of it: pást áll
Grásp Gód, thróned behínd
Death with a sovereignty that heeds but hides, bodes but abides;°

33

With a mercy that outrides°
The all of water, an ark
For the listener; for the lingerer with a love glides°
Lower than death and the dark;   260
A vein for the visiting of the past-prayer, pent in prison,°
The-last-breath penitent spirits—the uttermost mark°
Our passion-plungèd giant risen,
The Christ of the Father compassionate, fetched in the storm of his
    strides.

34

Now burn, new born to the world,
Double-naturèd name,
The heaven-flung, heart-fleshed, maiden-furled
Miracle-in-Mary-of-flame,°
Mid-numberèd he in three of the thunder-throne!
Not a dooms-day dazzle in his coming nor dark as he came;   270
Kind, but royally reclaiming his own;
A released shówer, let flásh to the shíre, not a líghtning of fíre
    hard-húrled.°

35

Dame, at our door
Drówned, and among oúr shóals,
Remember us in the roads, the heaven-haven of the reward:
Our Kíng back, Oh, upon Énglish sóuls!
Let him easter in us, be a dayspring to the dimness of us, be a
    crimson-cresseted east,°

More brightening her, rare-dear Britain, as his reign rolls,°
    Pride, rose, prince, hero of us, high-priest,
Our héarts' charity's héarth's fíre, our thóughts' chivalry's thróng's
      Lórd.   280

The Silver Jubilee

to James First Bishop of Shrewsbury on the
25th Year of his Episcopate July 28 1876

Though no high-hung bells or din
Of braggart bugles cry it in—
    What is sound? Nature's round
Makes the Silver Jubilee.

Five and twenty years have run°
Since sacred fountains to the sun
    Sprang, that but now were shut,
Showering Silver Jubilee.

Feasts, when we shall fall asleep,
Shrewsbury may see others keep;   10
    None but you this her true,
This her Silver Jubilee.

Not today we need lament
Your wealth of life is some way spent:
    Toil has shed round your head
Silver but for Jubilee.

Then for her whose velvet vales
Should have pealed with welcome, Wales,
    Let the chime of a rhyme
Utter Silver Jubilee.   20

Ad Episcopum Salopiensem

Quòd festas luces juvat instaurare Beatis
  Natalesque suis mos cumulare rosis,
His, pater, indiciis et consuetudine laeti
  Hac, colimus lecto te, pia turba, die;
Quique tuam quamcumque alias foret aptus in horam
  Serus in hanc nobis est revolutus honor.
Venit enim quintus vegeto et vigesimus annus
  Ex quo sacra tuumst lamina nacta caput.
Ut reor, is numerus mortalia saecula quadrat:
  Saecla quadras, eadem dimidiare queas.   10
Si Pius ille Petri pertingit et amplius annos
  Est cui longaevi nempe Joannis erunt.
Haud tamen ista animis in tempora vertor aruspex:
  Unum ego qui nunc est auguror esse diem;
Qui felix—at enim est felix patriaeque tibique:
  Tu quod es, hoc ut sis, id putat illa suum.
Te pastore, Deo quod visumst, integer Angli
  Grex in divinum coepimus ire gregem.
Quin etiam alma tuis sic secum agit Anglia lustris:
  'Scilicet ex illo tempore sancta feror.   20
His mihi post tantas, immania saecula, clades,
  His mihi, prisca, viris tu recidiva, fides.
Ergo optatarum salvete exordia rerum,
  Vos in fortunis O elementa meis.
Hinc ego jam numeror; fastas ego candida vestris;
  Quae potui per vos sponsa placere Deo.'

Cywydd

annerch i'r tra pharcedig Dr. Thomas Brown esgob yr Amwythig, wedi
cyrhaedd o hono ei bummed flwyddyn ar hugain, yr hon a elwir y Jubil; a
chwyno y mae'r bardd fôd daiar a dŵr yn tystiolaethu yn fwy i hên grefydd
Gwynedd nag y bydd dỹn, a dywed hefyd mai gobeithia fod hyny i gael ei
gyfnewid o waith yr esgob.

    Y mae'n llewyn yma'n llon
    Â ffrydan llawer ffynon,

Gweddill gwyn gadwyd i ni
Gan Feuno a Gwenfrewi.
Wlaw neu wlith, ni chei wlâd braidd
Tan rôd sydd fal hon iraidd.
Gwan ddwfr a ddwg, nis dwg dŷn,
Dyst ffyddlon am ein dyffryn;
Hên ddaiar ddengys â'i gwêdd
Ran drag'wyddawl o rinwedd;   10
Ni ddiffyg ond naws ddyniol,
Dŷn sydd yn unig yn ôl.
Dâd, o dy law di ela
Tardd a lîf â'r hardd brîf dda;
Tydi a ddygi trwy ffŷdd
Croyw feddygiaeth, maeth crefydd;
A gwela Gwalia'r awr hon
Gwîr saint, glân îr gwyryfon.

Brân Maenefa a'i cant
Ebrill y pedwerydd ar hugain
1876

Moonrise June 19
1876

I awoke in the midsummer not-to-call night, | in the white and the walk
    of the morning:
The móon, dwíndled and thínned to the frínge | of a fíngernail héld to
    the cándle,
Or páring of páradisáïcal frúit, | lóvely in wáning but lústreless,°
Stepped from the stool, drew back from the barrow, | of dark Maenefa
    the mountain;°
A cusp still clasped him, a fluke yet fanged him, | entangled him, not quit
    utterly.°
This was the prized, the desirable sight, | unsought, presented so easily,°
Parted me leaf and leaf, divided me, | eyelid and eyelid of slumber.

The Woodlark

Teevo cheevo cheevio chee:°
O where, what can thát be?
Weedio-weedio: there again!
So tiny a trickle of sóng-strain
And all round not to be found
For brier, bough, furrow, or gréen ground
Before or behind or far or at hand
Either left either right
Anywhere in the súnlight.
Well, after all! Ah but hark—   10
'I am the little wóodlark.

Today the sky is two and two
With white strokes and strains of the blue

Round a ring, around a ring
And while I sail (must listen) I sing

The skylark is my cousin and he
Is known to men more than me

                ... when the cry within
Says Go on then I go on
Till the longing is less and the good gone   20

But down drop, if it says Stop,
To the all-a-leaf of the tréetop
And after that off the bough

I ám so véry, O só very glád
That I dó thínk there is not to be had

The blue wheat-acre is underneath°
And the corn is corded and shoulders its sheaf,°
The ear in milk, lush the sash,
And crush-silk poppies aflash,
The blood-gush blade-gash   30
Flame-rash rudred
Bud shelling or broad-shed

Tatter-tangled and dingle-a-danglèd
Dandy-hung dainty head.

And down . . . the furrow dry
Sunspurge and oxeye°
And lace-leaved lovely
Foam-tuft fumitory°

Through the velvety wind V-winged
To the nest's nook I balance and buoy   40
With a sweet joy of a sweet joy,
Sweet, of a sweet, of a sweet joy
Of a sweet—a sweet—sweet—joy.'

In Theclam Virginem

Longa victa die, cum multo pulvere rerum,
    Deterior virtus ut queat esse queror;
Quod lateat niveae cunctos ita gloria Theclae
    Et post Mariam fama secunda meam.
Ducitur antiquis Pauli praeconis ab annis,
    Ducitur Eoo carmen ab Iconio.
Bellerophontëam monstrabat fabula Tarson
    At nunc excussus non male Paulus equis.
Finitima Iconio Tarsus, Cilicemque sequuntur
    Rite suae Paulum proxima fata Theclae.   10
Sederat in patulis longe pulcerrima tectis
    Forte et in apricas verterat ora vias,
Virgineo insignis cultu, sed sponsa, fereque
    Jam matronalis nactaque Thecla virum.
Mollis in his aetas se temperat arte severa
    Castaque composita membra quiete tenet.

Penmaen Pool:

for the Visitors' Book at the Inn

Who long for rest, who look for pleasure
Away from counter, court, or school
O where live well your lease of leisure
But here at, here at Penmaen Pool?

You'll dare the Alp? you'll dart the skiff?—
Each sport has here its tackle and tool:
Come, plant the staff by Cadair cliff;
Come, swing the sculls on Penmaen Pool.

What's yonder?—Grizzled Dyphwys dim:
The triple-hummocked Giant's Stool,°   10
Hoar messmate, hobs and nobs with him
To halve the bowl of Penmaen Pool.

And all the landscape under survey,
At tranquil turns, by nature's rule,
Rides repeated topsyturvy
In frank, in fairy Penmaen Pool.

And Charles's Wain, the wondrous seven,°
And sheep-flock clouds like worlds of wool,
For all they shine so, high in heaven,
Shew brighter shaken in Penmaen Pool.   20

The Mawddach, how she trips! though throttled
If floodtide teeming thrills her full,
And mazy sands all water-wattled
Waylay her at ebb, past Penmaen Pool.

But what's to see in stormy weather,
When grey showers gather and gusts are cool?—
Why, raindrop-roundels looped together
That lace the face of Penmaen Pool.

Then even in weariest wintry hour
Of New Year's Month or surly Yule   30
Furred snows, charged tuft above tuft, tower
From darksome darksome Penmaen Pool.°

And ever, if bound here hardest home,
You've parlour-pastime left and (who'll°
Not honour it?) ale like goldy foam
That frocks an oar in Penmaen Pool.

Then come who pine for peace or pleasure
Away from counter, court, or school,
Spend here your measure of time and treasure
And taste the treats of Penmaen Pool.   40

Ochenaid Sant Francis Xavier, Apostol yr Indiaid.

Nid, am i Ti fy ngwared i,
    Y'th garaf, Duw, yn lân,
Nac, am mai'r rhai na'th garant Di,
    Y berni am fyth i dân.

Ti, ti a'm hymgofleidiaist oll,
    Fy Jesu, ar y Groes;
Gan wayw, hoelion, enllib mawr,
    Goddefaist ddirfawr loes;

Aneirif ddolur darfu it',
    A phoen, a chwŷs eu dwyn,   10
Hyn erofi pechadur oll,
    Hyd farw er fy mwyn.

Gan hyny, 'r hygar Jesu, pam
    Na'th garwn yn ddilyth?
Nid er cael gennyt nef na phwyth,
    Na rhag fy mhoeni byth;

Ond megis Ti a'm ceraist i,
    A'th garaf, garu'r wyf,
Yn unig am Dy fod yn Dduw,
    A'th fod i mi yn Rhwyf.   20

(Margaret Clitheroe)

I

God's counsel cólumnar-severe°
But chaptered in the chief of bliss°

Had always doomed her down to this—
Pressed to death. He plants the year;
The weighty weeks without hands grow,
Heaved drum on drum; but hands alsó
Must deal with Margaret Clitheroe.

2

The very victim would prepare.
Like water soon to be sucked in
Will crisp itself or settle and spin°   10
So she: one sees that here and there
She mends the way she means to go.
The last thing Margaret's fingers sew
Is a shroud for Margaret Clitheroe.

3

The Christ-ed beauty of her mind
Her mould of features mated well.
She was admired. The spirit of hell
Being to her virtue clinching-blind°
No wonder therefore was not slow
To the bargain of its hate to throw   20
The body of Margaret Clitheroe.

Fawning fawning crocodiles
Days and days came round about
With tears to put her candle out;
They wound their winch of wicked smiles
To take her; while their tongues would go
God lighten your dark heart—but no,°
Christ lived in Margaret Clitheroe.

She caught the crying of those Three,
The Immortals of the eternal ring,   30
The Utterer, Utterèd, Uttering,°
And witness in her place would she.
She not considered whether or no
She pleased the Queen and Council. So
To the death with Margaret Clitheroe!

She was a woman upright, outright;
Her will was bent at God. For that
Word went she should be crushed out flat

Within her womb the child was quick.
Small matter of that then! Let him smother   40
And wreck in ruins of his mother

Great Thecla, the plumed passionflower,°
Next Mary mother of maid and nun,
— — — — — — —
And every saint of bloody hour
And breath immortal thronged that show;
Heaven turned its starlight eyes below
To the murder of Margaret Clitheroe.

She held her hands to, like in prayer;
They had them out and laid them wide
(Just like Jesus crucified);   50
They brought their hundredweights to bear.°
Jews killed Jesus long ago
God's son; these (they did not know)
God's daughter Margaret Clitheroe.

When she felt the kill-weights crush
She told His name times-over three;
I suffer this she said *for Thee.*
After that in perfect hush
For a quarter of an hour or so
She was with the choke of woe.—   60
It is over, Margaret Clitheroe.

'Hope holds to Christ'

Hope holds to Christ the mind's own mirror out°
To take His lovely likeness more and more.
It will not well, so she would bring about
A growing burnish brighter than before

And turns to wash it from her welling eyes
And breathes the blots off all with sighs on sighs.

Her glass is blest but she as good as blind°
Holds till hand aches and wonders what is there;
Her glass drinks light, she darkles down behind,
All of her glorious gainings unaware.°   10
I told you that she turned her mirror dim
Betweenwhiles, but she sees herself not Him.

God's Grandeur

The world is charged with the grandeur of God.°
    It will flame out, like shining from shook foil;°
    It gathers to a greatness, like the ooze of oil°
Crushed. Why do men then now not reck his rod?
Generations have trod, have trod, have trod;
    And all is seared with trade; bleared, smeared with toil;
    And wears man's smudge and shares man's smell: the soil
Is bare now, nor can foot feel, being shod.

And, for all this, nature is never spent;
    There lives the dearest freshness deep down things;   10
And though the last lights off the black West went
    Oh, morning, at the brown brink eastwards, springs—
Because the Holy Ghost over the bent°
    World broods with warm breast and with ah! bright wings.

The Starlight Night

Look at the stars! look, look up at the skies!
    O look at all the fire-folk sitting in the air!°
    The bright boroughs, the circle-citadels there!°
Down in dim woods the diamond delves! the elves'-eyes!°
The grey lawns cold where gold, where quickgold lies!°
    Wind-beat whitebeam! airy abeles set on a flare!°
    Flake-doves sent floating forth at a farmyard scare!—°
Ah well! it is all a purchase, all is a prize.°

Buy then! bid then!—What?—Prayer, patience, alms, vows.
Look, look: a May-mess, like on orchard boughs!°   10
  Look! March-bloom, like on mealed-with-yellow sallows!
These are indeed the barn; withindoors house°
The shocks. This piece-bright paling shuts the spouse°
  Christ home, Christ and his mother and all his hallows.

'The dark-out Lucifer'

The dark-out Lucifer detesting this°
  Self-trellises the touch-tree in live green twines°
  And loops the fruity boughs with beauty-bines

'As kingfishers catch fire'

As kingfishers catch fire, dragonflies draw flame;°
  As tumbled over rim in roundy wells
  Stones ring; like each tucked string tells, each hung bell's°
Bow swung finds tongue to fling out broad its name;
Each mortal thing does one thing and the same:
  Deals out that being indoors each one dwells;
  Selves—goes its self; *myself* it speaks and spells,
Crying *What I do is me: for that I came.*

I say more: the just man justices;°
  Keeps grace: that keeps all his goings graces;°   10
Acts in God's eye what in God's eye he is—°
  Christ. For Christ plays in ten thousand places,
Lovely in limbs, and lovely in eyes not his
  To the Father through the features of men's faces.

Ad Reverendum Patrem Fratrem Thomam Burke O.P. Collegium S. Beunonis invisentem

Ignotum spatiari horto, discumbere mensis,
  Et nova mirabar sacra litare virum.
Simplicibus propior quam nos candore columbis
  Ille erat et qualis veste referret ovem.

Mox ut quaesivi: Monacho quod nomen et ordo
    Qui velit ad nostros unicus esse lares;
Pura caput tonsum cui velat lana cucullo
    Et cadit ad medium cui toga pura pedem,
Nescio quod duplex a tergo, a pectore peplum est,
    Atque terit laevum magna corona latus?   10
Respondent: Haec vox toto clamantis in orbe
    Perque hominum Domino corda parantis iter.
Huic fuit Oceanus submissis utilis undis;
    Audiit occidua hunc, hunc oriente plaga.
Sed monachus non est verum est ex Fratribus unus,
    Quem pater agnoscit stelliger ille suum;°
Doctus Aquinatis reserare oracula Thomae,
    Si tamen est illo nunc quod in ore latet,
Quem tam Gudinus, Godatus, tamque Gonetus,°
    Tam Cajetanus perspicuum esse jubent,   20
Jamdudum innumeri patientem interpretis et quem
    Torqueat in sensus, nec mora, quisque suos.
Praeterea teneris fuit hic tironibus olim°
    Ductor et insuetae candida norma viae.
At non omnis in his, vel, si placet, omnis in his est,
    Sic tamen in magnis ut levis esse queat,
Intermiscet enim cum sacris ludicra curis,
    Nec vox nec facies constet ut una viro.
Haec et plura monent atque addunt nomen, at illud
    Non tulit aut aegre nostra Camoena tulit.   30
Talem ego nunc hominem multum salvere juberem
    Ancipitem sed me scrupulus unus habet;
Num sese velit ille a me laudarier Anglo,°
    Toto qui cives sternit in orbe meos.
Quidquid erit, passim mea dat Guenefrida salutem:°
    Huic det et aversum solvat amore sinum.
Quodque etiam possit plebi prodesse fideli,
    Muneris id nostro debeat ille solo.

Spring

Nothing is so beautiful as Spring—
    When weeds, in wheels, shoot long and lovely and lush;
    Thrush's eggs look little low heavens, and thrush°
Through the echoing timber does so rinse and wring

The ear, it strikes like lightnings to hear him sing;
  The glassy peartree leaves and blooms, they brush°
  The descending blue; that blue is all in a rush
With richness; the racing lambs too have fair their fling.

What is all this juice and all this joy?
  A strain of the earth's sweet being in the beginning   10
In Eden garden.—Have, get before it cloy,

  Before it cloud, Christ, lord, and sour with sinning,
Innocent mind and Mayday in girl and boy,
  Most, O maid's child, thy choice and worthy the winning.

The Sea and the Skylark

On ear and ear two noises too old to end
  Trench—right, the tide that ramps against the shore;°
  With a flood or a fall, low lull-off or all roar,°
Frequenting there while moon shall wear and wend.

Left hand, off land, I hear the lark ascend,
  His rash-fresh re-winded new-skeinèd score°
  In crisps of curl off wild winch whirl, and pour
And pelt music, till none's to spill nor spend.

How these two shame this shallow and frail town!°
  How ring right out our sordid turbid time,   10
Being pure! We, life's pride and cared-for crown,°

  Have lost that cheer and charm of earth's past prime:
Our make and making break, are breaking, down°
  To man's last dust, drain fast towards man's first slime.°

In the Valley of the Elwy

  I remember a house where all were good
    To me, God knows, deserving no such thing:
    Comforting smell breathed at very entering,
  Fetched fresh, as I suppose, off some sweet wood.

That cordial air made those kind people a hood°
  All over, as a bevy of eggs the mothering wing
  Will, or mild nights the new morsels of Spring:
Why, it séemed of coúrse; séemed of ríght it shóuld.

Lovely the woods, waters, meadows, combes, vales,
All the air things wear that build this world of Wales;   10
  Only the inmate does not correspond:

God, lover of souls, swaying considerate scales,
Complete thy creature dear O where it fails,
  Being mighty a master, being a father and fond.

The Windhover:

to Christ our Lord

I caught this morning morning's minion, king-
  dom of daylight's dauphin, dapple-dáwn-drawn Falcon, in his riding
  Of the rólling level úndernéath him steady áir, and stríding
High there, how he rung upon the rein of a wimpling wing°
In his ecstacy! then off, off forth on swing,°
  As a skate's heel sweeps smooth on a bow-bend: the hurl and gliding
  Rebuffed the big wind. My heart in hiding°
Stirred for a bird,—the achiève of, the mastery of the thing!°

Brute beauty and valour and act, oh, air, pride, plume, here°
  Buckle! AND the fire that breaks from thee then, a billion°   10
Times told lovelier, more dangerous, O my chevalier!°

  No wónder of it: shéer plód makes plóugh down síllion°
Shine, and blue-bleak embers, ah my dear,
  Fall, gáll themsélves, and gásh góld-vermílion.°

Pied Beauty

Glory be to God for dappled things—
  For skies of couple-colour as a brinded cow;°

For rose-moles all in stipple upon trout that swim;°
Fresh-firecoal chestnut-falls; finches' wings;°
   Landscape plotted and pieced—fold, fallow, and plough;
   And áll trades, their gear and tackle and trim.

All things counter, original, spáre, strange;°
   Whatever is fickle, frecklèd (who knows how?)°
   With swíft, slów; sweet, sóur; adázzle, dím;
He fathers-forth whose beauty is pást change:°   10
                Práise hím.

The Caged Skylark

As a dare-gale skylark scanted in a dull cage,
   Man's mounting spirit in his bone-house, mean house, dwells—°
   That bird beyond the remembering hís free fells;°
This in drudgery, day-labouring-out life's age.

Though aloft on turf or perch or poor low stage°
   Both sing sometímes the sweetest, sweetest spells,
   Yet both droop deadly sómetimes in their cells
Or wring their barriers in bursts of fear or rage.

Not that the sweet-fowl, song-fowl, needs no rest—°
Why, hear him, hear him babble and drop down to his nest,°   10
   But his own nest, wild nest, no prison.

Man's spirit will be flesh-bound, when found at best,°
But úncúmberèd: meadow-dówn is nót distréssed
   For a ráinbow fóoting it nor hé for his bónes rísen.

'To him who ever thought'

To him who ever thought with love of me
Or ever did for my sake some good deed
I will appear, looking such charity
And kind compassion, at his life's last need
That he will out of hand and heartily
Repent he sinned and all his sins be freed.°

Hurrahing in Harvest

Summer ends now; now, barbarous in beauty, the stooks rise
Around; up above, what wind-walks! what lovely behaviour°
Of silk-sack clouds! has wilder, wilful-wavier°
Meal-drift moulded ever and melted across skies?

I wálk, I líft up, Í lift úp heart, éyes,
Down all that glory in the heavens to glean our Saviour;
And, éyes, heárt, what looks, what lips yet gáve you a°
Rapturous love's greeting of realer, of rounder replies?

And the azurous hung hills are his world-wielding shoulder
Majestic—as a stallion stalwart, very-violet-sweet!—   10
These things, these things were here and but the beholder°
Wánting; whích two whén they ónce méet,
The heart rears wings bold and bolder
And hurls for him, O half hurls earth for him off under his feet.

The Lantern out of Doors

Sometimes a lantern moves along the night.
    That interests our eyes. And who goes there?
    I think; where from and bound, I wonder, where,
With, all down darkness wide, his wading light?

Men go by me, whom either beauty bright
    In mould or mind or what not else makes rare:
    They rain against our much-thick and marsh air
Rich beams, till death or distance buys them quite.°

Death or distance soon consumes them: wind,°
    What most I may eye after, be in at the end   10
I cannot, and out of sight is out of mind.

Christ minds: Christ's interest, what to avow or amend°
    There, éyes them, heart wánts, care háunts, foot fóllows kínd,
    Their ránsom, théir rescue, ánd first, fást, last fríend.°

The Loss of the Eurydice

foundered March 24 1878

The Eurydice—it concerned thee, O Lord:
Three hundred souls, O alas! on board,
    Some asleep unawakened, all un-
Warned, eleven fathoms fallen

Where she foundered! One stroke
Felled and furled them, the hearts of oak!°
    And flockbells off the aerial
Downs' forefalls beat to the burial.°

For did she pride her, freighted fully, on°
Bounden bales or a hoard of bullion?—   10
    Precious passing measure,
Lads and men her lade and treasure.

She had come from a cruise, training seamen—
Men, boldboys soon to be men:
    Must it, worst weather,
Blast bole and bloom together?

No Atlantic squall overwrought her
Or rearing billow of the Biscay water:
    Home was hard at hand
And the blow bore from land.   20

And you were a liar, O blue March day.
Bright sun lanced fire in the heavenly bay;°
    But what black Boreas wrecked her? he°
Came equipped, deadly-electric,

A beetling baldbright cloud thorough England
Riding: there did storms not mingle? and
    Hailropes hustle and grind their
Heavengravel? wolfsnow, worlds of it, wind there?

Now Carisbrook keep goes under in gloom;°
Now it overvaults Appledurcombe;   30
    Now near by Ventnor town
It hurls, hurls off Boniface Down.

Too proud, too proud, what a press she bore!°
Royal, and all her royals wore.°
    Sharp with her, shorten sail!
Too late; lost; gone with the gale.

This was that fell capsize.
As half she had righted and hoped to rise
    Death teeming in by her portholes
Raced down decks, round messes of mortals.   40

Then a lurch forward, frigate and men;
'All hands for themselves' the cry ran then;
    But she who had housed them thither
Was around them, bound them or wound them with her.

Marcus Hare, high her captain,
Kept to her—care-drowned and wrapped in
    Cheer's death, would follow°
His charge through the champ-white water-in-a-wallow,

All under Channel to bury in a beach her
Cheeks: Right, rude of feature,   50
    He thought he heard say
'Her commander! and thou too, and thou this way.'

It is even seen, time's something server,°
In mankind's medley a duty-swerver,
    At downright 'No or Yes?'
Doffs all, drives full for righteousness.

Sydney Fletcher, Bristol-bred,
(Low lie his mates now on watery bed)
    Takes to the seas and snows
As sheer down the ship goes.   60

Now her afterdraught gullies him too down;
Now he wrings for breath with the deathgush brown;
    Till a lifebelt and God's will
Lend him a lift from the sea-swill.

Now he shoots short up to the round air;
Now he gasps, now he gazes everywhere;
    But his eye no cliff, no coast or
Mark makes in the rivelling snowstorm.

Him, after an hour of wintry waves,
A schooner sights, with another, and saves,   70
    And he boards her in Oh! such joy
He has lost count what came next, poor boy.—

They say who saw one sea-corpse cold
He was all of lovely manly mould,
    Every inch a tar,
Of the best we boast our sailors are.

Look, foot to forelock, how all things suit! he
Is strung by duty, is strained to beauty,
    And brown-as-dawning-skinned
With brine and shine and whirling wind.   80

O his nimble finger, his gnarled grip!
Leagues, leagues of seamanship
    Slumber in these forsaken
Bones, this sinew, and will not waken.

He was but one like thousands more.
Day and night I deplore
    My people and born own nation,
Fast foundering own generation.

I might let bygones be—our curse°
Of ruinous shrine no hand or, worse,   90
    Robbery's hand is busy to
Dress, hoar-hallowèd shrines unvisited;

Only the breathing temple and fleet
Life, this wildworth blown so sweet,°
  These daredeaths, ay this crew, in
Unchrist, all rolled in ruin—

Deeply surely I need to deplore it,
Wondering why my master bore it,°
  The riving off that race°
So at home, time was, to his truth and grace   100

That a starlight-wender of ours would say
The marvellous Milk was Walsingham Way°
  And one—but let be, let be:°
More, more than was will yet be.—

O well wept, mother have lost son;
Wept, wife; wept, sweetheart would be one:
  Though grief yield them no good
Yet shed what tears sad truelove should.

But to Christ lord of thunder
Crouch; lay knee by earth low under: 110
  'Holiest, loveliest, bravest,
Save my hero, O Hero savest.°

And the prayer thou hearst me making°
Have, at the awful overtaking,
  Heard; have heard and granted
Grace that day grace was wanted.'

Not that hell knows redeeming,
But for souls sunk in seeming
  Fresh, till doomfire burn all,
Prayer shall fetch pity eternal.   120

The May Magnificat

May is Mary's month, and I
Muse at that and wonder why:
    Her feasts follow reason,
    Dated due to season—

Candlemas, Lady Day;°
But the Lady Month, May,
    Why fasten that upon her,
    With a feasting in her honour?

Is it only its being brighter
Than the most are must delight her?   10
    Is it opportunest
    And flowers finds soonest?

Ask of her, the mighty mother:
Her reply puts this other
    Question: What is Spring?—
    Growth in everything—

Flesh and fleece, fur and feather,
Grass and greenworld all together;
    Star-eyed strawberry-breasted
    Throstle above her nested   20

Cluster of bugle blue eggs thin°
Forms and warms the life within;
    And bird and blossom swell
    In sod or sheath or shell.

All things rising, all things sizing°
Mary sees, sympathising
    With that world of good,
    Nature's motherhood.

Their magnifying of each its kind
With delight calls to mind   30
    How she did in her stored
    Magnify the Lord.

Well but there was more than this:
Spring's universal bliss
    Much, had much to say
    To offering Mary May.

When drop-of-blood-and-foam-dapple°
Bloom lights the orchard-apple
    And thicket and thorp are merry
    With silver-surfèd cherry   40

And azuring-over greybell makes°
Wood banks and brakes wash wet like lakes
    And magic cuckoocall°
    Caps, clears, and clinches all—

This ecstacy all through mothering earth
Tells Mary her mirth till Christ's birth
    To remember and exultation
    In God who was her salvation.

'O where is it, the wilderness'

O where is it, the wilderness,
The wildness of the wilderness?
Where is it, the wilderness?
–   –   –   –   –   –

And wander in the wilderness;
In the weedy wilderness,
Wander in the wilderness.

'Denis'

Denis,
Whose motionable, alert, most vaulting wit
Caps occasion with an intellectual fit.
    Yet Arthur is a Bowman: his three-heeled timber'll hit°
The bald and bold blinking gold when all's done°
Right rooting in the bare butt's wincing navel in the sight of the sun.

'The furl of fresh-leaved dogrose'

The furl of fresh-leaved dogrose down°
His cheeks the forth-and-flaunting sun
Had swarthed about with lion-brown°
    Before the Spring was done.

His locks like all a ravel-rope's-end,
    With hempen strands in spray—°
Fallow, foam-fallow, hanks—fall'n off their ranks,
    Swung down at a disarray.

Or like a juicy and jostling shock°
    Of bluebells sheaved in May   10
Or wind-long fleeces on the flock°
    A day off shearing day.

Then over his turnèd temples—here—
    Was a rose, or, failing that,
Rough-Robin or five-lipped campion clear
    For a beauty-bow to his hat,
And the sunlight sidled, like dewdrops, like dandled diamonds
    Through the sieve of the straw of the plait.

'He mightbe slow'

He mightbe slow and something feckless first,°
Not feck at first, and here no harm,
But earnest, always earnest, *there* the charm

'What being in rank-old nature'

What being in rank-old nature should earlier have that breath been°
That here personal tells off these heart-song powerful peals?—
A bush-browed beetle-browed billow is it?°
With a south-westerly wind blustering, with a tide rolls reels
Of crumbling, fore-foundering, thundering all-surfy seas in; seen
Underneath, their glassy barrel, of a fairy green.

Or a jaunting vaunting vaulting assaulting trumpet telling.

Duns Scotus's Oxford

Towery city and branchy between towers;°
Cuckoo-echoing, bell-swarmèd, lark-charmèd, rook-racked,
   river-rounded;
The dapple-eared lily below thee; that country and town did°
Once encounter in, here coped and poisèd powers;

Thou hast a base and brickish skirt there, sours°
That neighbour-nature thy grey beauty is grounded
Best in; graceless growth, thou hast confounded
Rural rural keeping—folk, flocks, and flowers.°

Yet ah! this air I gather and I release
He lived on; these weeds and waters, these walls are what°   10
He haunted who of all men most sways my spirits to peace;°

Of realty the rarest-veinèd unraveller; a not°
Rivalled insight, be rival Italy or Greece;
Who fíred Fránce for Máry withóut spót.°

Binsey Poplars

felled 1879

My aspens dear, whose airy cages quelled,°
Quelled or quenched in leaves the leaping sun,
    Áll félled, félled, are áll félled;
    Of a fresh and following folded rank
        Not spared, not one
        That dandled a sandalled°
    Shadow that swam or sank
On meadow and river and wind-wandering weed-winding bank.

O if we but knew what we do
      When we delve or hew—   10
    Hack and rack the growing green!
      Since country is so tender
      To tóuch, her béing só slénder,

That, like this sleek and seeing ball
But a prick will make no eye at all,
Where we, even where we mean
                    To mend her we end her,
        When we hew or delve:
    After-comers cannot guess the beauty been.
            Ten or twelve, only ten or twelve   20
            Strokes of havoc unselve
                The sweet especial scene,
            Rural scene, a rural scene,
            Sweet especial rural scene.

Henry Purcell

The poet wishes well to the divine genius of Purcell and praises him that,
whereas other musicians have given utterance to the moods of man's mind, he
has, beyond that, uttered in notes the very make and species of man as created
both in him and in all men generally

Have fáir fállen, O fáir, fáir have fállen, so déar
To me, so arch-especial a spirit as heaves in Henry Purcell,
An age is now since passed, since parted; with the reversal
Of the outward sentence low lays him, listed to a heresy, here.

Not mood in him nor meaning, proud fire or sacred fear,
Or love, or pity, or all that sweet notes not his might nursle:°
It is the forgèd feature finds me; it is the rehearsal°
Of own, of abrupt self there so thrusts on, so throngs the ear.°

Let him oh! with his air of angels then lift me, lay me! only I'll
Have an eye to the sakes of him, quaint moonmarks, to his pelted
    plumage under   10
Wings: so some great stormfowl, whenever he has walked his while

The thunder-purple seabeach, plumèd purple-of-thunder,
If a wuthering of his palmy snow-pinions scatter a colossal smile
Off him, but meaning motion fans fresh our wits with wonder.

'Repeat that, repeat'

Repeat that, repeat,
Cuckoo, bird, and open ear wells, heart-springs, delightfully sweet,
With a ballad, with a ballad, a rebound
Off trundled timber and scoops of the hillside ground, hollow hollow
    hollow ground:
The whole landscape flushes on a sudden at a sound

The Candle Indoors

Some candle clear burns somewhere I come by.
I muse at how its being puts blissful back
With yellowy moisture mild night's blear-all black°
Or to-fro tender trambeams truckle at the eye.°

By that window what task what fingers ply,
I plod wondering, a-wanting, just for lack
Of answer the eagerer a-wanting Jessy or Jack
There/God to aggrándise, God to glorify.—°

Come you indoors, come home; your fading fire°
Mend first and vital candle in close heart's vault;   10
You there are master, do your own desire;

What hinders? Are you beam-blind, yet to a fault°
In a neighbour deft-handed? Are you that liar°
And, cast by conscience out, spendsavour salt?°

The Handsome Heart:

at a gracious answer

'But tell me, child, your choice, | your fancy; what to buy
You?'—'Father, what you buy me | I shall like the best'.
With the sweetest earnest air | his purport, once expressed,
Ever he swung to, push | what plea I might and ply°

Him. Áh, whát the heart is! | Like carriers let fly—°
Doff darkness: homing nature, | nature knows the rest—°
Heart to its own fine function, | wild and self-instressed,°
Falls as light as, life-long, | schooled to what and why.

Heárt mánnerly | is more than handsome face,°
Beauty's bearing or | muse of mounting vein;   10
And whát when, as ín this cáse, | bathed in high hallowing grace?—

Of heaven then now what boon | to buy you, boy, or gain°
Not granted? None but this, | all your road your race°
To match and more than match | its sweet forestalling strain.

'How all is one way wrought'

How all is one way wrought!°
How all things suit and sit!
Then ah! the tune that thought
Trod to that fancied it.

Nor angel insight can
Learn how the heart is hence:
Since all the make of man°
Is law's indifference.

Who built these walls made known°
The music of his mind,   10
Yet here he has but shewn
His ruder-rounded rind.

Not free in this because
His powers seemed free to play:°
He swept what scope he was
To sweep and must obey.

Though down his being's bent
Like air he changed in choice,
That was an instrument
Which overvaulted voice.°   20

Therefóre this masterhood,
This piece of perfect song,°
This fault-not-found-with good,°
Is neither right nor wrong.

No more than red and blue,
No more than Re and Mi,
Or sweet the golden glue
That's built for by the bee.°

What makes the man and what
The man within that makes:°   30
Ask whom he serves or not
Serves and what side he takes.

For good grows wild and wide,
Has shades, is nowhere none;
But right must seek a side°
And choose for chieftain one.

Cheery Beggar

Beyond Magdalen and by the Bridge, on a place called there the Plain,
        In Summer, in a burst of summertime
        Following falls and falls of rain,°
When the air was sweet-and-sour of the flown fineflour of°
Those goldnails and their gaylinks that hang along a lime;°

        The motion of that man's heart is fine
        Whom want could not make pine, pine
That struggling should not sear him, a gift should cheer him
Like that poor pocket of pence, poor pence of mine.

The Bugler's First Communion

A bugler boy from barrack (it is over the hill
There)—boy bugler, born, he tells me, of Irish
        Mother to an English sire (he
Shares their best gifts surely, fall how things will),

This very very day came down to us after a boon he on°
My late being there begged of me, overflowing
      Boon in my bestowing,
Came, I say, this day to it—to a First Communion.

Here he knelt then in regimental red.
Forth Christ from cupboard fetched, how fain I of feet   10
      To his youngster take his treat!
Low-latched in leaf-light housel his too huge godhead.°

There! and your sweetest sendings, ah divine,
By it, heavens, befall him! as a heart Christ's darling, dauntless;
      Tongue true, vaunt- and tauntless;
Breathing bloom of a chastity in mansex fine.

Frowning and forefending angel-warder°
Squander the hell-rook ranks sally to molest him;°
      March, kind comrade, abreast him;
Dress his days to a dexterous and starlight order.   20

How it does my heart good, visiting at that bleak hill,
When limber liquid youth, that to all I teach
      Yields ténder as a púshed peách,
Hies headstrong to its wellbeing of a self-wise self-will!°

Then though Í should tréad túfts of consolation°
Dáys áfter, só I in a sort deserve to
      And do serve God to serve to
Just such slips of soldiery Christ's royal ration.

Nothing else is like it, no, not all so strains°
Us—freshyouth fretted in a bloomfall all portending°   30
      That sweet's sweeter ending;
Realm both Christ is heir to and there reigns.

O now well work that sealing sacred ointment!
O for now charms, arms, what bans off bad
      And locks love ever in a lad!
Let mé though sée no more of him, and not disappointment

Those sweet hopes quell whose least me quickenings lift,°
In scarlet or somewhere of some day seeing
     That brow and bead of being,
An our day's God's own Galahad. Though this child's drift   40

Seems bý a divíne doom chánnelled, nor do I cry
Disaster there; but may he not rankle and roam
     In backwheels, though bound home?—
That left to the Lord of the Eucharist, I here lie by;

Recorded only, I have put my lips on pleas
Would brandle adamantine heaven with ride and jar, did°
     Prayer go disregarded:
Forward-like, but however, and like favourable heaven heard these.°

Andromeda

Now Time's Andromeda on this rock rude,
  With not her either beauty's equal or
  Her injury's, looks off by both horns of shore,
Her flower, her piece of being, doomed dragon food.

Time past she has been attempted and pursued
  By many blows and banes; but now hears roar
  A wilder beast from West than all were, more
Rife in her wrongs, more lawless, and more lewd.

Her Perseus linger and leave her tó her extremes?—
  Pillowy air he treads a time and hangs   10
  His thoughts on her, forsaken that she seems,

All while her patience, morselled into pangs,
  Mounts; then to alight disarming, no one dreams,
  With Gorgon's gear and barebill/thongs and fangs.

Morning, Midday, and Evening Sacrifice

    The dappled die-away°
    Cheek and the wimpled lip,°

The gold-wisp, the airy-grey
Eye, all in fellowship—
This, all this beauty blooming,
This, all this freshness fuming,°
Give God while worth consuming.

Both thought and thew now bolder
And told by Nature: Tower;
Head, heart, hand, heel, and shoulder   10
That beat and breathe in power—
This pride of prime's enjoyment
Take as for tool, not toy meant°
And hold at Christ's employment.

The vault and scope and schooling
And mastery in the mind,
In silk-ash kept from cooling°
And ripest under rind—°
What death half lifts the latch of,
What hell hopes soon the snatch of,   20
Your offering, with despatch, of!°

Peace

When will you ever, Peace, wild wooddove, shy wings shut,
Your round me roaming end, and under be my boughs?°
When, when, Peace, will you, Peace?—I'll not play hypocrite

To own my heart: I yield you do come sometimes; but
That piecemeal peace is poor peace. What pure peace allows
Alarms of wars, the daunting, wars, the death of it?

O surely, reaving Peace, my Lord should leave in lieu°
Some good! And so he does leave Patience exquisite,
That plumes to Peace thereafter. And when Peace here does house°
He comes with work to do, he does not come to coo,   10
                He comes to brood and sit.

At the Wedding March

God with honour hang your head,
Groom, and grace you, bride, your bed
With lissome scions, sweet scions,°
Out of hallowed bodies bred.

Each be other's comfort kind:
Deep, deeper than divined,
Divine charity, dear charity,
Fast you ever, fast bind.

Then let the March tread our ears:
I to him turn with tears°   10
Who to wedlock, his wonder wedlock,
Deals triumph and immortal years.

Felix Randal

Félix Rándal the fárrier, O is he déad then? my dúty all énded,
Who have watched his mould of man, big-boned and hardy-handsome
Pining, pining, till time when reason rambled in it and some
Fatal four disorders, fleshed there, all contended?

Sickness broke him. Impatient, he cursed at first, but mended
Being anointed and all; though a heavenlier heart began some°
Mónths éarlier, since Í had our swéet repriéve and ránsom°
Téndered to him. Áh well, God rést him áll road éver he offénded!°

This séeing the síck endéars them tó us, us tóo it endéars.
My tongue had taught thee comfort, touch had quenched thy tears,   10
Thy tears that touched my heart, child, Felix, poor Felix Randal;

How far from then forethought of, all thy more boisterous years,
When thou at the random grim forge, powerful amidst peers,
Didst fettle for the great grey drayhorse his bright and battering sandal!°

Brothers

How lovely the elder brother's
Life all laced in the other's,
Love-laced!—what once I well
Witnessed; so fortune fell.
When Shrovetide, two years gone,
Our boys' plays brought on
Part was picked for John,
Young John: then fear, then joy
Ran revel in the elder boy.
Now the night come, all
Our company thronged the hall.
Henry, by the wall,
Beckoned me beside him.
I came where called and eyed him
By meanwhiles; making my play
Turn most on tender by-play.
For, wrung all on love's rack,
My lad, and lost in Jack,
Smiled, blushed, and bit his lip,
Or drove, with a diver's dip,
Clutched hands through claspèd knees;
And many a mark like these
Told tales with what heart's stress
He hung on the imp's success.
Now the óther was bráss-bóld:°
He had no work to hold
His heart up at the strain;
Nay, roguish ran the vein.
Two tedious acts were past;°
Jack's call and cue at last;
When Henry, heart-forsook,
Dropped eyes and dared not look.
There! the hall rung;
Dog, he did give tongue!
Oh, Harry,—in his hands he has flung
His tear-tricked cheeks of flame
For fond love and for shame.—

10

20

30

Ah Nature, framed in fault,
There's comfort then, there's salt!
Nature bad, base, and blind,   40
Dearly thou canst be kind;
There deárly thén, deárly,
Dearly thou canst be kind.

Spring and Fall:

to a Young Child

Margaret, are you grieving
Over Goldengrove unleaving?°
Leaves, like the things of man, you
With your fresh thoughts care for, can you?
Ah! as the heart grows older
It will come to such sights colder
By and by, nor spare a sigh
Though worlds of wanwood leafmeal lie;°
And yet you *will* weep and know why.
Now no matter, child, the name: 10
Sorrow's springs are the same.
Nor mouth had, no nor mind, expressed
What heárt heárd of, ghóst guéssed:°
It is the blight man was born for,
It is Margaret you mourn for.

Milton

(translated from Dryden)

Ævo diversi tres et regione poetae
   Hellados, Ausoniae sunt Britonumque decus.
Ardor in hoc animi, majestas praestat in illo,
   Tertius ingenio junxit untrumque suo.
Scilicet inventrix cedens Natura labori
   'Quidquid erant isti' dixerat 'unus eris.'

Inversnaid

This dárksome búrn, hórseback brówn,
His rollrock highroad roaring down,
In coop and in comb the fleece of his foam°
Flutes and low to the lake falls home.°

A wíndpuff-bónnet of fáwn-fróth
Turns and twindles over the broth°
Of a póol so pitchblack, féll-fró wning,
It rounds and rounds Despair to drowning.

Degged with dew, dappled with dew°
Are the groins of the braes that the brook treads through,   10
Wiry heathpacks, flitches of fern,°
And the beadbonny ash that sits over the burn.°

What would the world be, once bereft°
Of wet and of wildness? Let them be left,
O let them be left, wildness and wet;
Long live the weeds and the wilderness yet.

Angelus ad virginem

Gabriel, from heaven's king
Sent to the maiden sweet,
Brought to her blissfúl tidíng°
And fair 'gan her to greet.
'Hail be thou, full of grace aright!
For so God's Son, the heaven's light,
Loves man, that He | a man will be | and take
Flesh óf thee, maiden bright,
Mankind free for to make
Of sin and devil's might.'   10

Gently tó him gave answér
The gentle maiden then:
'And in what wise should I bear
Child, that know not man?'

The angel said: 'O dread thee nought.
'Tis through the Holy Ghost that wrought
Shall be this thing | whereof tidíng | I bring:
Lost mánkind shall be bought°
By thy sweet childbearíng,
And back from sorrow brought.'   20

When the maiden understood
And the angel's words had heard,
Mildly, of her own mild mood,
The angel she answéred:
'Our Lord His handmaiden, I wis,
I am, that here above us is:
As touching me | fulfillèd be | thy saw;
That I, since His will is,
Be, out of nature's law°
A maid with mother's bliss.'   30

The angel went away thereon
And parted from her sight
And straightway she conceived a Son
Through th' Holy Ghost His might.
In her was Christ contained anon,
True God, true man, in flesh and bone;
Born óf her too | when time was due; | who then
Redeemed us for His own,°
And bought us out of pain,
And died for us t'atone.   40

Fillèd full of charity,
Thou matchless maiden-mother,°
Pray for us to him that He
For thý love above other,
Away our sin and guilt should take,
And clean of every stain us make
And heaven's bliss, | when our time is | to die,
Would give us for thy sake;
With grace to serve him by
Till He us to him take. Amen.   50

The Leaden Echo and the Golden Echo

(Maidens' song from *St. Winefred's Well*)

The Leaden Echo—

How to keep—is there ány any, is there none such, nowhere known
    some, bow or brooch or braid or brace, lace, latch or catch or key to
    keep
Back beauty, keep it, beauty, beauty, beauty, . . . from vanishing away?
Ó is there no frowning of these wrinkles, rankèd wrinkles deep,°
Down? no waving off of these most mournful messengers, still messen-
    gers, sad and stealing messengers of grey?—
No there's none, there's none, O no there's none,
Nor can you long be, what you now are, called fair,
Do what you may do, what, do what you may,
And wisdom is early to despair:°
Be beginning; since, no, nothing can be done
To keep at bay   10
Age and age's evils, hoar hair,
Ruck and wrinkle, drooping, dying, death's worst, winding sheets, tombs
    and worms and tumbling to decay;
So be beginning, be beginning to despair.
O there's none; no no no there's none:
Be beginning to despair, to despair,
Despair, despair, despair, despair.

The Golden Echo—   Spare!°
There is one, yes I have one (Hush there!);
Only not within seeing of the sun.
Not within the singeing of the strong sun,
Tall sun's tingeing, or treacherous the tainting of the earth's air,
Somewhere elsewhere there is ah well where! one,
One. Yes I can tell such a key, I do know such a place,
Where whatever's prizèd and passes of us, everything that's fresh and
    fast flying of us, seems to us sweet of us and swiftly away with, done
    away with, undone,
Undone, done with, soon done with, and yet dearly and dangerously
    sweet
Of us, the wimpledwater-dimpled, not-by-morning-matchèd face,°   10
The flower of beauty, fleece of beauty, too too apt to, ah! to fleet,°
Never fleets more, fastened with the tenderest truth

To its own best being and its loveliness of youth: it is an everlastingness
    of, O it is an all youth!
Cóme then, your ways and airs and looks, locks, maidengear, gallantry
    and gaiety and grace,
Winning ways, airs innocent, maidenmanners, sweet looks, loose locks,
    long locks, lovelocks, gaygear, going gallant, girlgrace—
Resign them, sign them, seal them, send them, motion them with breath,
And with sighs soaring, soaring sighs, deliver
Them; beauty-in-the-ghost, deliver it, early now, long before death
Give beauty back, beauty, beauty, beauty, back to God
    beauty's self and beauty's giver.
See; not a hair is, not an eyelash, not the least lash lost; every hair   20
Is, hair of the head, numbéred.°
Nay, what we had lighthanded left in surly the mere mould°
Will have waked and have waxed and have walked with the wind what
    while we slept,
This side, that side hurling a heavy-headed hundredfold
What while we, while we slumbered.
O then, weary then whý should we tread? O why are we so haggard at the
    heart, so care-coiled, care-killed, so fagged, so fashed, so cogged, so
    cumbered,°
When the thing we freely fórfeit is kept with fonder a care,°
Fonder a care kept than we could have kept it, kept
Far with fonder a care (and we, we should have lost it) finer, fonder
A care kept.—Where kept? do but tell us where kept, where.—   30
Yonder.—What high as that! We follow, now we follow.—Yonder, yes
    yonder, yonder,
Yonder.

Ribblesdale

    Earth, sweet Earth, sweet landscape, with leavès throng°
    And louchèd low grass, heaven that dost appeal°
    To with no tongue to plead, no heart to feel;
    That canst but only be, but dost that long—

    Thou canst but be, but that thou well dost; strong
    Thy plea with him who dealt, nay does now deal,°
    Thy lovely dale down thus and thus bids reel
    Thy river, and o'er gives all to rack or wrong.

And what is Earth's eye, tongue, or heart else, where
Else, but in dear and dogged man? Ah, the heir   10
To his own selfbent so bound, so tied to his turn,°

To thriftless reave both our rich round world bare
And none reck of world after, this bids wear°
Earth brows of such care, care and dear concern.

A Trio of Triolets

No. 1—λέγεταί τι καινόν;°

'No news in the *Times* to-day,'
Each man tells his next-door neighbour.
He, to see if what they say,
'No news in the *Times* to-day'
Is correct, must plough his way
Through that: after three hours' labour,
'No news in the *Times* to-day,'
Each man tells his next-door neighbour.

No. 2—Cockle's Antibilious Pills   10

'When you ask for Cockle's Pills,
Beware of spurious imitations.'
Yes, when you ask for every ill's
Cure, when you ask for Cockle's Pills,
Some hollow counterfeit that kills
Would fain mock that which heals the nations.
Oh, when you ask for Cockle's Pills
Beware of heartless imitations.

No. 3—'The Child is Father to the Man'
                    (Wordsworth)

'The child is father to the man.'   20
How can he be? The words are wild.
Suck any sense from that who can:
'The child is father to the man.'

No; what the poet did write ran,
'The man is father to the child.'
'The child is father to the man!'
How *can* he be? The words are wild.

The Blessed Virgin compared to the Air we Breathe

Wild air, world-mothering air,
Nestling me everywhere,
That each eyelash or hair
Girdles; goes home betwixt
The fleeciest, frailest-flixed°
Snowflake; that's fairly mixed
With, riddles, and is rife
In every least thing's life;
This needful, never spent,
And nursing element;   10
My more than meat and drink,
My meal at every wink;
This air, which, by life's law,
My lung must draw and draw
Now but to breathe its praise,
Minds me in many ways
Of her who nót only
Gave God's infinity
Dwindled to infancy
Welcome in womb and breast,   20
Birth, milk, and all the rest
But mothers each new grace°
That does now reach our race—
Mary Immaculate,
Merely a woman, yet
Whose presence, power is
Great as no goddess's
Was deemèd, dreamèd; who
This one work has to do—
Let all God's glory through,   30
God's glory which would go
Through her and from her flow
Off, and no way but so.

I say that we are wound
With mercy round and round
As if with air: the same
Is Mary, more by name.°
She, wild web, wondrous robe,
Mantles the guilty globe,
Since God has let dispense°   40
Her prayers his providence:
Nay, more than almoner,
The sweet alms' self is her
And men are meant to share
Her life as life does air.
    If I have understood,°
She holds high motherhood
Towards all our ghostly good
And plays in grace her part
About man's beating heart,   50
Laying, like air's fine flood,
The deathdance in his blood;
Yet no part but what will°
Be Christ our Saviour still.
Of her flesh he took flesh:
He does take fresh and fresh,
Though much the mystery how,
Not flesh but spirit now
And makes, O marvellous!
New Nazareths in us,°   60
Where she shall yet conceive
Him, morning, noon, and eve;
New Bethlems, and he born
There, evening, noon, and morn—
Bethlem or Nazareth,
Men here may draw like breath
More Christ and baffle death;
Who, born so, comes to be
New self and nobler me
In each one and each one   70
More makes, when all is done,
Both God's and Mary's Son.
    Again, look overhead
How air is azurèd;

O how! Nay do but stand°
Where you can lift your hand
Skywards: rich, rich it laps
Round the four fingergaps.
Yet such a sapphire-shot,
Charged, steepèd sky will not   80
Stain light. Yea, mark you this:
It does no prejudice.
The glass-blue days are those
When every colour glows,
Each shape and shadow shows.
Blue be it: this blue heaven
The seven or seven times seven
Hued sunbeam will transmit
Perfect, not alter it.
Or if there does some soft,   90
On things aloof, aloft,
Bloom breathe, that one breath more
Earth is the fairer for.
Whereas did air not make
This bath of blue and slake
His fire, the sun would shake,
A blear and blinding ball
With blackness bound, and all
The thick stars round him roll
Flashing like flecks of coal,   100
Quartz-fret, or sparks of salt,
In grimy vasty vault.
    So God was god of old:°
A mother came to mould°
Those limbs like ours which are
What must make our daystar
Much dearer to mankind;
Whose glory bare would blind
Or less would win man's mind.
Through her we may see him   110
Made sweeter, not made dim,
And her hand leaves his light
Sifted to suit our sight.
    Be thou then, O thou dear
Mother, my atmosphere;

My happier world, wherein
To wend and meet no sin;
Above me, round me lie
Fronting my froward eye
With sweet and scarless sky;   120
Stir in my ears, speak there
Of God's love, O live air,
Of patience, penance, prayer:
World-mothering air, air wild,
Wound with thee, in thee isled,
Fold home, fast fold thy child.

'The times are nightfall'

The times are nightfall, look, their light grows less;°
The times are winter, watch, a world undone:
They waste, they wither worse; they as they run
Or bring more or more blazon man's distress.°
And I not help. Nor word now of success:°
All is from wreck, here, there, to rescue one—
Work which to see scarce so much as begun
Makes welcome death, does dear forgetfulness.°
Or what is else? There is your world within.°
There rid the dragons, root out there the sin.°   10
Your will is law in that small commonweal.

From *St. Winefred's Well*

Act I. Sc. 1

Enter Teryth from riding, Winefred following.

T.   What is it, Gwen, my girl? | why do you hover and haunt me?
W.   You came by Caerwys, sir? |
T.   I came by Caerwys.
W.   There°
Some messenger there might have | met you from my uncle.
T.   Your uncle met the messenger— | met me; and this the message:
Lord Beuno comes tonight. |
W.   Tonight, sir!

T.   Soon, now: therefore
        Have all things ready in his room. |
W.   There needs but little doing.
T.   Let what there needs be done. | Stay! with him one companion,
        His deacon, Dirvan. Warm | twice over must the welcome be,
        But both will share one cell,— | This was good news,
            Gwenvrewi.°
W.   Áh, yes!
T.   Why, get thee gone then; | tell thy mother I want her.
            Exit Winefred.   10
        No man has such a daughter. | The fathers of the world
        Call no such maiden 'mine'. | The deeper grows her dearness
        And more and móre times laces | round and round my heart,
        The more some monstrous hand | gropes with clammy fingers
            there,
        Támpering with those sweet bines, | draws them out, strains
            them, strains them;
        Meantime some tongue cries 'Whát, Teryth! | what, thou poor
            fond father!
        How when this bloom, this honeysuckle, | that rides the air so
            rich about thee,
        Is all, all sheared away, | thus!' Then I sweat for fear.
        Or else a funeral | and yet 'tis nót a funeral,
        Some pageant which takes tears | and I must foot with feeling
            that   20
        Alive or dead my girl | is carried in it, endlessly
        Goes marching thro' my mind. | What sense is this? It has none.
        This is too much the father; | nay the mother. Fanciful!
        I here forbid my thoughts | to fóol themselves with fears.
            Enter Gwenlo.

                .   .   .   .

Act II.—*Scene, a wood, ending in a steep bank over a dry dean. Winefred
having been murdered within, re-enter Caradoc with a bloody sword.*

C.   My héart, where have we been? | What have we séen, my mind?
        What stroke has Caradoc's right arm dealt? | what done? Head of
            a rebel
        Struck óff it has; written | upon lovely limbs,
        In bloody letters, lessons | of earnest, of revenge;
        Mónuments of my earnest, | récords of my revenge,

On one that went against me ǀ whereas I had warned her—
Wárned her! well she knew ǀ I warned her of this work.
What work? what harm's done? There is ǀ no harm done, none
    yet;
Perháps we struck no blow, ǀ Gwenvrewi lives perhaps;
To mákebelieve my mood was— ǀ móck. O I might think so 10
But here, here is a workman ǀ from his day's task swéats.
Wiped I am sure this was; ǀ it seems, not well; for still,
Still the scarlet swings ǀ and dances on the blade.
So be it. Thou steel, thou butcher,
I cán scour thee, fresh burnish thee, ǀ sheathe thee in thy dark
    lair; these drops
Never, never, never ǀ in their blue banks again.
The wóeful, Cradock, O ǀ the woeful word! Then what,°
What have we seen? Her head, ǀ sheared from her shoulders, fall,
And lapped in shining hair, ǀ róll to the bank's edge; then
Down the beetling banks, ǀ like water in waterfalls,   20
It stooped and flashed and fell ǀ and ran like water away.
Her eyes, oh and her eyes!
In all her beauty, and sunlight ǀ to it is a pit, den, darkness,
Foamfalling is not fresh to it, ǀ rainbow by it not beaming,
In all her body, I say, ǀ no place was like her eyes,
No piece matched those eyes ǀ kept most part much cast down
But, being lífted, ímmortal, ǀ of immórtal brightness.
Several times I saw them, ǀ thrice or four times turning;
Round and round they came ǀ and flashed towards heaven: O
    there,
There they did appeal. ǀ Therefore airy vengeances   30
Are afoot; heaven-vault fast purpling ǀ portends, and what first
    lightning
Any instant falls means me. ǀ And I do not repent;
I do not and I will not ǀ repent, not repent.
The blame bear who aroused me. ǀ What Í have dóne violent
I have like a lion done, ǀ lionlike done,
Honouring an uncontrolled ǀ royal wrathful nature,
Mantling passion in a grandeur, ǀ crimson grandeur.
Now be my pride then perfect, ǀ all one piece. Henceforth
In a wide world of defiance ǀ Caradoc lives alone,
Loyal to his own soul, laying his ǀ own law down, no law nor 40
Lord now curb him for ever. ǀ O daring! O deep insight!
What is virtue? Valour; ǀ only the heart valiant.

And right? Only resolution; | will, his will unwavering
Who, like me, knowing his nature | to the heart home, nature's
    business,
Despatches with no flinching. | But will flesh, O can flesh
Second this fiery strain? | Not always; O no no!
We cannot live this life out; | sometimes we must weary
And in this darksome world | what comfort can I find?
Down this darksome world | comfort where can Í find
When 'ts light I quenched; its rose, | time's one rich rose, my
    hand,   50
By her bloom, fast by | her fresh, her fleecèd bloom,
Hideous dashed down, leaving | earth a winter withering
With no now, nó Gwenvrewi. | I must miss her most
That might have spared her were it | but for passion-sake. Yes,
To hunger and not have, yet | hope on for, to storm and strive and
Be at every assault fresh foiled, | worse flung, deeper dis-
    appointed,
The turmoil and the torment, | it has, I swear, a sweetness
Keeps a kind of joy in it, | a zest, an edge, an ecstasy,
Next after sweet success. | I am not left even this;
I all my being have hacked | in half with her neck: one part, 60
Reason, selfdisposal, | choice of better or worse way,
Is corpse now, cannot change; | my other self, this soul,
Life's quick, this kínd, this kéen self-feeling,
With dreadful distillation | of thoughts sour as blood,
Must all day long taste murder. | What do now then? Do? Nay,
Deed-bound I am; one deed treads all down here | cramps all
    doing. What do? Not yield,
Not hope, not pray; despair; | ay, that: brazen despair out,
Brave all, and take what comes— | as here this rabble is come,
Whose bloods I reck no more of, | no more rank with hers
Than sewers with sacred oils. | Mankind, that mob, comes.
    Come!   70
Enter a crowd, among them Teryth, Gwenlo, Beuno, etc.

.   .   .   .

(C.) *After Winefred's raising from the dead and the breaking out of the
fountain.*

Beuno.   O now while skies are blue, | now while seas are salt,
    While rushy rains shall fall | or brooks shall fleet from fountains,

While sick men shall cast sighs, | of sweet health all despairing,
While blind men's eyes shall thírst after | daylight, draughts of
    daylight,
Or deaf ears shall desire that | lípmusic that's lóst upon them,
While cripples aré, while lepers, | dancers in dismal limb-dance,
Fallers in dreadful frothpits, | waterfearers wild,
Stone, palsy, cancer, cough, | lung-wasting, womb-not-bearing,
Rupture, running sores, | what more? in brief, in burden,
As long as men are mortal | and God merciful,   10
So long to this sweet spot, | this leafy lean-over,
This Dry Dean, now no longer dry | nor dumb, but moist and
    musical
With the uproll and the downcarol | of day and night delivering
Water, which keeps thy name, | (for not in róck written,
But in pale water, fráil water, | wild rash and reeling water,
That will not wear a print, | that will not stain a pen,
Thy venerable record, | virgin, is recorded)
Here to this holy well | shall pilgrimages be,
And not from purple Wáles only | nor from elmy England,
But from beyond seas, Erin, | France and Flanders,
    everywhere,   20
Pilgrims, still pilgrims, móre | pilgrims, still more poor pilgrims

   .   .   .   .

What sights shall be when some | that swung, wretches, on
    crutches
Their crutches shall cast from them, | on heels of air departing,
Or they go rich as roseleaves | hence that loathsome cáme hither!
Not now to náme even
Those dearer, more divine | boons whose haven the heart is.

   .   .   .   .

As sure as what is most sure, | sure as that spring primroses
Shall new-dapple next year, | sure as tomorrow morning,
Amóngst come-back-agáin things, | thíngs with a revival, things
    with a recovery,
Thy name . . .   30

   .   .   .   .

'To seem the stranger'

To seem the stranger lies my lot, my life
Among strangers. Father and mother dear,°
Brothers and sisters are in Christ not near
And he my peace/my parting, sword and strife.°

England, whose honour O all my heart woos, wife°
To my creating thought, would neither hear°
Me, were I pleading, plead nor do I: weár-
Y of idle a being but by where wars are rife.°

I am in Ireland now; now I am at a thírd°
Remove. Not but in all removes I can   10
Kind love both give and get. Only what word°

Wisest my heart breeds dark heaven's baffling ban
Bars or hell's spell thwarts. This to hoard unheard,
Heard unheeded, leaves me a lonely began.°

'I wake and feel'

I wake and feel the fell of dark, not day.°
What hours, O what black hours we have spent°
This night! what sights you, heart, saw; ways you went!
And more must, in yet longer light's delay.

With witness I speak this. But where I say
Hours I mean years, mean life. And my lament
Is cries countless, cries like dead letters sent°
To dearest him that lives alas! away.

I am gall, I am heartburn. God's most deep decree°
Bitter would have me taste: my taste was me;°   10
Bones built in me, flesh filled, blood brimmed the curse.

Selfyeast of spirit a dull dough sours. I see°
The lost are like this, and their scourge to be
As I am mine, their sweating selves; but worse.

'Strike, churl'

Strike, churl; hurl, cheerless wind, then; heltering hail
May's beauty massacre and wispèd wild clouds grow
Out on the giant air; tell Summer No,
Bid joy back, have at the harvest, keep Hope pale.°

'No worst'

No worst, there is none. Pitched past pitch of grief,°
More pangs will, schooled at forepangs, wilder wring.°
Comforter, where, where is your comforting?°
Mary, mother of us, where is your relief?
My cries heave, herds-long; huddle in a main, a chief-
Woe, wórld-sorrow; on an áge-old ánvil wínce and síng—°
Then lull, then leave off. Fury had shrieked 'No ling-
Ering! Let me be fell: force I must be brief.'°
O the mind, mind has mountains; cliffs of fall
Frightful, sheer, no-man-fathomed. Hold them cheap°   10
May who ne'er hung there. Nor does long our small
Durance deal with that steep or deep. Here! creep,°
Wretch, under a comfort serves in a whirlwind: all°
Life death does end and each day dies with sleep.

To what serves Mortal Beauty?

To what serves mortal beauty— | dangerous; does set danc-
Ing blood—the O-seal-that-so | feature, flung prouder form°
Than Purcell tune lets tread to? | See: it does this: keeps warm°
Men's wit to the things that are; | to what good means—where a glance°
Master more may than gaze, | gaze out of countenance.
Those lovely lads once, wet-fresh | windfalls of war's storm,°
How then should Gregory, a father, | have gleanèd else from swarm-
Èd Rome? But God to a nation | dealt that day's dear chance.
To man, that once would worship | block or barren stone,°
Our law says / love what are | love's worthiest, were all known;   10
World's loveliest—men's selves. Self | flashes off frame and face.
What do then? how meet beauty? | Merely meet it; own,°
Home at heart, heaven's sweet gift; | then leave, let that alone.°
Yea, wish that though, wish all, | God's better beauty, grace.°

(*Carrion Comfort*)

Not, I'll not, carrion comfort, Despair, not feast on thee;°
Not untwist—slack they may be—these last strands of man
In me ór, most weary, cry *I can no more*. I can;
Can something, hope, wish day come, not choose not to be.°

But ah, but O thou terrible, why wouldst thou rude on me°
Thy wring-earth right foot rock? lay a lionlimb against me? scan
With darksome devouring eyes my bruisèd bones? and fan,°
O in turns of tempest, me heaped there; me frantic to avoid thee and
  flee?°

Why? That my chaff might fly; my grain lie, sheer and clear.°
Nay in all that toil, that coil, since (seems) I kissed the rod,°   10
Hand rather, my heart lo! lapped strength, stole joy, would laugh, cheer.°

Cheer whóm though? The héro whose héaven-handling flúng me, fóot
  tród°
Me? or mé that fóught him? O whích one? is it eách one? That níght, that
  year
Of now done darkness I wretch lay wrestling with (my God!) my God.

(*The Soldier*)

Yes. Whý do we áll, séeing of a ǀ soldier, bless him? bléss
Our redcoats, our tars? Both ǀ thése being, the greater part,
But frail clay, nay but foul clay. ǀ Hére it is:. the heart,°
Since, proud, it calls the calling ǀ manly, gives a guess
That, hopes that, mákesbelieve, ǀ the men must be no less;
It fancies, feigns, deems, déars ǀ the artist after his art;°
And fain will find as sterling ǀ all as all is smart
And scarlet wéar the spirit of war thére express.
Mark Christ our King. He knows war, served this soldiering through;°
He of all can reave a rope best. ǀ There he bides in bliss   10
Now, and seeing somewhere some man ǀ do all that man can do,
For love he léans forth, needs ǀ his neck must fall on, kiss,
And cry 'O Christ-done deed! ǀ So God-made-flesh does too:°
Were I come o'er again' cries ǀ Christ 'it should be this.'

'Thee, God, I come from'

Thee, God, I come from, to thee go,
All dáy long I like fountain flow
From thy hand out, swayed about
Mote-like in thy mighty glow.

What I know of thee I bless,
As acknowledging thy stress°
On my being and as seeing
Something of thy holiness.

Once I turned from thee and hid,
Bound on what thou hadst forbid;   10
Sow the wind I would; I sinned:
I repent of what I did.

Bad I am, but yet thy child.
Father, be thou reconciled.
Spare thou me, since I see
With thy might that thou art mild.

I have life left with me still
And thy purpose to fulfil;
Yea a debt to pay thee yet:
Help me, sir, and so I will.   20

But thou bidst, and just thou art,
Me shew mercy from my heart
Towards my brother, every other
Man my mate and counterpart.

.   .   .   .   .

Jesus Christ sacrificed
On the cross

Moulded, he, in maiden's womb,
Lived and died and from the tomb
Rose in power and is our
Judge that comes to deal our doom.   30

'Patience, hard thing'

Patience, hard thing! the hard thing but to pray,
But bid for, patience is! Patience who asks°
Wants war, wants wounds; weary his times, his tasks;
To do without, take tosses, and obey.

Rare patience roots in these, and, these away,
No-where. Natural heart's-ivy Patience masks
Our ruins of wrecked past purpose. There she basks
Purple eyes and seas of liquid leaves all day.°

We hear our hearts grate on themselves: it kills°
To bruise them dearer. Yet the rebellious wills°   10
Of us wé do bid God bend to him even so.

And where is he who more and more distills
Delicious kindness?—He is patient. Patience fills
His crisp combs, and that comes those ways we know.°

'My own heart'

My own heart let me more have pity on; let
Me live to my sad self hereafter kind,
Charitable; not live this tormented mind
With this tormented mind tormenting yet.

I cast for comfort I can no more get
By groping round my comfortless than blind°
Eyes in their dark can day or thirst can find
Thirst's all-in-all in all a world of wet.°

Soul, self; come, poor Jackself, I do advise°
You, jaded, lét be; call off thoughts awhile   10
Elsewhere; leave comfort root-room; let joy size°

At God knows when to God knows what; whose smile
'S not wrung, see you; unforseentimes rather—as skies
Betweenpie mountains—lights a lovely mile.°

To his Watch

Mortal my mate, bearing my rock-a-heart,
Warm beat with cold beat company, shall I°
Earlier or you fail at our force and lie°
The ruins of, rifled, once a world of art?°
The telling time our task is; time's some part,°
Not all, but we were framed to fail and die—°
One spell and well that one. There, ah thereby°
Is sweetest comfort's carol or worst woe's smart.°

Field-flown, the departed day no morning brings°
Saying 'This was yours' with her, but new one, worse,   10
And then that last and shortest.

Songs from Shakespeare, in Latin and Greek

(i)

'Come unto these yellow sands'
(*The Tempest*, I. ii)

Ocius O flavas has, ocius O ad arenas
    Manusque manibus jungite,
Post Salve dictum, post oscula, dum neque venti
    Ferum neque obstrepit mare.
Tum pede sic agiles terram pulsabitis et sic
    Pulsabitis terram pede.
Vos, dulces nymphae, spectabitis interea; quin
    Plausu modos signabitis.
Lascivae latrare; ita plaudere. At hoc juvat: ergo
    Nos Hecuba et Hecubae nos canes°   10
Adlatrant. Gallus sed enim occinit, occinat: aequumst°
    Cantare gallos temperi.

(ii)

'Full fathom five'
(*The Tempest*, I. ii)

Occidit, O juvenis, pater et sub syrtibus his est,
    Ossaque concretum paene coralium habet,

Quique fuere oculi vertunt in iaspidas undae:
    In rem Nereidum et Tethyos omnis abit.
Quidquid enim poterat corrumpi corpore in illo
    Malunt aequoream fata subire vicem.
Exsequias, quod tu miraberis, illi Phorcys°
    Delphinis ducunt Oceanusque suis.
Fallor an ipsa vadis haec nenia redditur imis?
    Glauci mortalem flet, mihi crede, chorus.   10

(iii)

'While you here do snoring lie'
(The Tempest, II. i)

Vos dum stertitis ore sic supino
Grandes insidiae parantur estque
Fraus quod optat adesse nacta tempus,
Extremis digitis levis minaxque.
Qui, somnum nisi vultis hunc supremum,
Nostra voce nimis periculoso
Expergiscimini, viri, sopore.

(iv)

'Tell me where is Fancy bred'
(The Merchant of Venice, III. ii)

Rogo vos Amor unde sit, Camenae.
Quis illum genuit? quis educavit?
Qua vel parte oriundus ille nostra
Sit frontis mage pectorisne alumnus
Consultae memorabitis, sorores.
Amorem teneri creant ocelli;
Pascunt qui peperere; mox eumdem
Aversi patiuntur interire.
Nam cunas abiisse ita in feretrum!
Amorem tamen efferamus omnes,   10
Quem salvere jubemus et valere
Sic, O vos pueri atque vos puellae:
Eheu heu, Amor, ilicet, valeto.
Eheu heu, Amor, ilicet, valeto.

(v)
'Tell me where is Fancy bred'

στροφή·
χορευτὴς α΄]   τίς ἔρωτος, τίς ποτ᾽ ἄρ᾽ ἁ πατρὶς ἦν;
τίς δέ νιν τίκτει, τίς ἔθρεψεν, ἀνδρῶν ἢ θεῶν;
πότερ᾽ αὑτὸν καρδίαν ἢ κεφαλᾶς ἐτήτυμον εἴπω°
τὸν καὶ πάλαι ὡς ἐπιστρωφῶντα μᾶλλον
τόπον; οὐ γάρ, οὐκ ἔχω πᾶ τάδε θεὶς δὴ τύχοιμ᾽ ἄν.

ἀντιστροφή·
χορευτὴς β΄]   τὸν ἔρωτ᾽ ἄρ᾽ οὐχ ἑλικοβλεφάροις
ὡς ἐν ὀφθαλμοῖσι τραφέντ᾽ ἀκούεις παῖδα μὲν
συνέφαβον δ᾽ ἱμέρου καὶ χάριτος τέως νεοθαλοῦς
τηλαυγέσιν ἐν προσώπου τοῖς θεάτροις
τέλος ἐκπεσόντα φροῦδον, θανάτῳ φροῦδον ἔρρειν;   10

ἐπῳδός· κορυφαῖος]   φροῦδος ἔρως, φροῦδος ἡμῖν.
ἡμιχόριον α] ἀλλ᾽ αἴλινον αἴλινον εἴπωμεν, ἄνδρες.
ἡμιχόριον β] αἴλινον γὰρ αἴλινον εἴπωμεν.
χορός]   αἰαῖ,
φροῦδος ἔρως τὸ λοιπόν, φροῦδος ἡμῖν ἔρως.

(vi)
'Orpheus with his lute made trees'
(*King Henry VIII*, III. i)

Orpheus fertur et arbores canendo
Et pigros nive concitasse montes.
Si quid luserat ille, vitis uvas
Extemplo referebat, herba flores.
Diceres Zephyros eoque Phoebum
Conspirasse diem in sereniorem
Et ver continuare sempiternum.
Tum venti posuere, tum resedit
Omnis fluctus ab obsequente ponto.
Est hoc imperium artis atque Musae:
Importunior aegriorque nuper
Cura quae fuerat, loquente plectro
Conticescere vel mori necessest.

(vii)

'Orpheus with his lute made trees'

στροφή]   λόγος Ὀρφέως λύραν καὶ δένδρεσιν χοραγεῖν
καὶ νιφοκτύπων ὄρεων κορυφαῖσιν θαμά, δαμείσαις πόθῳ,
κελαδοῦντι δ᾽ εὐθὺς ἀνθῆσαι ῥόδοισίν θ᾽ ἁλίου τε γᾶν καὶ
ψακάδος οὐρανίου βλαστήμασι καλλικάρποις

ἀντιστροφή]   χιόνος κρύος μεσούσας. πόντιον δὲ κῦμα
τῶν τ᾽ ἐριβρόμων ἀίοντ᾽ ἀνέμων πνεύματα γαλάνα πέσεν.
κιθάρᾳ δὲ ταῖς τε Μούσαις ὡς ἔνεστ᾽ εἰπεῖν τὸ παυσίλυπον
ἀδύνατον· κατεκοίμασ᾽ αὐτίκα πάντα λάθα.

(viii)

Unfinished rendering of 'When icicles hang by the wall'
(*Love's Labour's Lost*, V. ii)

Institit acris hiemps: glacies simul imbrices ad imas
    Promissa passim ut horret haec! Camillus
Pastor, primores quotiens miser afflat ore in ungues,
    Ut ore, rore, vix fovet rigentes!
Grandia ligna foco fert Marcipor uvidis struendo
    Vestigiis in atrium secutus
Aut stupet, e tepido quod presserat ubere ipse, mulctris
    Haesisse tam liquore posse nullo.

Robert Bridges: *'In all things beautiful, I cannot see'*

Incomplete Latin Version

Nempe ea formosa est: adeo omne quod aut facit aut fit
    Cynthia continuo fomes amoris adest.
Stat, sedet, incedit: quantumst modo pulchra quod instat,
    Haec modo res! sequitur pulchrior illa tamen.
Nec mora nec modus est: nam quod mihi saepe negavi
    Suavius illam unquam posse placere placet.
Quid? tacet. At taceat. Jam vera fatebor: ut illud,°
    Ut vincit vestros, Musa, tacere choros!
Si quis in ulla volet perpellere verba silentem
    Vexet marmoreos improbus ille deos.
Hunc in Olympiaca post tot fore saecla sereno   10
    Intempestivum non pudet aede Jovi.

· · · · ·

Postremo si qua jam de re disputat, his et
Ipsa velit Virtus dicere et ipsa Fides;
Aurea non alio sunt saecula more locuta;
Astraeam his usam vocibus esse reor.

Spelt from Sibyl's Leaves

Earnest, earthless, equal, attuneable, | vaulty, voluminous, . . .
    stupendous°
Evening strains to be tíme's vást, | womb-of-all, home-of-all, hearse-of-
    all night.°
Her fond yellow hornlight wound to the west, | her wild hollow hoarlight
    hung to the height°
Waste; her earliest stars, earlstars, | stars principal, overbend us,°
Fíre-féaturing héaven. For éarth | her béing has unbóund; her dápple is
    at énd, as-
Tray or aswarm, all throughther, in throngs; | self ín self stéepèd and
    páshed—qúite°
Disremembering, dismembering | all now. Heart, you round me right°
With: Óur évening is óver us; óur night | whélms, whélms, ánd will énd
    us.
Only the beakleaved boughs dragonish | damask the tool-smooth bleak
    light; black,°
Ever so black on it. Óur tale, O óur oracle! | Lét life, wáned, ah lét life
    wínd   10
Off hér once skéined stained véined varíety | upon, áll on twó spools;
    párt, pen, páck°
Now her áll in twó flocks, twó folds—bláck, white; | ríght, wrong; réckon
    but, réck but, mínd
But thése two; wáre of a wórld where bút these | twó tell, éach off the
    óther; of a ráck°
Where, selfwrúng, selfstrung, sheathe- and shelterless, | thoúghts
    agáinst thoughts ín groans grínd.

On the Portrait of Two Beautiful Young People

a Brother and Sister

O I admire and sorrow! The heart's eye grieves
Discovering you, dark tramplers, tyrant years.
A juice rides rich through bluebells, in vine leaves,°
And beauty's dearest veriest vein is tears.

Happy the father, mother of these! Too fast:
Not that, but thus far, all with frailty, blest
In one fair fall; but, for time's aftercast,°
Creatures all heft, hope, hazard, interest.°

And are they thus? The fine, the fingering beams
Their young delightful hour do feature down   10
That fleeted else like day-dissolvèd dreams
Or ringlet-race on burling Barrow brown.°

She leans on him with such contentment fond
As well the sister sits, would well the wife;
His looks, the soul's own letters, see beyond,
Gaze on, and fall directly forth on life.

But ah, bright forelock, cluster that you are
Of favoured make and mind and health and youth,
Where lies your landmark, seamark, or soul's star?
There's none but truth can stead you. Christ is truth.   20

There's none but good can bé good, both for you
And what sways with you, maybe this sweet maid;
None good but God—a warning wavèd to°
One once that was found wanting when Good weighed.

Man lives that list, that leaning in the will°
No wisdom can forecast by gauge or guess,
The selfless self of self, most strange, most still,
Fast furled and all foredrawn to No or Yes.

Your feast of; that most in you earnest eye°
May but call on your banes to more carouse.°   30
Worst will the best. What worm was here, we cry,
To have havoc-pocked so, see, the hung-heavenward boughs?

Enough: corruption was the world's first woe.
What need I strain my heart beyond my ken?
O but I bear my burning witness though
Against the wild and wanton work of men.

Harry Ploughman

Hard as hurdle arms, with a broth of goldish flue°
Breathed round; the rack of ribs; the scooped flank; lank°
Rope-over thigh; knee-nave; and barrelled shank—°
  Head and foot, shouldér and shank—
By a grey eye's heed steered well, one crew, fall to;
Stand at stress. Each limb's barrowy brawn, his thew
That onewhere curded, onewhere sucked or sank—°
  Soared ór sank—,
Though as a beechbole firm, finds his, as at a rollcall, rank°
And features, in flesh, what deed he each must do—   10
  His sinew-service where do.
He leans to it, Harry bends, look. Back, elbow, and liquid waist
In him, all quáil to the wallowing o' the plough. 'S cheek crímsons; curls
Wag or crossbridle, in a wind lifted, windlaced—°
  Wind-lilylocks-laced;°
Churlsgrace too, chíld of Amansstrength, how it hángs or hurls°
Them—broad in bluff hide his frowning feet lashed! raced°
With, along them, cragiron under and cold furls—
  With-a-fountain's shining-shot furls.°

(Ashboughs)

Not of all my eyes see, wándering on the world,
Is anything a milk to the mind so, só sighs déep
Poetry to it, as a tree whose boughs break in the sky.
Say it is áshboughs: whether on a December day and furled

Fast or they in clammyish láshtender combs creep°
Apart wide and new-nestle at heaven most high.
They touch heaven, tabour on it; how their talons sweep°
The smouldering enormous winter welkin! May
Mells blue and snowwhite through them, a fringe and fray°
Of greenery: it is old earth's groping towards the steep   10
                Heaven whom she childs us by.

(Second version from l. 7)

They touch, they tabour on it, hover on it; here, there hurled,
                With talons sweep
The smouldering enormous winter welkin. Eye,
                But more cheer is when May   10
Mells blue with snowwhite through their fringe and fray
Of greenery and old earth gropes for, grasps at steep
                Heaven with it whom she childs things by.

Tom's Garland:

upon the Unemployed

Tom—garlanded with squat and surly steel
Tom; then Tom's fallowbootfellow piles pick°
By him and rips out rockfire homeforth—sturdy Dick;
Tom Heart-at-ease, Tom Navvy: he is all for his meal
Sure, 's bed now. Low be it: lustily he his low lot (feel
That ne'er need hunger, Tom; Tom seldom sick,
Seldomer heartsóre; that treads through, prickproof, thick
Thousands of thorns, thoughts) swings though. Commonweal
Little Í reck ho! lacklevel in, if all had bread:
What! Country is honour enough in all us—lordly head,   10
With heaven's lights high hung round, or, mother-ground
That mammocks, mighty foot. But nó way sped,
Nor mind nor mainstrength; gold go garlanded
With, perilous, O nó; nor yet plod safe shod sound;
                Undenizened, beyond bound
Of earth's glory, earth's ease, all; no-one, nowhere,
In wide the world's weal; rare gold, bold steel, bare
                In both; care, but share care—
This, by Despair, bred Hangdog dull; by Rage,
Manwolf, worse; and their packs infest the age.   20

Epithalamion

Hark, hearer, hear what I do; lend a thought now, make believe
We are leaf-whelmed somewhere with the hood
Of some branchy bunchy bushybowered wood,
Southern dean or Lancashire clough or Devon cleave,
That leans along the loins of hills, where a candycoloured, where a
    gluegold-brown
Marbled river, boisterously beautiful, between
Roots and rocks is danced and dandled, all in froth and waterblowballs,
    down.
We are there, when we hear a shout
That the hanging honeysuck, the dogeared hazels in the cover
Makes dither, makes hover   10
And the riot of a rout
Of, it must be, boys from the town
Bathing: it is summer's sovereign good.
By there comes a listless stranger: beckoned by the noise
He drops towards the river: unseen
Sees the bevy of them, how the boys
With dare and with downdolfinry and bellbright bodies huddling out,
Are earthworld, airworld, waterworld thorough hurled, all by turn and
    turn about.
This garland of their gambol flashes in his breast
Into such a sudden zest   20
Of summertime joys
That he hies to a pool neighbouring; sees it is the best
There; sweetest, freshest, shadowiest;
Fairyland; silk-beech, scrolled ash, packed sycamore, wild wychelm,
    hornbeam fretty overstood
By. Rafts and rafts of flake leaves light, dealt so, painted on the air,
Hang as still as hawk or hawkmoth, as the stars or as the angels there,
Like the thing that never knew the earth, never off roots
Rose. Here he feasts: lovely all is! Nó more: off with—down he dings
His bleachèd both and woolwoven wear:
Careless these in coloured wisp   30
All lie tumbled-to; then with loop-locks
Forward falling, forehead frowning, lips crisp
Over fingerteasing task, his twiny boots
Fast he opens, last he off wrings

Till walk the world he can with bare his feet
And come where lies a coffer, burly all of blocks
Built of chancequarrièd, selfquainèd, hoar-huskèd rocks°
And the water warbles over into, filleted | with glassy grassy quicksilvery
    shivès and shoots°
And with heavenfallen freshness down from moorland still brims,
Dark or daylight on and on. Here he will then, here he will the fleet 40
Flinty kindcold element let break across his limbs
Long. Where we leave him, froliclavish, while he looks about him,
    laughs, swims.

Enough now; since the sacred matter that I mean
I should be wronging longer leaving it to float
Upon this only gambolling and echoing-of-earth note
What is   the delightful dean?
Wedlock. What the water? Spousal love

                        to Everard, as I surmise,
Sparkled first in Amy's eyes
                turns
Father, mother, brothers, sisters, friends
Into fairy trees, wildflowers, woodferns
Rankèd round the bower

'The sea took pity'

    The sea took pity: it interposed with doom:
    'I have tall daughters dear that heed my hand:
    Let Winter wed one, sow them in her womb,
    And she shall child them on the New-world strand.'

That Nature is a Heraclitean Fire and of the comfort of the Resurrection

Cloud-puffball, torn tufts, tossed pillows | flaunt forth, then chevy on an
    air—°
Built thoroughfare: heaven-roysterers, in gay-gangs | they throng; they
    glitter in marches.

Down roughcast, down dazzling whitewash, | wherever an elm arches,°
Shivelights and shadowtackle in long | lashes lace, lance, and pair.°
Delightfully the bright wind boisterous | ropes, wrestles, beats earth
      bare°
Of yestertempest's creases; in pool and rutpeel parches
Squandering ooze to squeezed | dough, crust, dust; stanches, starches°
Squadroned masks and manmarks | treadmire toil there
Footfretted in it. Million-fuelèd, | nature's bonfire burns on.°
But quench her bonniest, dearest | to her, her clearest-selvèd
      spark°   10
Man, how fast his firedint, | his mark on mind, is gone!
Both are in an únfathomable, all is in an enormous dark
Drowned. O pity and indig | nation! Manshape, that shone°
Sheer off, disseveral, a star, | death blots black out; nor mark°
        Is any of him at all so stark
But vastness blurs and time | beats level. Enough! the Resurrection,
A heart's-clarion! Away grief's gasping, | joyless days, dejection.
        Across my foundering deck shone
A beacon, an eternal beam. | Flesh fade, and mortal trash°
Fall to the residuary worm; | world's wildfire, leave but ash: 20
        In a flash, at a trumpet crash,°
I am all at once what Christ is, | since he was what I am, and
This Jack, joke, poor potsherd, | patch, matchwood, immortal diamond,
        Is immortal diamond.

'What shall I do for the land that bred me'

What shall I do for the land that bred me,
Her homes and fields that folded and fed me?—
Be under her banner and live for her honour:
Under her banner I'll live for her honour.
    CHORUS. Under her banner [we] live for her honour.

Not the pleasure, the pay, the plunder,
But country and flag, the flag I am under—
There is the shilling that finds me willing
To follow a banner and fight for honour.   10
    CH. We follow her banner, we fight for her honour.

Call me England's fame's fond lover,
Her fame to keep, her fame to recover.
Spend me or end me what God shall send me,
But under her banner I live for her honour.
 CH. Under her banner we march for her honour.

Where is the field I must play the man on?
O welcome there their steel or cannon.
Immortal beauty is death with duty,
If under her banner I fall for her honour.
 CH. Under her banner we fall for her honour.   20

In honour of
St. Alphonsus Rodriguez

Laybrother of the Society of Jesus upon the
first falling of his feast after his canonisation
For the College of Palma in the Island
of Majorca, where the saint lived for 40 years as
Hall porter

Glory is a flame off exploit, so we say,°
And those fell strokes that once scarred flesh, scored shield,
Should tongue that time now, trumpet now that field,
Record, and on the fighter forge the day.
On Christ they do, they on the martyr may;
But where war is within, what sword we wield
Not seen, the heroic breast not outward-steeled,
Earth hears no hurtle then from fiercest fray.
Yet, he that hews out mountain, continent,
Earth, all, at last; who, with fine increment°   10
Trickling, veins violets and tall trees makes more
Could crowd career with conquest while there went
Those years and years by of world without event
That in Majorca Alfonso watched the door.

Justus quidem tu es, Domine,

si disputem tecum; verumtamen justa loquar ad te: quare via impiorum
prosperatur? etc (Jerem. xii 1.)

Thou art indeed just, Lord, if I contend
With thee; but, sir, so what I plead is just.
Why do sinners' ways prosper? and why must
Disappointment all I endeavour end?

Wert thou my enemy, O thou my friend,
How wouldst thou worse, I wonder, than thou dost
Defeat, thwart me? Oh, the sots and thralls of lust
Do in spare hours more thrive than I that spend,

Sir, life upon thy cause. See, banks and brakes°
Now, leavèd how thick! lacèd they are again   10
With fretty chervil, look, and fresh wind shakes°

Them; birds build—but not I build; no, but strain,°
Time's eunuch, and not breed one work that wakes.
Mine, O thou lord of life, send my roots rain.

'The shepherd's brow'

The shepherd's brow, fronting forked lightning, owns°
The horror and the havoc and the glory
Of it. Angels fall, they are towers, from heaven—a story°
Of just, majestical, and giant groans.
But man—we, scaffold of score brittle bones;
Who breathe, from groundlong babyhood to hoary
Age gasp; whose breath is our *memento mori*—
What bass is *our* viol for tragic tones?°
He! Hand to mouth he lives, and voids with shame;
And, blazoned in however bold the name,   10
Man Jack the man is, just; his mate a hussy.°
And I that die these deaths, that feed this flame,
That . . . in smooth spoons spy life's masque mirrored: tame°
My tempests there, my fire and fever fussy.

To R. B.

The fine delight that fathers thought; the strong
Spur, live and lancing like the blowpipe flame,°
Breathes once and, quenchèd faster than it came,
Leaves yet the mind a mother of immortal song.

Nine months she then, nay years, nine years she long°
Within her wears, bears, cares and combs the same:°
The widow of an insight lost she lives, with aim
Now known and hand at work now never wrong.

Sweet fire the sire of muse, my soul needs this;
I want the one rapture of an inspiration.   10
O then if in my lagging lines you miss

The roll, the rise, the carol, the creation,°
My winter world, that scarcely breathes that bliss°
Now, yields you, with some sighs, our explanation.

EARLY DIARIES

[Oxford, 1863]

Flick, fillip, flip, fleck, flake.°

Flick means to touch or strike lightly as with the end of a whip, a finger etc. To *fleck* is the next tone above flick, still meaning to touch or strike lightly (and leave a mark of the touch or stroke) but in a broader less slight manner. Hence substantively a *fleck* is a piece of light, colour, substance etc. looking as though shaped or produced by such touches. *Flake* is a broad and decided *fleck*, a thin plate of something, the tone above it. Their connection is more clearly seen in the applications of the words to natural objects than in explanations. It would seem that *fillip* generally pronounced *flip* is a variation of *flick*, which however seems connected with *fly*, *flee*, *flit*, meaning to make fly off. Key to meaning of *flick*, *fleck* and *flake* is that of striking or cutting off the surface of a thing; in *flick* (as to flick off a fly) something little or light from the surface, while *flake* is a thin scale of surface. *Flay* is therefore connected, perhaps *flitch*.

[1863–4]

Whitby Abbey.° I have not seen any parallel to this kind of tracery in French or Italian Gothic. The style did not last long I think and seems to me to have been more capable of grand development than any other. The bars split at the ends, which connect the bights or recesses of the four-sided openings with other parts of the tracery are at a distance, and in effect, straight and yet harmonize completely. This is the only successful manner of introducing them in Decorated windows that I know, for those in early geometrical are poor and the instance in Merton choir erected in finest style and in company with other windows of exquisite tracery is quite unworthy of the others and a failure. The above window I have restored as far as possible from a photograph by Uncle George. There was probably no circle or other opening within the four-sided ones. The mouldings I have not given. The whole rough.

[August–September 1864]

The poetical language lowest.° To use that, which poetasters, and indeed almost everyone, can do, is no more necessarily to be uttering poetry than striking the keys of piano is playing a tune. Only, when the tune is played it is on the keys. So when poetry is uttered it is in this language. Next, Parnassian. Can only be used by real poets. Can be written without inspiration Good instance in Enoch Arden's island.° Common in professedly descriptive pieces. Much of it in *Paradise Lost* and *Regained*. Nearly all *The Faery Queen*. It is the effect of fine age to enable ordinary people to write something very near it.—Third and highest, poetry proper, language of inspiration. Explain inspiration. On first reading a strange poet his merest Parnassian seems inspired. This is because then first we perceive genius. But when we have read more of him and are accustomed to the genius we shall see distinctly the inspirations and much that would have struck us with great pleasure at first loses much of its charm and becomes Parnassian.—Castalian, highest sort of Parnassian. e.g. 'Yet despair Touches me not, though pensive as a bird Whose vernal coverts winter hath laid bare'. Or 'On roses for the flush of youth etc.'. Real Parnassian only written by poets and is as impossible for others as poetry, as practically it is as hard to reach the moon as the stars, but something very like it may be. Much Parnassian takes down a poet's reputation, lowers his average, as it were. Pope and all artificial schools great writers of Parnassian. This is the real meaning of an artificial poet.—The poetical language may be called language of the sacred *Plain*, Delphic. There is seemingly much Parnassian music. Same thing no doubt exists in painting.

Prose scenes from 'Floris in Italy'°

Giulia writing. Fool jumps up and seats himself in window.

F. Madam.

G. You startled me.

F. Madam, what are you doing?

G. Fool, writing a letter.

F. I thought it was your will. I approve your care; but indeed it is better to have a lawyer at once. For my part I never send a loveletter without an attorney. I would not bid anyone to dinner without taking legal opinion.

G. This is not a loveletter nor an invitation. It is to my cousin. I can make nothing of it. Dictate to me now, Fool.

F. Truth or untruth?

G. Truth.

F. And will you set down whatever I read you?

G. Why, truth, they say, is not expedient to speak at all times.

F. Do you defend lying, Madam?

G. You know what I mean. It is better to conceal at times.

F. There are some ladies who conceal all things at all times. Crystal sincerity hath found no shelter but in a fool's cap; I have long found it so.

It loves the innocent tinkle of the bells, and only speaks by the mouths of men of my profession. But to the letter. Whether when it is set down you will send it or no, you shall decide. If you do not send it, I shall despair of your judgment. But it shall be as you will. Now will you promise to set down what I read you?

G. If it be truth.

F. You must forfeit a gold piece, if you refuse.

G. Very well, I will forfeit a gold piece.

F. Lay it down on the table.

G. Can you not trust me?

F. No, Madam, not a woman; least of all in matters of money.

G. Then you shall not have it at all.

F. I said so. Madam, you stand convicted. You must ever pack with your sex.

G. Then there it is. (*Laying it down.*)

F. A hostage. Now, truth, you say?

G. Why, would you have me write lies?

F. Madam, if you follow me, I will take care it be nothing but truth. If at any place you refuse to write you forfeit. Is it agreed?

G. As long as you keep to truth, it is.

F. Thus then. *Cousin,—*

G. Why what a boorish opening is that!

September 9, 1864

... Do you suppose I assail my cousin with such martial peremptory salutations? I say *dearest Cousin* or *dear Cousin*.

F. But she is neither *dearest Cousin* nor *dear Cousin* now. And you have forfeited your gold piece.

G. No, I have put it down. Go on.

F. Cousin, Neither wish to deceive me, for you shall never put out my eyes; nor

G. Why,—

F. Madam, beware for your forfeit. *Neither wish to deceive me, for you shall never put out my eyes; nor think that I can be silent on what I see. You are doing that thing a woman can never forgive, and which, in your way of doing it, is a very shame to a woman to do.*

G. What is all this?

F. Madam! *That I love desperately you know well: that you love at all I much doubt: that I am not loved is my misery: that you are loved is the fear that graces my Lent of lovelessness with the diet of gall and the mortifying of tears.* You are not writing ...

Scene. A cave in a quarry. Evening. Gabriel comes to ask the advice of the hermit, who has however died. He is half-mad. He runs out and finds some night-shade berries which he eats. These make him delirious. A shepherd and his wife take refuge in the cave from the violence of the rain; she crouching in the corner, he standing at the door. Re-enter Gabriel.

G. Can you remember why he set me this penance? What has happened with me? Have I wronged any man's wife? I can call none to mind.—Who are you?

S. What do you want with me?

G. Are you married?

S. Who are you that ask me these questions?

G. What, do you think I am the only man that has been shamed in his bed? Get into the wet. There is nightshade about. Out, out, cuckoo. Out of the nest. (Thrusting him out.)

S. Keep back. (Strikes him with his heavy stick. Gabriel falls with a cry.)

G. O Maurice, you have hurt me. You have struck me, Maurice. I have not wronged your wife, nor any mans wife. You are handsome and strong and my friend: there is not such another in the court, but you strike too hard.

W. Nay, John, you have hurt him. He bleeds. Now see here, John; 'tis a thousand pities if you have hurt him. There's a face to be sure.

G. Gabrielle! I know you. But you are under a cloud. Ay, they say so: 'tis the talk of the whole court. Yes, I know your husband; good but weak. They say he still loves her very, very, very much. Oh the misery. It is a weakness. The last time I saw him he lay in a quarry bleeding. I am cold: cover me up.

W. It is wicked to laugh, but he does talk wild. Dear, dear, poor soul. There put your hands down.

G. See, it rains blood. The moon shall be turned into blood.—Why if all the jealous husbands run their horns at us as you did, shepherd, there'll be no gallantry left in these latter days.

S. Best leave him. We can do nought for him. He is clean mad.

W. Now John, how can be so hard-hearted. Come, I'll not stir; so you may do as you like.

G. No, never leave her. And yet I have been bitterly, horribly, horribly wronged.—Well the tale runs thus. The husband went away, his friend committed adultery with his wife, the husband comes back, does nothing, but goes as near madness as the scalp is to the scull,* and the devil has a good find of souls.—Well 'tis the story of Launcelot and

Guinevere again. Some call her Guinevera, some Guinevere, but the story is the same.

January 23, 1866

For Lent.° No pudding on Sundays. No tea except if to keep me awake and then without sugar. Meat only once a day. No verses in Passion Week or on Fridays. No lunch or meat on Fridays. Not to sit in armchair except can work in no other way. Ash Wednesday and Good Friday bread and water.

* *Thus in MS.*

JOURNAL

[1866]

July 17. Dull, curds-and-whey clouds faintly at times.—It was this night I believe but possibly the next that I saw clearly the impossibility of staying in the Church of England, but resolved to say nothing to anyone till three months are over, that is the end of the Long, and then of course to take no step till after my Degree.°

[1867]

Aug. 22. Bright.—Walked to Finchley and turned down a lane to a field where I sketched an appletree. Their sprays against the sky are gracefully curved and the leaves looping over edge them, as it looks, with rows of scales.° In something the same way I saw some tall young slender wych-elms of thin growth the leaves of which enclosed the light in successive eyebrows. From the spot where I sketched—under an oak, beyond a brook, and reached by the above green lane between a park-ground and a pretty field—there was a charming view, the field, lying then on the right of the lane, being a close-shaven smoothly-rounded shield of bright green ended near the high road by a row of viol-headed or flask-shaped elms—not rounded merely but squared—of much beauty—dense leafing, rich dark colour, ribs and spandrils of timber garlanded with leaf between tree and tree. But what most struck me was a pair of ashes in going up the lane again. The further one was the finer—a globeish just-sided head with one launching-out member on the right; the nearer one was more naked and horny. By taking a few steps one could pass the further behind the nearer or make the stems close, either coincidingly, so far as disagreeing outlines will coincide, or allowing a slit on either side, or again on either side making a broader stem than either would make alone. It was this which was so beautiful—making a noble shaft and base to the double tree, which was crested by the horns of the nearer ash and shaped on the right by the bosom of the hinder one with its springing bough. The outline of the double stem was beautiful to whichever of the two sides you slid the hinder tree—in one (not, I think, in both) shaft-like and narrowing at the ground. Besides I saw how great the richness and subtlety is of the curves in the clusters, both in the forward bow mentioned before and in some most graceful hangers on the other side: it combines somewhat-slanted outward strokes with rounding, but I cannot very well characterise it now.—Elm-leaves:—

they shine much in the sun—bright green when near from underneath but higher up they look olive: their shapelessness in the flat is from their being made, διὰ τὸ πεφυκέναι,° to be dimpled and dog's-eared: their leaf-growth is in this point more rudimentary than that of oak, ash, beech, etc that the leaves lie in long rows and do not subdivide or have central knots but tooth or cog their woody twigs.

For July 6, 1866° I have a note on elm-leaves, that they sit crisp, dark, glossy, and saddle-shaped along their twigs, on which at that time an inner frill of soft juicy young leaves had just been run; they chip the sky, and where their waved edge turns downwards they gleam and blaze like an underlip sometimes will when seen against the light.

Aug. 23. Fine and cloudless; fiery sunset.—Some wych-elms seem to have leaves smaller, others bigger, than the common elm: see Sept. 1.

Papa, Mamma, and Milicent went off to Brittany. I went down to call on Mrs. Cunliffe, who was out, and walked a little in Hyde Park, where I noticed a fine oblate chestnut-tree with noble long ramping boughs more like an oak. Then to the chapel of the poor Clares, where I made my resolution 'if it is better',° but now, Sept. 4, nothing is decided. For the evening to Aunt Kate's. See *infra* May 2 and 11.

Aug. 30. Fair; in afternoon fine; the clouds had a good deal of crisping and mottling.—A round by Plumley.—Stands of ash in a copse: they consisted of two or three rods most gracefully leaved, for each wing or comb finally curled inwards, that is upwards.—Putting my hand up against the sky whilst we lay on the grass I saw more richness and beauty in the blue than I had known of before, not brilliance but glow and colour. It was not transparent and sapphire-like but turquoise-like, swarming and blushing round the edge of the hand and in the pieces clipped in by the fingers,° the flesh being sometimes sunlit, sometimes glassy with reflected light, sometimes lightly shadowed in that violet one makes with cobalt and Indian red.

[The Oratory, near Birmingham]

Sept. 13. Fine. There was an eclipse at night of the moon, and some of the Fathers told me that from the golder colour she had had at first she became, at the eclipse and while it was going on, intensely silver, while the stars did not brighten but became yellowish green.

[1868]

[Retreat at Roehampton]

May 2. Fine, with some haze, and warm.

This day, I think, I resolved.° See *supra* last 23rd of August and *infra* May 11.

May 5. Cold.

Resolved to be a religious.°

May 6. Fine but rather thick and with a very cold N.E. wind.

May 7. Warm; misty morning; then beautiful turquoise sky.

Home [Hampstead], after having decided to be a priest and religious but still doubtful between St. Benedict and St. Ignatius.°

May 8. Dim sunlight; wind not cold, yet East.

May 9. Sultry and, I believe, dull.

May 10. Thick, but fine evening.

May 11. Dull; afternoon fine.

Slaughter of the innocents.° See above, the 2nd.

[Holiday in Switzerland]

July 11. Fine.

We took a guide° up the Wylerhorn but the top being clouded dismissed him and stayed up the mountain, lunching by a waterfall. Presently after long climbing—for there was a good chance of a clearance—we nearly reached the top, when a cloud coming on thick frightened me back: had we gone on we should have had the view, for it cleared quite. Still we saw the neighbouring mountains well. The snow is often cross-harrowed and lies too in the straightest paths as though artificial, which again comes from the planing. In the sheet it glistens yellow to the sun. How fond of and warped to the mountains it would be easy to become! For every cliff and limb and edge and jutty has its own nobility.—Two boys came down the mountain yodelling.—We saw the snow in the hollows for the first time. In one the surface was crisped across the direction of the cleft and the other way, that is across the broader crisping and down the stream, combed: the stream ran below and smoke came from the hollow: the edge of the snow hewn in curves as if by moulding-planes.—Crowd of mountain flowers—gentians; gentianellas; blood-red lucerne; a deep blue glossy spiked flower like plantain, flowering gradually up the spike, so that at the top it looks like clover or honeysuckle; rich big harebells glistening black like the cases of our veins when dry and heated from without; and others. All the herbage enthronged with every fingered or fretted leaf.—Firs very tall, with the swell of the branching on the outer side of the slope so that the peaks seem to point inwards to the mountain peak, like the lines of the Parthenon, and the outline melodious and moving on many focuses.—I wore my pagharee° and turned it with harebells below and gentians in two rows above like double pan-pipes.—In coming down we lost our way and each had a dangerous slide down the long wet grass of a steep slope.

Waterfalls not only skeined but silky too—one saw it from the inn across the meadows: at one quain of the rock the water glistened above and took shadow below, and the rock was reddened a little way each side with the wet, which sets off the silkiness.

Goat-flocks, each goat with its bell.

Ashes here are often pollarded and look different from ours and they give off their sprays at the outline in marked parallels justifying the Italian painters.

July 15. Showers; little sun.

Walked to the Hôtel Bellevue on the Little Scheidegg.

The mountains and in particular the Silberhorn are shaped and nippled like the sand in an hourglass and the Silberhorn has a subsidiary pyramidal peak naped sharply down the sides. Then one of their beauties is in nearly vertical places the fine pleatings of the snow running to or from one another, like the newness of lawn in an alb and sometimes cut off short as crisp as celery.

There are round one of the heights of the Jungfrau two ends or falls of a glacier. If you took the skin of a white tiger or the deep fell of some other animal and swung it tossing high in the air and then cast it out before you it would fall and so clasp and lap round anything in its way just as this glacier does and the fleece would part in the same rifts: you must suppose a lazuli under-flix to appear. The spraying out of one end I tried to catch but it would have taken hours: it was this which first made me think of a tiger-skin, and it ends in tongues and points like the tail and claws: indeed the ends of the glaciers are knotted or knuckled like talons. Above, in a plane nearly parallel to the eye, becoming thus foreshortened, it forms saddle-curves with dips and swells.

The view was not good: a few times we saw the Silberhorn but the Eiger never clearly and the Jungfrau itself scarcely or not at all.

It is curious how blue the glimpses of the mountain-sides and valleys look through the lifting cloud.

July 18. Up a little after sunrise but rain was falling and nothing worth seeing but the orange slit in the E. Then it became fine; dull afternoon; fine evening.

In coming down the Faulhorn saw the Finster Aarhorn at last, lonely, standing like a high-gabled steeple. The two heights of the Viescherhörner group are striking too, rising like thorns (as many *hörner* are like thorns or talons) from the level ridged wall which forms the theatre of the Grindelwald glacier.

Rushing streams may be described as inscaped ordinarily in pillows— and upturned troughs.

We lunched at the Baths of Rosenlaui° and walked on to Meyringen down the valley of the Reichenbach in torrent. Sycomores grew on the slopes of the valley, scantily leaved, sharply quained and accidented by perhaps the valley winds, and often most gracefully inscaped.—On the wall of the cliff bounding the valley on the further side of the river was a bright silver-tackled waterfall parted into slender shanks.

When Meyringen came in sight in the broader thwart valley below we sat on a little bridge over the Reichenbach there narrowed in, all leaved over, and rushing from one cornered rocky chamber to another. I saw that below the cuffs or long lips of lather that form there descended a webby space of foamy water. On the one side of the bridge an ash rose with eye-taking sky-clusters, the leaves making the outlines of their two sides smartly cross and recross and so giving the disputed bat's-wing.

Then we saw the three falls of the Reichenbach. The upper one is the biggest. At the take-off it falls in discharges of rice or meal but each cluster as it descends sharpens and tapers, and from halfway down the whole cascade is inscaped in fretted falling vandykes in each of which the frets or points, just like the startings of a just-lit lucifer match, keep shooting in races, one beyond the other, to the bottom. The vapour which beats up from the impact of the falling water makes little feeder rills down the rocks and these catching and running in drops along the sharp ledges in the rock are shaken and delayed and chased along them and even cut off and blown upwards by the blast of the vapour as it rises: saw the same thing at Handeck too.—In the second fall when facing the great limbs in which the water is packed saw well how they are tretted like open sponge or light bread-crumb where the yeast has supped in the texture in big and little holes.

July 19, 1868. Sunday, but no Catholics, I found, at Meyringen. The day fine.

Walked up the valley of the Aar, sallow-coloured and torrent, to the Grimsel. The heights bounding the valley soon became a mingle of lilac and green, the first the colour of the rock, the other the grass crestings, and seemed to group above in crops and rounded buttresses, yet to be cut sharp in horizontal or leaning planes below.

We came up with a guide who reminded me of F. John. He took E.B.'s knapsack and on finding the reason why I would not let him take mine said 'Le bon Dieu n'est pas comme ça.' The man probably was a rational Protestant; if a Catholic at least he rationalised gracefully, as they do in Switzerland.

At a turn in the road the foam-cuffs in the river, looked down upon, were of the crispiest endive spraying.

We lunched at Guttannen, where there was that strange party of Americans.

I was arguing about the planing of rocks and made a sketch of two in the Aar, and after that it was strange, for Nature became Nemesis, so precise they were, and E.B. himself pointed out two which looked, he said, as if they had been sawn. And of the hills themselves it could sometimes be seen, but on the other hand the sides of the valley often descended in trending sweeps of vertical section and so met at the bottom.

At times the valley opened in *cirques*, amphitheatres, enclosing levels of plain, and the river then ran between flaky flat-fish isles made of cindery lily-white stones.—In or near one of these openings the guide cries out 'Voulez-vous une Alp-rose?' and up he springs the side of the hill and brings us each bunches of flowers down.

In one place over a smooth table of rock came slipping down a blade of water looking like and as evenly crisped as fruitnets let drop and falling slack.

We saw Handeck waterfall. It is in fact the meeting of two waters, the right the Aar sallow and jade-coloured, the left a smaller stream of clear lilac foam. It is the greatest fall we have seen. The lower half is hidden in spray. I watched the great bushes of foam-water, the texture of branch-ings and water-spandrils which makes them up. At their outsides nearest the rock they gave off showers of drops strung together into little quills which sprang out in fans.

On crossing the Aar again there was as good a fall as some we have paid to see, all in jostling foam-bags.

Across the valley too we saw the fall of the Gelmer—like milk chasing round blocks of coal; or a girdle or long purse of white weighted with irregular black rubies, carelessly thrown aside and lying in jutty bends, with a black clasp of the same stone at the top—for those were the biggest blocks, squared, and built up, as it happened, in lessening stories, and the cascade enclosed them on the right and left hand with its foam; or once more like the skin of a white snake square-pied with black.

July 26. Sunday. There was no church nearer than Valtournanches, but there was to be mass said in a little chapel for the guides going up with Tyndal° at two o'clock in the morning and so I got up for this, my burnt face in a dreadful state and running. We went down with lanterns. It was an odd scene: two of the guides or porters served; the noise of a torrent outside accompanied the priest. Then to bed again.

Day fine. We did not get a completely clear view of the Matterhorn from this side.

In the afternoon we walked down the valley, which is beautiful, to Valtournanches.—We passed a gorge at the end of which it was curious to see a tree rubbing and ruffling with the water at the neck just above a fall.—Then we saw a grotto, that is deep and partly covered chambers of rock through which the torrent river runs.—A little beyond, I think, was a wayside chapel with a woman kneeling at a window a long time.—Further, across the valley a pretty village, the houses white, deep-eaved, pierced with small square windows at effective distances, and crossed with balconies, and above, a grove of ash or sycomore or both, sprayed all one way like water-weed beds in a running stream, very English-looking.—Beyond again, in midst of a slope of meadow slightly pulled like an unsteady and swelling surface of water, some ashes growing in a beautifully clustered 'bouquet', the skeleton as below—the inward bend of the left-hand stem being partly real, partly apparent and helped by τύχη τέχνην στεργούσῃ.°—Dim mountains down the valley red in the sunset.

July 31.
In the afternoon we took the train for Paris and passed through a country of pale grey rocky hills of a strong and simple outscape covered with fields of wormy green vines.

Aug. 1. Through Paris to Dieppe and by Newhaven home [Hampstead].

Day bright. Sea calm, with little walking wavelets edged with fine eyebrow crispings, and later nothing but a netting or chain-work on the surface, and even that went, so that the smoothness was marbly and perfect and, between the just-corded near sides of the waves rising like fishes' backs and breaking with darker blue the pale blue of the general field, in the very sleek hollows came out golden crumbs of reflections from the chalk cliffs.—Peach-coloured sundown and above some simple gilded messes of cloud, which later became finer, smaller, and scattering all away.—Here the sunlight had been dim.

The fields are burnt white, the heat has gone on.

[1869]

[The Novitiate, Roehampton]
Jan. 24. One day at the end of the year some heavy rain changed into snow which melted as it touched the ground. Else there has been no

snow this winter. It was mild—sun and rain—till the 20th or 21st I think, when there were for sunrise webs of rosy cloud and afterwards ranks of sharply edged crops or slices and all day delicate clouding: this red did not mean rain, but frost followed till the 25th, on which day it was giving; the next it was gone. Since then mild weather, more and more remarkably mild, with sun, gales, and much rain. Feb. 5 and 6 were almost hot. Daffodils have been in bloom for some days. A weeping-willow here is all green. The elms have long been in red bloom and yesterday (the 11th) I saw small leaves on the brushwood at their roots. Some primroses out. But a penance° which I was doing from Jan. 25 to July 25 prevented my seeing much that half-year.

April 30. Br. Wm. Kerr told me some days ago that in Australia(?) the English trees introduced had driven out the natives, mostly different kinds of gum-trees, and that he had seen a park planted with them, which were dying or dead. In particular our furze, which thrives wonderfully and grows into great hedges, has driven the native vegetation before it.

A cold May, and in fact no such hot weather as we had in April till the beginning of June and the haymaking, and then again cold winds.

Br. Wells calls a grindstone a *grindlestone*.

To *lead* north-country for to *carry* (a field of hay etc). *Geet* north-country preterite of *get*: 'he geet agate agoing'.

Trees sold 'top and lop': Br. Rickaby told me and suggests *top* is the higher, outer, and lighter wood good for firing only, *lop* the stem and bigger boughs when the rest has been lopped off used for timber.

Br. Wells calls white bryony Dead Creepers, because it kills what it entwines.

Fr. Casano's pronunciation of Latin instructive. (He is a Sicilian but has spent many years in Spain.) *Quod* he calls *c'od* and *quae hora* becomes almost *c'ora*—the *u* disappearing in a slight apostrophe; *Deus* sounds like *da-us* or *do-us*, the *e* being kept quite open; *meis* is almost a diphthong—like *mace*; *m* in *omnis* and, if I am not mistaken, final *ms* less strongly he gives the metallic nasal sound and the first syllable of *sanctus* he calls as if it were French.— Feb. 4, '70. Fr. Goldie gives long *e* like short *e* merely lengthened or even opener (the broad vowel between broad *a* and our closed *a*, the substitute for *e*, *i*, or *u* followed by *r*). Fr. Morris gives long *u* very full (*Luca*); he emphasises the semi-consonant and the vowel before it where two vowels meet—*Pio* becomes *Pī-jo* and *tuam tū-vam* (that is *pee-yo* and *too-wam*)— but in *tuum* the vowel is simply repeated. This morning I noticed Fr. Sangalli saying mass give the *ms* very slightly or bluntly.

The sunset June 20 was wine-coloured, with pencillings of purple, and next day there was rain.

June 27. The weather turned warm again two or three days ago and today is warmer still. Before that there had been cold, rain, and gloom.

Br. Sidgreaves has heard the high ridges of a field called *folds* and the hollow between the *drip*.

June 28. The cuckoo *has* changed his tune: the two notes can scarcely be told apart, that is their pitch is almost the same.

Nov. 20—Two large planets, the one an evening star, the other distant today from it as in the diagram, both nearly of an altitude and of one size—such counterparts that each seems the reflection of the other in opposite bays of the sky and not two distinct things.

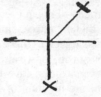

[1869–70]

Dec. 23—Yesterday morning I was dreaming I was with George
Consciousness of Simcox and was considering how to get away in time
dreaming to ring the bells here which as porter I had to ring (I was made porter on the 12th of the month, I think, and had the office for a little more than two months). I knew that I was dreaming and made this odd dilemma in my dream: either I am not really with Simcox and then it does not matter what I do, or if I am, waking will carry me off without my needing to do anything—and with this I was satisfied.

Another day in the evening after Litanies as Father Rector was giving the points for meditation I shut my eyes, being very tired, and without ceasing to hear him began to dream. The dream-images seemed to rise and overlie those which belonged to what he was saying and I saw one of the Apostles—he was talking about the Apostles—as if pressed against by a piece of wood about half a yard long and a few inches across, like a long box with two of the long sides cut off. Even then I could not understand what the piece of wood did encumbering the apostle. Now this piece of wood I had often seen in an outhouse and being that week 'A Secretis'° I had seen it longer together and had been that day wondering what it was: in reality it is used to hold a little heap of cinders against the wall which keep from the frost a piece of earthenware pipe which there comes out and goes in again making a projection in the wall. It is just the things which produce dead impressions, which the mind, either because you cannot make them out or because they were perceived across other more engrossing thoughts, has made nothing of and brought into no scaping,° that force themselves up in this way afterwards.—It seems true what Ed. Bond said, that you can trace your dreams to something or other in your waking life, especially of things that have been lately—I would not say this universally however. But the connection may be capricious, almost punning: I remember in one case to have detected a real pun but what it was I forget.

The dream-images also appear to have little or no projection, to be flat like pictures, and often one seems to be holding one's eyes close to them—I mean even while dreaming. This probably due to a difference still felt between images brought by ordinary use of function of sight and those seen as these are 'between our eyelids and our eyes'—though this is not all, for we also see the colours, brothy motes and figures, and at all events the positive darkness, made by the shut eyelids by the ordinary use of the function of sight, but these images are brought upon that dark field, as I imagine, by a reverse action of the visual nerves (the same will hold of the sounds, sensations of touch, etc of dreams)—or by other nerves, but it seems reasonable to suppose impressions of sight belong to the organ of sight—and once lodged there are stalled by the mind like other images: only you cannot make them at will when awake, for the very effort and advertence would be destructive to them, since the eye in its sane waking office kens only impressions brought from without, that is to say either from beyond the body or from the body itself produced upon the dark field of the eyelids. Nevertheless I have seen in favourable moments the images brought from within lying there like others: if I am not mistaken they are coarser and simpler and something like the spectra made by bright things looked hard at. I can therefore believe what Chandler told E.B., that at waking he could see—which is a step beyond seeing them on the field of the eyelids—the images of his dream upon the wall of his room.

It is not in reality harder for the mind to have ken at the same time of what the eye sees and also of the belonging images of our thoughts without ever or almost ever confounding them than it is for it to multiply the pictures brought by the two eyes into one without ever or almost ever separating them (March 23, '70).

One day towards the end of that year ... I passed a music shop somewhere in the outskirts of Notting Hill and in the window my eye was caught by 'the Disraeli Walz'. Some days before I had been trying unsuccessfully to recall Mr. Maclaren's Débutante Walz (in reality I think it is a polka). A few steps further on I found myself humming it.

One day in the Long Retreat (which ended on Xmas Day) they were reading in the refectory Sister Emmerich's account of the Agony in the Garden and I suddenly began to cry and sob and could not stop. I put it down for this reason, that if I had been asked a minute beforehand I should have said that nothing of the sort was going to happen and even when it did I stood in a manner wondering at myself not seeing in my reason the traces of an adequate cause for such strong emotion—the traces of it I say because of course the cause in itself is adequate for the sorrow of a lifetime. I remember much the same thing on Maundy Thursday when the presanctified Host was carried to the sacristy. But neither the weight nor the stress of sorrow, that is to say of the thing which should cause sorrow, by themselves move us or bring the tears as a sharp knife does not cut for being pressed as long as it is pressed without any shaking of the hand but there is always one touch, something striking sideways and unlooked for, which in both cases undoes resistance and pierces, and this may be so delicate that the pathos seems to have gone directly to the body and cleared the understanding in its passage. On the other hand the pathetic touch by itself, as in dramatic pathos, will only draw slight tears if its matter is not important or not of import to us, the strong emotion coming from a force which was gathered before it was discharged: in this way a knife may pierce the flesh which it had happened only to graze and only grazing will go no deeper.

The winter was called severe. There were three spells of frost with skating, the third beginning on Feb. 9. No snow to speak of till that day. Some days before Feb. 7 I saw catkins hanging. On the 9th there was snow but not lying on the roads. On the grass it became a crust lifted on the heads of the blades. As we went down a field near Caesar's Camp° I noticed it before me *squalentem*, coat below coat, sketched in intersecting edges bearing 'idiom', all down the slope:—I have no other word yet for that which takes the eye or mind in a bold hand or effective sketching or in marked features or again in graphic writing, which not being beauty nor true inscape yet gives interest and makes ugliness even better than meaninglessness.—On the Common the snow was channelled all in parallels by the sharp driving wind and upon the tufts of grass (where by the dark colour shewing through it looked greyish) it came to turret-like clusters or like broken shafts of basalt.—In the Park in the afternoon the wind was driving little clouds of snow-dust which caught the sun as they rose and delightfully took the eyes: flying up the slopes they looked like breaks of sunlight fallen through ravelled cloud upon the hills and again like deep flossy velvet blown to the root by breath which passed all along. Nearer at hand along the road it was gliding over the ground in white wisps that between trailing and flying shifted and wimpled like so many silvery worms to and from one another.

The squirrel was about in our trees all the winter. For instance about Jan 2 I often saw it.

April 4°—In taking off my jersey of knitted wool in the dark with an accidental stroke of my finger down the stuff I drew a flash of electric light. This explains the crackling I had often heard.

On March 27 I asked the Brentford boys about a ghost story they had told me before that. At Norris's market gardens by Sion Lane there is a place where according to tradition two men (and some boys, I think) were ploughing with four horses: in bringing the plough round at the headland they fell into a covered well which they did not see and were killed. And now if you lean your ear against a wall at the place you can hear the horses going and the men singing at their work.—There are other ghosts belonging to Sion House. E.g. there is an image (of our Lady, if I remember) in a stained window which every year is broken by an unseen hand and invisibly mended again. . .

Br. Byrne:—Hockey and football are much played in Ireland and the great day is Shrove Tuesday, on which the 'merits' are awarded. A player who had greatly distinguished himself at football was that day going home when in a lonely field a ball came rolling to his feet: he kicked it, it was kicked back, and soon he found himself playing the game with a fieldfull of fairies and in a place which was strange to him. The fairies would not let him go but they did their best to amuse him, they danced and wrestled before him so that he should never want for entertainment, but they could not get him to eat, for knowing that if he eat what they gave him they would have a claim upon him he preferred to starve and they for fear he should die on their hands at last put him on the right road home. On reaching home he found a pot of stirabout on the fire and had only had time to taste a ladlefull when the fairies were in upon him and began to drag him away again. He caught hold of the doorpost and called on the saints but when he came to our Lady's name they let go and troubled him no more.

[Stonyhurst, Lancashire]

Sept. 24—First saw the Northern Lights. My eye was caught by beams of light and dark very like the crown of horny rays the sun makes behind a cloud. At first I thought of silvery cloud until I saw that these were more luminous and did not dim the clearness of the stars in the Bear. They rose slightly radiating thrown out from the earthline. Then I saw soft pulses of light one after another rise and pass upwards arched in shape but waveringly and with the arch broken. They seemed to float, not following the warp of the sphere as falling stars look to do but free though concentrical with it. This busy working of nature wholly independent of the earth and seeming to go on in a strain of time not reckoned by our reckoning of days and years but simpler and as if correcting the preoccupation of the world by being preoccupied with and appealing to and dated to the day of judgment was like a new witness to God and filled me with delightful fear

Oct. 20—Laus Deo—the river today and yesterday. Yesterday it was a sallow glassy gold at Hodder Roughs and by watching hard the banks began to sail upstream, the scaping unfolded, the river was all in tumult but not running, only the lateral motions were perceived, and the curls of froth where the waves overlap shaped and turned easily and idly.—I meant to have written more.—Today the river was wild, very full, glossy brown with mud, furrowed in permanent billows through which from head to head the water swung with a great down and up again. These heads were scalped with rags of jumping foam. But at the Roughs the sight was the burly water-backs which heave after heave kept tumbling up from the broken foam and their plump heap turning open in ropes of velvet

[1871]

The spring weather began with March about

I have been watching clouds this spring and evaporation, for instance over our Lenten chocolate. It seems as if the heat by *aestus*, throes/one after another threw films of vapour off as boiling water throws off steam under films of water, that is bubbles. One query then is whether these films contain gas or no. The film seems to be set with tiny bubbles which gives it a grey and grained look. By throes perhaps which represent the moments at which the evener stress of the heat has overcome the resistance of the surface or of the whole liquid. It would be reasonable then to consider the films as the shell of gas-bubbles and the grain on them as a network of bubbles condensed by the air as the gas rises.— Candle smoke goes by just the same laws, the visible film being here of unconsumed substance, not hollow bubbles. The throes can be perceived/like the thrills of a candle in the socket: this is precisely to *reech*, whence *reek*. They may by a breath of air be laid again and then shew like grey wisps on the surface—which shews their part-solidity. They seem to be drawn off the chocolate as you might take up a napkin between your fingers that covered something, not so much from here or there as from the whole surface at once reech, so that the film is perceived at the edges and makes in fact a collar or ring just within the walls all round the cup; it then draws together in a cowl like a candleflame but not regularly or without a break: the question is why.

Perhaps in perfect stillness it would not but the air breathing it aside entangles it with itself. The film seems to rise not quite simultaneously but to peel off as if you were tearing cloth; then giving an end forward like the corner of a handkerchief and beginning to coil it makes a long wavy hose you may sometimes look down, as a ribbon or a carpenter's shaving may be made to do. Higher running into frets and silvering in the sun with the endless coiling, the soft bound of the general motion and yet the side lurches sliding into some particular pitch it makes a baffling and charming sight.—Clouds however solid they may look far off are I think wholly made of film in the sheet or in the tuft. The bright woolpacks that pelt before a gale in a clear sky are in the tuft and you can see the wind unravelling and rending them finer than any sponge till within one easy reach overhead they are morselled to nothing and consumed—it depends of course on their size. Possibly each tuft in forepitch or in origin is quained and a crystal. Rarer and wilder packs have sometimes film in the sheet, which may be caught as it turns on the edge of the cloud  like an outlying eyebrow. The one in which I saw this was in a north-east wind, solid but not crisp, white like the white of egg, and bloated-looking

What you look hard at seems to look hard at you, hence the true and the false instress of nature. One day early in March when long streamers were rising from over Kemble End one large flake loop-shaped, not a streamer but belonging to the string, moving too slowly to be seen, seemed to cap and fill the zenith with a white shire of cloud. I looked long up at it till the tall height and the beauty of the scaping—regularly curled knots springing if I remember from fine stems, like foliation in wood or stone—had strongly grown on me. It changed beautiful changes, growing more into ribs and one stretch of running into branching like coral. Unless you refresh the mind from time to time you cannot always remember or believe how deep the inscape in things is

March 14—Bright morning, pied skies, hail. In the afternoon the wind was from the N., very cold; long bows of soft grey cloud straining the whole heaven but spanning the skyline with a slow entasis which left a strip of cold porcelain blue. The long ribs or girders were as rollers/ across the wind, not in it, but across them there lay fine grass-ends, sided off down the perspective, as if locks of vapour blown free from the main ribs down the wind. Next day and next snow. Then in walking I saw the water-runs in the sand of unusual delicacy and the broken blots of snow in the dead bents of the hedge-banks I could find a square scaping in which helped the eye over another hitherto disordered field of things.

(And if you look well at big pack-clouds overhead you will soon find a strong large quaining and squaring in them which makes each pack impressive and whole.) Pendle was beautiful: the face of snow on it and the tracks or gulleys which streaked and parted this well shaped out its roundness and boss and marked the slow tune of its long shoulder. One time it lay above a near hill of green field which, with the lands in it lined and plated by snow, was striped like a zebra: this Pendle repeated finer and dimmer

March 17—In the morning clouds chalky and milk-coloured, with remarkable oyster-shell moulding. (From a rough pencil sketch)

Between eleven and twelve at night a shock of earthquake

End of March and beginning of April—This is the time to study inscape in the spraying of trees, for the swelling buds carry them to a pitch which the eye could not else gather—for out of much much more, out of little not much, out of nothing nothing: in these sprays at all events there is a new world of inscape. The male ashes are very boldly jotted with the heads of the bloom which tuft the outer ends of the branches. The staff of each of these branches is closely knotted with the places where buds are or have been, so that it is something like a finger which has been tied up with string and keeps the marks. They are in knops of a pair, one on each side, and the knops are set alternately, at crosses with the knops above and the knops below, the bud of course is a short smoke-black pointed nail-head or beak pieced of four lids or nippers. Below it, like the hollow below the eye or the piece between the knuckle and the root of the nail, is a half-moon-shaped sill as if once chipped from the wood and this gives the twig its quaining in the outline. When the bud breaks at first it shews a heap of fruity purplish anthers looking something like unripe elder-berries but these push open into richly-branched tree-pieces coloured buff and brown, shaking out loads of pollen, and drawing the tuft as a whole into peaked quains—mainly four, I think, two bigger and two smaller

The bushes in the woods and hedgerows are spanned over and twisted upon by the woody cords of the honeysuckle: the cloves of leaf these bear are some purple, some grave green. But the young green of the briars is gay and neat and smooth as if cut in ivory.—One bay or hollow of Hodder Wood is curled all over with bright green garlic

The sycomores are quite the earliest trees out: some have been fully out some days (April 15). The behaviour of the opening clusters is very beautiful and when fully opened not the single leaves but the whole tuft is strongly templed like the belly of a drum or bell

The half-opened wood-sorrel leaves, the centre or spring of the leaflets rising foremost and the leaflets dropping báck like ears leaving straight-chipped clefts between them, look like some green lettering and cut as sharp as dice

The white violets are broader and smell; the blue, scentless and finer made, have a sharper whelking and a more winged recoil in the leaves

Take a *few* primroses in a glass and the instress of—brilliancy, sort of starriness: I have not the right word—so simple a flower gives is remarkable. It is, I think, due to the strong swell given by the deeper yellow middle

'The young lambs bound As to the tabour's sound'.°
They toss and toss: it is as if it were the earth that flung them, not themselves. It is the pitch of graceful agility when we think that.—April 16—Sometimes they rest a little space on the hind legs and the forefeet drop curling in on the breast, not so liquidly as we see it in the limbs of foals though

Bright afternoon; clear distances; Pendle dappled with tufted shadow; west wind; interesting clouding, flat and lying in the warp of the heaven but the pieces with rounded outline and dolphin-backs shewing in places and all was at odds and at Z's, one piece with another. Later beautifully delicate crisping. Later rippling as in the drawing°

April 21—We have had other such afternoons, one today—the sky a beautiful grained blue, silky lingering clouds in flat-bottomed loaves, others a little browner in ropes or in burly-shouldered ridges swany and lustrous, more in the Zenith stray packs of a sort of violet paleness. White-rose cloud formed fast, not in the same density—some caked and swimming in a wan whiteness, the rest soaked with the blue and like the leaf of a flower held against the light and diapered out by the worm or veining of deeper blue between rosette and rosette. Later/moulding, which brought rain: in perspective it was vaulted in very regular ribs with fretting between: but these are not ribs; they are a 'wracking' install made of these two realities—the frets, which are scarves of rotten cloud bellying upwards and drooping at their ends and shaded darkest at the brow or tropic where they double to the eye, and the whiter field of sky shewing between: the illusion looking down the 'waggon' is complete. These swaths of fretted cloud move in rank, not in file

April 22—But such a lovely damasking in the sky as today I never felt before. The blue was charged with simple instress, the higher, zenith sky earnest and frowning, lower more light and sweet. High up again, breathing through woolly coats of cloud or on the quains and branches of the flying pieces it was the true exchange of crimson, nearer the earth/ against the sun/it was turquoise, and in the opposite south-western bay below the sun it was like clear oil but just as full of colour, shaken over with slanted flashing 'travellers', all in flight, stepping one behind the other, their edges tossed with bright ravelling, as if white napkins were thrown up in the sun but not quite at the same moment so that they were all in a scale down the air falling one after the other to the ground

April 27—Went to see Sauley Abbey (Cistercian): there is little to see

Mesmerised a duck with chalk lines drawn from her beak sometimes level and sometimes forwards on a black table. They explain that the bird keeping the abiding offscape of the hand grasping her neck fancies she is still held down and cannot lift her head as long as she looks at the chalk line, which she associates with the power that holds her. This duck lifted her head at once when I put it down on the table without chalk. But this seems inadequate. It is most likely the fascinating instress of the straight white stroke

May 9—A simple behaviour of the cloudscape I have not realised before. Before a N.E. wind great bars or rafters of cloud all the morning and in a manner all the day marching across the sky in regular rank and with equal spaces between. They seem prism-shaped, flat-bottomed and banked up to a ridge: their make is like light tufty snow in coats

This day and May 11 the bluebells in the little wood between the College and the highroad and in one of the Hurst Green cloughs. In the little wood/ opposite the light/ they stood in blackish spreads or sheddings like the spots on a snake. The heads are then like thongs and solemn in grain and grape-colour. But in the clough/ through the light/ they came in falls of sky-colour washing the brows and slacks of the ground with vein-blue, thickening at the double, vertical themselves and the young grass and brake fern combed vertical, but the brake struck the upright of all this with light winged transomes. It was a lovely sight.— The bluebells in your hand baffle you with their inscape, made to every sense: if you draw your fingers through them they are lodged and struggle/ with a shock of wet heads; the long stalks rub and click and flatten to a fan on one another like your fingers themselves would when you passed the palms hard across one another, making a brittle rub and jostle like the noise of a hurdle strained by leaning against; then there is the faint honey smell and in the mouth the sweet gum when you bite them. But this is easy, it is the eye they baffle. They give one a fancy of panpipes and of some wind instrument with stops—a trombone perhaps. The overhung necks—for growing they are little more than a staff with a simple crook but in water, where they stiffen, they take stronger turns, in the head like sheephooks or, when more waved throughout, like the waves riding through a whip that is being smacked—what with these overhung necks and what with the crisped ruffled bells dropping mostly on one side and the gloss these have at their footstalks they have an air of the knights at chess. Then the knot or 'knoop' of buds some shut, some just gaping, which makes the pencil of the whole spike, should be noticed: the inscape of the flower most finely carried out in the siding of the axes, each striking a greater and greater slant, is finished in these clustered buds, which for the most part are not straightened but rise to the end like a tongue and this and their tapering and a little flattening they have make them look like the heads of snakes

May 17 etc—I have several times seen the peacock with train spread lately. It has a very regular warp, like a shell, in which the bird embays himself, the bulge being inwards below but the hollow inwards above, cooping him in and only opening towards the brim, where the feathers are beginning to rive apart. The eyes, which lie alternately when the train is shut, like scales or gadroons, fall into irregular rows when it is opened, and then it thins and darkens against the light, it loses the moistness and satin it has when in the pack but takes another/grave and expressive splendour, and the outermost eyes, detached and singled, give with their corner fringes the suggestion of that inscape of the flowing cusped trefoil which is often effective in art. He shivers it when he first rears it and then again at intervals and when this happens the rest blurs and the eyes start forward.—I have thought it looks like a tray or green basket or fresh-cut willow hurdle set all over with Paradise fruits cut through—first through a beard of golden fibre and then through wet flesh greener than greengages or purpler than grapes—or say that the knife had caught a tatter or flag of the skin and laid it flat across the flesh—and then within all a sluggish corner drop of black or purple oil

On Whit Monday (May 29) went to Preston to see the procession. Though not very splendid it moved me. But just as it was beginning we heard the news of the murder of the hostages by the Commune at the entry of the Government troops into Paris—64 in all, including the Archbishop, Mgr. Maret bishop of Sura, the Curé of the Madeleine, and Fr. Olivain with four other of our Fathers. It was at the same time the burning of the Tuileries and the other public buildings was carried out

---

Lancashire—'of all the wind instruments big droôm fots me best'.— Old Wells directing someone how to set a wedge in a tree told him that if he would put it so and so he would 'fot it agate a riving'.—The omission of *the* is I think an extension of the way in which we say 'Father', 'government' etc: they use it when there is a relative/in order to define.—They say *frae* and *aboon*

June 13—A beautiful instance of inscape sided on the slide, that is/ successive sidings of one inscape, is seen in the behaviour of the flag flower from the shut bud to the full blowing: each term you can distinguish is beautiful in itself and of course if the whole 'behaviour' were gathered up and so stalled it would have a beauty of all the higher degree

July 8—After much rain, some thunder, and no summer as yet, the river swollen and golden and, where charged with air, like ropes and hills of melting candy, there was this day a thunderstorm on a greater scale— huge rocky clouds lit with livid light, hail and rain that flooded the garden, and thunder ringing and echoing round like brass, so that there is in a manner earwitness to the χαλκεον οὐρανόν.° The lightning seemed to me white like a flash from a lookingglass but Mr. Lentaigne in the afternoon noticed it rose-coloured and lilac. I noticed two kinds of flash but I am not sure that sometimes there were not the two together from different points of the same cloud or starting from the same point different ways—one a straight stroke, broad like a stroke with chalk and liquid, as if the blade of an oar just stripped open a ribbon scar in smooth water and it caught the light; the other narrow and wire-like, like the splitting of a rock and danced down-along in a thousand jags. I noticed this too, that there was a perceptible interval between the blaze and first inset of the flash and its score in the sky and that that seemed to be first of all laid in a bright confusion and then uttered by a tongue of brightness (what is strange) running up from the ground to the cloud, not the other way

July ?—At eight o'clock about sunset hanging due opposite the house in the east the greatest stack of cloud, to call it one cloud, I ever can recall seeing. Singled by the eye and taken up by itself it was shining white but taken with the sky, which was a strong hard blue, it was anointed with warm brassy glow: only near the earth it was stunned with purplish shadow. The instress of its size came from comparison not with what was visible but with the remembrance of other clouds: like the Monte Rosa range from the Gorner Grat its burliness forced out everything else and loaded the eyesight. It was in two limbs fairly level above and below but not equal in breadth—as 2 to 3 or 3 to 4 perhaps—, like two waggons or loaded trucks. The left was rawly made, a fleece parcelled in wavy locks flowing open upwards, with shady gutturs° between, like the ringlets of a ram's fleece blowing; the right was shapely, roped like a heavy cable being slowly paid and by its weight settling into gross coils and beautifully plotted with tortoise-shell squares of shading—indeed much as a snake is plotted, and this one rose steep up like an immeasurable cliff

[The two rocks on which Dumbarton Castle in the Clyde stands reminded me of this cloud Aug. 28]

[While I am writing, Aug. 12, in a room in the Old Magazine at the College I hear every now and then the deathwatch ticking. It goes for a few seconds at a time. Several of us have heard it]

July 24—Robert says the first grass from the scythe is the *swathe*, then comes the *strow* (tedding), then *rowing*, then the footcocks, then *breaking*, then the *hubrows*, which are gathered into *hubs*, then sometimes another break and *turning*, then *rickles*, the biggest of all the cocks, which are run together into *placks*, the shapeless heaps from which the hay is carted.

Dec. 17–18 at night—Rescued a little kitten that was perched in the sill of the round window at the sink over the gasjet and dared not jump down. I heard her mew a piteous long time till I could bear it no longer; but I make a note of it because of her gratitude after I had taken her down, which made her follow me about and at each turn of the stairs as I went down leading her to the kitchen run back a few steps and try to get up to lick me through the banisters from the flight above

———

Some events of interest of 1871 and the end of '70 (see that date) partly from my memoranda and partly got out of Whitaker's Almanack by Cyril and Uncle John

Recapture of Orleans by the Germans Dec. 5, 1870

Death of Prim and landing of King Amadeo in Spain Dec. 30

Le Mans occupied by the Germans (after several days of fighting) Jan. 12, 1871

Surrender of Paris Jan. 28

Signing of peace Feb. 26

Paris seized by the Commune March 18

First defeat of the Communal troops and death of Flourens April 4(?)

Death of the Archbishop etc May 24

Suppression of the insurrection May 28

Exercise of royal prerogative to carry the Army Purchase bill against the House of Lords Aug. 17

[1872]

July 19 ... Stepped into a barn of ours, a great shadowy barn, where the hay had been stacked on either side, and looking at the great rudely arched timberframes—principals(?) and tie-beams, which make them look like bold big *A*s with the cross-bar high up—I thought how sadly beauty of inscape was unknown and buried away from simple people and yet how near at hand it was if they had eyes to see it and it could be called out everywhere again

This month here and all over the country many great thunderstorms. Cyril, in bed I think, at Liverpool after a simultaneous flash and crash felt a shock like one from a galvanic battery and for some time one of his arms went numbed. At Roehampton Fr. Williams was doubled up and another Father had his breviary struck out of his hand. Here a tree was struck near the boys' cricketfield and a cow was ripped up

[Holiday in the Isle of Man]

After the examinations we went for our holiday out to Douglas in the Isle of Man Aug. 3. At this time I had first begun to get hold of the copy of Scotus on the Sentences in the Baddely library° and was flush with a new stroke of enthusiasm. It may come to nothing or it may be a mercy from God. But just then when I took in any inscape of the sky or sea I thought of Scotus

Aug. 4—Kirk Onchan church is modern and if you looked to anything but the steeple very poor but the steeple is so strongly and boldly designed that it quite deceived me and I took it for old work well preserved. In the churchyard is an old engraved cross with knotwork such as on those at Whalley

Aug. 8—Walked to Ramsey, and back by steamer. From the highroad I saw how the sea, dark blue with violet cloud-shadows, was warped to the round of the world like a coat upon a ball and often later I marked that perspective. I had many beautiful sights of it, sometimes to the foot of the cliff, where it was of a strong smouldering green over the sunken rocks—these rocks, which are coated with small limpets, discolour the coast all along with a fringe of yellow at the tide-mark and under water reflect light and make themselves felt where the smooth black ones would not shew—, but farther out blue shadowed with gusts from the shore; at other times with the brinks hidden by the fall of the hill, packing the land in/it was not seen how far, and then you see best how it is drawn up to a brow at the skyline and stoops away on either side, tumbling over towards the eye in the broad smooth fall of a lakish apron of water, which seems bound over or lashed to land below by a splay of dark and light braids: they are the gusts of wind all along the perspective with which all the sea that day was dressed.

And it is common for the sea looked down upon, where the sheety spread is well seen but the depth and mass unfelt, to sway and follow the wind like the tumbled canvas of a loose sail

The flowers in the island are plentiful and strongly coloured. On the sides of the cliff above our house, Derby Castle, the brambles were often doubled. The flower was bigger, purplish pink, and the five petals changed for a multitude of small strap leaves as in daisies and auriculas

The country is bare and you see the valleys and fell-sides plotted and painted with the squares of the fields and their hedges far and wide. But the trees are rich and thickly leaved where they grow. I remember one little square house cushioned up in a thatched grove of green like a man with an earache. These groves are stunted and shaped by the sea breeze but plighted thick together and cast a deep green shade. Often the cage of boughs is bare and ragged but thick tufts at the top. The ashes thrive and the combs are not wiry and straight but rich and beautifully curled. The climate varies little and is said to have a higher average of heat in winter than Rhodez or Milan. Fuchsias, strawberry trees, and tamarisks do well

On the way we went aside to see the Laxey waterwheel 72 ft. in diameter, said to be the biggest in the world. It is on the side of a hill up Laxey glen, the water is delivered a little below its highest point and turns the wheel towards itself, acting mainly no doubt by its weight in the buckets but perhaps also by its flow, and I do not know whether it should be called a breastwheel or overshot. It is geared by a long timber shaft or beam or piece running by little wheels on a rail to an oscillating head carrying at one side a makeweight and at the other a connecting rod working a pump. The wheel is used to pump to lead mine. It turns once in 25 seconds

Aug. 10—I was looking at high waves. The breakers always are parallel to the coast and shape themselves to it except where the curve is sharp however the wind blows. They are rolled out by the shallowing shore just as a piece of putty between the palms whatever its shape runs into a long roll. The slant ruck or crease one sees in them shows the way of the wind. The regularity of the barrels surprised and charmed the eye; the edge behind the comb or crest was as smooth and bright as glass. It may be noticed to be green behind and silver white in front: the silver marks where the air begins, the pure white is foam, the green/solid water. Then looked at to the right or left they are scrolled over like mouldboards or feathers or jibsails seen by the edge. It is pretty to see the hollow of the barrel disappearing as the white combs on each side run along the wave gaining ground till the two meet at a pitch and crush and overlap each other

About all the turns of the scaping from the break and flooding of wave to its run out again I have not yet satisfied myself. The shores are swimming and the eyes have before them a region of milky surf but it is hard for them to unpack the huddling and gnarls of the water and law out the shapes and the sequence of the running: I catch however the looped or forked wisp made by every big pebble the backwater runs over—if it were clear and smooth there would be a network from their overlapping, such as can in fact be seen on smooth sand after the tide is out—; then I saw it run browner, the foam dwindling and twitched into long chains of suds, while the strength of the back-draught shrugged the stones together and clocked them one against another

Looking from the cliff I saw well that work of dimpled foamlaps—strings of short loops or halfmoons—which I had studied at Freshwater years ago

It is pretty to see the dance and swagging of the light green tongues or ripples of waves in a place locked between rocks

[Lancashire]

Oct. 5—A goldencrested wren had got into my room at night and circled round dazzled by the gaslight on the white cieling;° when caught even and put out it would come in again. Ruffling the crest which is mounted over the crown and eyes like beetlebrows, I smoothed and fingered the little orange and yellow feathers which are hidden in it. Next morning I found many of these about the room and enclosed them in a letter to Cyril° on his wedding day

Oct. 27—Fr. Gallwey came up. Before night litanies he came to my room as I lay on my bed making my examen, for I had some fever, and sitting by the bedside took my hand within his and said some affectionate and most encouraging words

That fever came from a chill I caught one Blandyke° and the chill from weakness brought about by my old complaint,° which before and much more after the fever was worse than usual. Indeed then I lost so much blood that I hardly saw how I was to recover. Nevertheless it stopped suddenly, almost at the worst. This was why I came up to town at Christmas

Dec. 12—A Blandyke. Hard frost, bright sun, a sky of blue 'water'. On the fells with Mr. Lucas. Parlick Pike and that ridge ruddy with fern and evening light. Ground sheeted with taut tattered streaks of crisp gritty snow. Green-white tufts of long bleached grass like heads of hair or the crowns of heads of hair, each a whorl of slender curves, one tuft taking up another—however these I might have noticed any day. I saw the inscape though freshly, as if my eye were still growing, though with a companion the eye and the ear are for the most part shut and instress cannot come. We started pheasants and a grouse with flickering wings. On the slope of the far side under the trees the fern looked ginger-coloured over the snow. When there was no snow and dark greens about, as I saw it just over the stile at the top of the Forty-Acre the other day, it made bats and splinters of smooth caky road-rut-colour

Dec. 19—Under a dark sky walking by the river at Brockennook. There all was sad-coloured and the colour caught the eye, red and blue stones in the river beaches brought out by patches of white-blue snow, that is/snow quite white and dead but yet it seems as if some blue or lilac screen masked it somewhere between it and the eye: I have often noticed it. The swells and hillocks of the river sands and the fields were sketched and gilded out by frill upon frill of snow—they must be seen: this is only to shew which way the curve lies.

Where the snow lies as in a field the damasking of white light and silvery shade may be watched indeed till brightness and glare is all lost in a perplexity of shadow and in the whitest of things the sense of white is lost, but at a shorter gaze I see two degrees in it—the darker, facing the sky, and the lighter in the tiny cliffs or scarps where the snow is broken or raised into ridges, these catching the sun perhaps or at all event more directly hitting the eye and gilded with an arch brightness, like the sweat in the moist hollow between the eyebrows and the eyelids on a hot day or in the way the light of a taper Tommy was screening with his hand the other morning in the dark refectory struck out the same shells of the eyes and the cleft of the nostrils and flat of the chin and tufts on the cheeks in gay leaves of gold

[1873]

Feb. 24—In the snow flat-topped hillocks and shoulders outlined with wavy edges, ridge below ridge, very like the grain of wood in line and in projection like relief maps. These the wind makes I think and of course drifts, which are in fact snow waves. The sharp nape of a drift is sometimes broken by slant flutes or channels. I think this must be when the wind after shaping the drift first has changed and cast waves in the body of the wave itself. All the world is full of inscape and chance left free to act falls into an order as well as purpose: looking out of my window I caught in the random clods and broken heaps of snow made by the cast of a broom. The same of the path trenched by footsteps in ankledeep snow across the fields leading to Hodder wood through which we went to see the river. The sun was bright, the broken brambles and all boughs and banks limed and cloyed with white, the brook down the clough pulling its way by drops and by bubbles in turn under a shell of ice

In March there was much snow

May 11—Bluebells in Hodder wood, all hanging their heads one way. I caught as well as I could while my companions talked the Greek rightness of their beauty, the lovely/what people call/ 'gracious' bidding one to another or all one way, the level or stage or shire of colour they make hanging in the air a foot above the grass, and a notable glare the eye may abstract and sever from the blue colour/of light beating up from so many glassy heads, which like water is good to float their deeper instress in upon the mind

May 30—The swifts round and scurl under the clouds in the sky: light streamers were about; the swifts seemed rather to hang and be at rest and to fling these away row by row behind them like spokes of a lighthung wheel

June 5 etc—The turkey and hens will let a little chick mount their backs and sit between the wings

June 15—Sunday after Corpus Christi. Some of us went to Billington to join in their procession.° Mr. Lucas was with me. The day was very beautiful. A few streamer clouds and a grapy yellowing team moving along the horizon. At the ferry a man said 'Hāst a penny, Tom?'—the old ferry was below the rocks

June 16—Still brighter and warmer, southern-like. Shadows sharp in the quarry and on the shoulders of our two young white pigeons. There is some charm about a thing such as these pigeons or trees when they dapple their boles in wearing its own shadow. I was on the fells with Mr. Strappini. They were all melled and painted with colour and full of roaming scents, and winged silver slips of young brake rising against the light trim and symmetrical and gloried from within reminded me of I do not remember what detail of coats of arms, perhaps the lilies of Eton College. Meadows smeared yellow with buttercups and bright squares of rapefield in the landscape. Fine-weather bales of cloud. Napkin folds brought out on the Parlick ridge and capfulls of shadow in them. A cuckoo flew by with a little bird after it as we lay in the quarry at Kemble End

As I passed the stables later and stayed to look at the peacocks John Myerscough came out to shew me a brood of little peafowl (though it could not be found at that time) and the kindness touched my heart

I looked at the pigeons down in the kitchen yard and so on. They look like little gay jugs by shape when they walk, strutting and jod-jodding with their heads. The two young ones are all white and the pins of the folded wings, quill pleated over quill, are like crisp and shapely cuttle-shells found on the shore. The others are dull thundercolour or black-grape-colour except in the white pieings, the quills and tail, and in the shot of the neck. I saw one up on the eaves of the roof: as it moved its head a crush of satin green came and went, a wet or soft flaming of the light

July 22—Very hot, though the wind, which was south, dappled very sweetly on one's face and when I came out I seemed to put it on like a gown as a man puts on the shadow he walks into and hoods or hats himself with the shelter of a roof, a penthouse, or a copse of trees, I mean it rippled and fluttered like light linen, one could feel the folds and braids of it—and indeed a floating flag is like wind visible and what weeds are in a current; it gives it thew and fires it and bloods it in.—Thunderstorm in the evening, first booming in gong-sounds, as at Aosta, as if high up and so not reechoed from the hills; the lightning very slender and nimble and as if playing very near but after supper it was so bright and terrible some people said they had never seen its like. People were killed, but in other parts of the country it was more violent than with us. Flashes lacing two clouds above or the cloud and the earth started upon the eyes in live veins of rincing or riddling liquid white, inched and jagged as if it were the shivering of a bright riband string which had once been kept bound round a blade and danced back into its pleatings. Several strong thrills of light followed the flash but a grey smother of darkness blotted the eyes if they had seen the fork, also dull furry thickened scapes of it were left in them

[Holiday in the Isle of Man; Lancashire]

Aug. 16—We rose at four, when it was stormy and I saw duncoloured waves leaving trailing hoods of white breaking on the beach. Before going I took a last look at the breakers, wanting to make out how the comb is morselled so fine into string and tassel, as I have lately noticed it to be. I saw big smooth flinty waves, carved and scuppled in shallow grooves, much swelling when the wind freshened, burst on the rocky spurs of the cliff at the little cove and break into bushes of foam. In an enclosure of rocks the peaks of the water romped and wandered and a light crown of tufty scum standing high on the surface kept slowly turning round: chips of it blew off and gadded about without weight in the air. At eight we sailed for Liverpool in wind and rain. I think it is the salt that makes rain at sea sting so much. There was a good-looking young man on board that got drunk and sung 'I want to go home to Mamma'. I did not look much at the sea: the crests I saw ravelled up by the wind into the air in arching whips and straps of glassy spray and higher broken into clouds of white and blown away. Under the curl shone a bright juice of beautiful green. The foam exploding and smouldering under water makes a chrysoprase green. From Blackburn I walked: infinite stiles and sloppy fields, for there has been much rain. A few big shining drops hit us aslant as if they were blown off from eaves or leaves. Bright sunset: all the sky hung with tall tossed clouds, in the west with strong printing glass edges, westward lamping with tipsy bufflight, the colour of yellow roses. Parlick ridge like a pale goldish skin without body. The plain about Clitheroe was sponged out by a tall white storm of rain. The sun itself and the spot of 'session' dappled with big laps and flowers-in-damask of cloud. But we hurried too fast and it knocked me up. We went to the College, the seminary being wanted for the secular priests' retreat: almost no gas, for the retorts are being mended; therefore candles in bottles, things not ready, darkness and despair. In fact being unwell I was quite downcast: nature in all her parcels and faculties gaped and fell apart, *fatiscebat*, like a clod cleaving and holding only by strings of root. But this must often be

We found the German Divines from Ditton Hall with their rector and professors spending their villa° at the college . . .

Aug. 17—The Germans gave us a concert and again on the 19th, I think, and we a return with jokes of various kinds on the 21st

Aug. 22—We went back to the Seminary

Aug. 27—Farewell concert from the Germans, who went back to Ditton next day. They were kind, amiable, and edifying people. Some of us went down to Whalley with them and afterwards I walked with Herbert Lucas by the river and talked Scotism with him for the last time. In the evening I received orders to go to Roehampton° to teach rhetoric and started next morning early, by Preston, travelling to town with Vaughan and Considine, who were bound for Beaumont. At Manresa I caught the Provincial° who spoke most kindly and encouragingly

[Roehampton]

Sept. 18—At the Kensington Museum.° Bold masterly rudeness of the blue twelvemonth service of plates or platters by Luca Della Robbia—Giovanni's (1260) and Niccola (early in next century) Pisano's pulpits—Bronze gilt doors for Cathedral of Florence by°?—The cartoons° and a full sized chalk drawing from the Transfiguration—Standard portfolios of Indian architecture—also of Michael Angelo's paintings at the Vatican: the *might*, with which I was more deeply struck than ever before, though this was in the dark side courts and I could not see well, seems to come not merely from the simplifying and then amplifying or emphasising of parts but from a masterly realism in the simplification, both these things: there is the simplifying and strong emphasising of anatomy in Rubens, the emphasising and great simplifying in Raphael for instance, and on the other hand the realism in Velasquez, but here force came together from both sides—Thought more highly of Mulready° than ever before—Watts: Two sisters and a couple of Italian peasants with a yoke of oxen—instress of expression in the faces, as in other characteristic English work, Burne Jones', Mason's, Walker's etc—Musical instruments—harpsichords (English for clavecin); spinets (small portable harpsichords); virginals (square, differing from spinet, which is three-cornered like the harpsichord as

cottage piano from grand); dulcimers (this-shaped);

lutes; theorbos; (viols, I think, differ from lutes in having slacks, hollows, in the sides, so as to be the original of the violin); mandolas and mandolines (small lutes, I think); viol-de-gambois (held between the knees); citherns; panduras

Yes, the viol is the origin of the violin. It has been thought the parent of all the viol family is the Welsh crwth. The name looks against this. They are characterised by the bridge and the use of the bow. The viol has 5 strings. Another day at the Kensington I made some notes. The lute is round-bottomed and has frets—Fétis° says 10 and 11 strings, 9 of them double, 3 tuned in unison and 6 in octaves. The theorbo I have noted to have the neck very much put back, two sets of pegs (Fétis says it has 2 fingerboards, the smaller that of the lute, the other, much bigger, with 8 strings for the bass—but my note says it has no frets). The pandora again I have marked as round-bottomed: Fétis says it differs from the lute in being flat and having metal instead of catgut strings. The mandola is round-bottomed, with frets: Fétis says it has 4 strings 'tuned from 5ths to 4ths'. The mandoline he says is smaller and with a fingerboard like a guitar, played with a quill, and the strings tuned in unison with the violin. The cither is very like the guitar, flat-bottomed, with frets

I had a nightmare that night. I thought something or someone leapt onto me and held me quite fast: this I think woke me, so that after this I shall have had the use of reason. This first start is, I think, a nervous collapse of the same sort as when one is very tired and holding oneself at stress not to sleep yet/suddenly goes slack and seems to fall and wakes, only on a greater scale and with a loss of muscular control reaching more or less deep; this one to the chest and not further, so that I could speak, whispering at first, then louder—for the chest is the first and greatest centre of motion and action, the seat of θυμός.° I had lost all muscular stress elsewhere but not sensitive, feeling where each limb lay and thinking that I could recover myself if I could move my finger, I said, and then the arm and so the whole body. The feeling is terrible: the body no longer swayed as a piece by the nervous and muscular instress seems to fall in and hang like a dead weight on the chest. I cried on the holy name and by degrees recovered myself as I thought to do. It made me think that this was how the souls in hell would be imprisoned in their bodies as in prisons and of what St. Theresa says of the 'little press in the wall' where she felt herself to be in her vision

[1874]

May 23—Dark, very heavy, fine rain. The change this morning was not so much from temperate to warm as from cold to temperate, the weather has been so wintry: I even got chilblains again

I went one day to the Academy and again June 12, when Fr. Johnson (Superior in the absence of Fr. Porter, who is gone to take the waters at Carlsbad in Bohemia) kindly sent me to town with Br. Bampton for change. These are the notes on the two days— . . .

Briton Rivière's° Apollo (from Euripides)—Like a roughened boldened Leighton, very fine. Leopards shewing the flow and slow spraying of the streams of spots down from the backbone and making this flow word-in and <u>inscape</u> the whole animal and even the group of them; lion and lioness's paws outlined and threaded round by a touch of fur or what not, as one sees it in cats—very true broad realism; herd of stags between firtrees all giving one inscape in the moulding of their flanks and bodies and hollow shell of the horns . . .

Millais°—*Scotch Firs: 'The silence that is in the lonely woods'*—No such thing, <u>instress</u> absent, firtrunks ungrouped, four or so pairing but not markedly, true bold realism but quite a casual install of woodland with casual heathertufts, broom with black beanpods and so on, but the master shewn in the slouch and toss-up of the firtree-head in near background, in the tufts of fir-needles, and in everything. So too *Winter Fuel: 'Bare ruined choirs'*° etc—almost no sorrow of autumn; a rawness (though I felt this less the second time), unvelvety papery colouring, especially in raw silver and purple birchstems, crude rusty cartwheels, aimless mess or minglemangle of cut underwood in under-your-nose foreground; aimlessly posed truthful child on shaft of cart; but then most masterly Turner-like outline of craggy hill, silver-streaked with birch-trees, which fielded in an equally masterly rust-coloured young oak, with strong curl and seizure in the dead leaves. There were two scales of colour in this picture—browns running to scarlet (in the Red-Riding-Hood girl) and greys to blue (little girl's bow or something) and purple in the smoke on the hill, heather, birchwoods, and in foreground the deep mouldy purple of the stems; then for a gobetween a soft green meadow. There was a beautiful spray-off of the dead oak-scrolls against dark trees behind with flowing blue smoke above. Toss or dance of twig and light-wood hereabouts

July 14—To the House of Commons. The debate was on the Schools Endowment bill moved by Lord Sandon, who spoke well; so did, not *so* well, Mr. Forster in reply. We heard Newdigate. Gladstone was preparing to speak and writing fast but we could not stay to hear him. Lowe, who sat next him, looked something like an apple in the snow

July 23—To Beaumont: it was the rector's day. It was a lovely day: shires-long of pearled <u>cloud under cloud</u>, with a grey stroke underneath marking each row; beautiful blushing yellow in the straw of the uncut ryefields, the wheat looking white and all the ears making a delicate and very true crisping along the top and with just enough air stirring for them to come and go gently; then there were fields reaping. All this I would have looked at again in returning but during dinner I talked too freely and unkindly and had to do penance going home. One field I saw from the balcony of the house behind an elmtree, which it threw up, like a square of pale goldleaf, as it might be, catching the light

Our schools at Roehampton ended with two days of examination before St. Ignatius' feast the 31st. I was very tired and seemed deeply cast down till I had some kind words from the Provincial. Altogether perhaps my heart has never been so burdened and cast down as this year. The tax on my strength has been greater than I have felt before: at least now at Teignmouth I feel myself weak and can do little. But in all this our Lord goes His own way

[St Beuno's, Wales]

Oct. 8—Bright and beautiful day. Crests of snow could be seen on the mountains. Barraud and I walked over to Holywell and bathed at the well° and returned very joyously. The sight of the water in the well as clear as glass, greenish like beryl or aquamarine, trembling at the surface with the force of the springs, and shaping out the five foils of the well quite drew and held my eyes to it. Within a month or six weeks from this (I think Fr. di Pietro said) a young man from Liverpool, Arthur Kent (?), was cured of rupture/ in the water. The strong unfailing flow of the water and the chain of cures from year to year all these centuries took hold of my mind with wonder at the bounty of God in one of His saints, the sensible thing so naturally and gracefully uttering the spiritual reason of its being (which is all in true keeping with the story of St. Winefred's death and recovery) and the spring in place leading back the thoughts by its spring in time to its spring in eternity: even now the stress and buoyancy and abundance of the water is before my eyes

Oct. 12—The bp came, so we got a half holiday and I went with Rickaby to Cwm. We came back by the woods on the Rhuallt and the view was so like Ribblesdale from the fells that you might have thought you were there. The sky was iron grey and the valley, full of Welsh charm and graceful sadness, all in grave colours lay like a painted napkin

Oct. 19—I was there again with Purbrick, at the scaffolding which is left as a mark of the survey at the highest point. We climbed on this and looked round: it was a fresh and delightful sight. The day was rainy and a rolling wind; parts of the landscape, as the Orms' Heads, were blotted out by rain. The clouds westwards were a pied piece—sail-coloured brown and milky blue; a dun yellow tent of rays opened upon the skyline far off. Cobalt blue was poured on the hills bounding the valley of the Clwyd and far in the south spread a bluish damp, but all the nearer valley was showered with tapered diamond flakes of fields in purple and brown and green

Nov. 8—Walking with Wm. Splaine we saw a vast multitude of starlings making an unspeakable jangle. They would settle in a row of trees; then, one tree after another, rising at a signal they looked like a cloud of specks of black snuff or powder struck up from a brush or broom or shaken from a wig; then they would sweep round in whirlwinds—you could see the nearer and farther bow of the rings by the size and blackness; many would be in one phase at once, all narrow black flakes hurling round, then in another; then they would fall upon a field and so on. Splaine wanted a gun: then 'there it would rain meat' he said. I thought they must be full of enthusiasm and delight hearing their cries and stirring and cheering one another

LETTERS

My dear Father,—I must begin with a practical immediate point. The Church strictly forbids all communion in sacred things with non-Catholics. I have only just learnt this, but it prevents me going to chapel, and so yesterday I had to inform the Dean of Chapel. Today the Master° sent for me and said he cd. not grant me leave of absence without an application from you. As the College last term passed a resolution admitting Catholics and took a Catholic into residence it has no right to alter its principle in my case. I wish you therefore not to give yourself the pain of making this application, even if you were willing: I am of age moreover and am alone concerned. If you refuse to make the application, the Master explains that he shall lay my case before the common-room. In this case there is very little doubt indeed that the Fellows wd. take the reasonable course and give me leave of absence fr. chapel, and if not, I am quite contented: but in fact I am satisfied as to the course our Fellows will take and the Master will at the last hesitate to lay the matter before them perhaps even. I want you therefore to write at once, if you will,— not to the Master who has no right to ask what he does, but to me, with a refusal: no harm will follow.

The following is the position of things with me. You ask me to suspend my judgment for a long time, or at the very least more than half a year, in other words to stand still for a time. Now to stand still is not possible, thus: I must either obey the Church or disobey. If I disobey, I am not suspending judgment but deciding, namely to take backward steps fr. the grounds I have already come to. To stand still if it were possible might be justifiable, but to go back nothing can justify. I must therefore obey the Church by ceasing to attend any service of the Church of England. If I am to wait then I must either be altogether without services and sacraments, which you will of course know is impossible, or else I must attend the services of the Church—still being unreceived. But what can be more contradictory than, in order to avoid joining the Church, attending the services of that very Church? Three of my friends, whose conversions were later than mine, Garrett,° Addis,° and Wood,° have already been received, but this is by the way. Only one thing remains to be done: I cannot fight against God Who calls me to His Church: if I were to delay and die in the meantime I shd. have no plea why my soul was not forfeit. I have no power in fact to stir a finger: it is God Who makes the decision and not I.

But you do not understand what is involved in asking me to delay and how little good you wd. get from it. I shall hold as a Catholic what I have long held as an Anglican, that literal truth of our Lord's words by which I learn that the least fragment of the consecrated elements in the Blessed Sacrament of the Altar is the whole Body of Christ born of the Blessed Virgin, before which the whole host of saints and angels as it lies on the altar trembles with adoration. This belief once got is the life of the soul and when I doubted it I shd. become an atheist the next day. But, as Monsignor Eyre° says, it is a gross superstition unless guaranteed by infallibility. I cannot hold this doctrine confessedly except as a Tractarian or a Catholic: the Tractarian ground I have seen broken to pieces under my feet. What end then can be served by a delay in wh. I shd. go on believing this doctrine as long as I believed in God and shd. be by the fact of my belief drawn by a lasting strain towards the Catholic Church?

About my hastiness I wish to say this. If the question which is the Church of Christ? cd. only be settled by laborious search, a year and ten years and a lifetime are too little, when the vastness of the subject of theology is taken into account. But God must have made his Church such as to attract and convince the poor and unlearned as well as the learned. And surely it is true, though it will sound pride to say it, that the judgment of one who has seen both sides for a week is better than his who has seen only one for a lifetime. I am surprised you shd. say fancy and aesthetic tastes have led me to my present state of mind: these wd. be better satisfied in the Church of England, for bad taste is always meeting one in the accessories of Catholicism. My conversion is due to the following reasons mainly (I have put them down without order)—(i) simple and strictly drawn arguments partly my own, partly others', (ii) common sense, (iii) reading the Bible, especially the Holy Gospels, where texts like 'Thou art Peter' (the evasions proposed for this alone are enough to make one a Catholic) and the manifest position of St. Peter among the Apostles so pursued me that at one time I thought it best to stop thinking of them, (iv) an increasing knowledge of the Catholic system (at first under the form of Tractarianism, later in its genuine place), which only wants to be known in order to be loved—its consolations, its marvellous ideal of holiness, the faith and devotion of its children, its multiplicity, its array of saints and martyrs, its consistency and unity, its glowing prayers, the daring majesty of its claims, etc etc. You speak of the claims of the Church of England, but it is to me the strange thing that the Church of England makes no claims: it is true that Tractarians make them for her and find them faintly or only in a few instances borne out for them by her liturgy, and are strongly assailed for their extravagances while they do it. Then about applying to Mr. Liddon° and the Bp. of Oxford. Mr. Liddon writes begging me to pause: it wd. take too long to explain how I did not apply to him at first and why it wd. have been useless. If Dr. Pusey° is in Oxford tomorrow I will see him, if it is any satisfaction to you. The Bishop is too much engaged to listen to individual difficulties and those who do apply to him may get such answers as young Mr. Lane Fox did, who gave up £30,000 a year just lately to become a Catholic. He wrote back about a cob which he wanted to sell to the Dean of some place and wh. Lane Fox was to put his own price on and ride over for the Bishop to the place of sale. In fact Dr. Pusey and Mr. Liddon were the only two men in the world who cd. avail to detain me: the fact that they were Anglicans kept me one, for arguments for the Church of England I had long ago felt there were none that wd. hold water, and when that influence gave way everything was gone.

You are so kind as not to forbid me your house, to which I have no claim, on condition, if I understand, that I promise not to try to convert my brothers and sisters. Before I can promise this I must get permission, wh. I have no doubt will be given. Of course this promise will not apply after they come of age. Whether after my reception you will still speak as you do now I cannot tell.

You ask me if I have had no thought of the estrangement. I have had months to think of everything. Our Lord's last care on the cross was to commend His mother to His Church and His Church to His mother in the person of St. John. If even now you wd. put yourselves into that position wh. Christ so unmistakeably gives us and ask the Mother of sorrows to remember her three hours' compassion at the cross, the piercing of the sword prophecied by Simeon, and her seven dolours, and her spouse Joseph, the lily of chastity, to remember the flight into Egypt, the searching for his Foster-Son at twelve years old, and his last ecstacy with Christ at his death-bed, the prayers of this Holy Family wd. in a few days put an end to estrangements for ever. If you shrink fr. doing this, though the Gospels cry aloud to you to do it, at least for once—if you like, only once—approach Christ in a new way in which you will at all events feel that you are exactly in unison with me, that is, not vaguely, but casting yourselves into His sacred broken Heart and His five adorable Wounds. Those who do not pray to Him in His Passion pray to God but scarcely to Christ. I have the right to propose this, for I have tried both ways, and if you will not give one trial to this way you will see you are prolonging the estrangement and not I.

After saying this I feel lighter-hearted, though I still can by no means make my pen write what I shd. wish. I am your loving son,

Gerard M. Hopkins.

23 New Inn Hall Street, Oct. 17, 1866.

P.S. I am most anxious that you shd. not think of my future. It is likely that the positions you wd. like to see me in wd. have no attraction for me, and surely the happiness of my prospects depends on the happiness to me and not on intrinsic advantages. It is possible even to be very sad and very happy at once and the time that I was with Bridges,° when my anxiety came to its height, was I believe the happiest fortnight of my life. My only strong wish is to be independent.

If you are really willing to make the application to the Master, well and good; but I do not want you to put yourself to pain. I have written a remonstrance to him.

Many thanks to Arthur° for his letter.

To his mother

St. Beuno's. Christmas Eve, 1875.

My dearest Mother,—Many thanks to you for your loving letter and presents and a very happy Christmas to you all. In particular thank Kate° for her letter. I also return Grace's° paper but I am persuaded that if she did write on the subject she would express herself far more simply and intelligibly than Mr. Nicholas Breakspear,° the effect of whose style on me is not, in his own words, 'the impartance of Emotional Pleasure': it is a great babble and he cannot say a plain thing in a plain way (I believe musical people never can). (I have just found that his name is Eustace, not Nicholas, by George.) When I next write to her I will add what little remains to be said. The nasty oilstain on the outside sheet was made in the post: this is the third paper here I have seen in that condition, but the others were worse.

I am obliged for the cuttings, nevertheless you made two oversights. You sent two duplicates, for one thing, and the other was that you omitted the most interesting piece of all, the account of the actual shipwreck°: fortunately I had read it but still I should have been glad to have had it by me to refer to again, for I am writing something on this wreck, which may perhaps appear but it depends on how I am speeded. It made a deep impression on me, more than any other wreck or accident I ever read of.

My gas does flicker but I have ceased to care for it or notice it. My neighbour has got a new burner, lucky for him: it does not perceptibly lessen my light. On the other hand he has lost eight teeth.

Do you know if anything is said of a comet?° I have seen one three nights. It appears to be in Cancer. It is small and pale but quite visible. If it is not a comet it must be a nebula and then it is strange I should not have noticed it before but its appearance is in all respects that of a comet. At ten o'clock it is well visible in the northeast, not high; later it would be higher.

Have you guessed the charade in the Xmas *Illustrated?* I have.

Where is Aunt Anne spending her Christmas? and where Aunt Kate?

With the best Christmas wishes to all I remain your loving son

Gerard M. Hopkins S.J.

To Bridges

St. Beuno's, St. Asaph. Aug. 21 1877.

Dearest Bridges,—Your letter cannot amuse Father Provincial, for he is on the unfathering deeps outward bound to Jamaica: I shd. not think of telling you anything about his reverence's goings and comings if it were it° not that I know this fact has been chronicled in the Catholic papers.

Enough that it amuses me, especially the story about Wooldridge° and the Wagnerite, wh. is very good.

Your parody reassures me about your understanding the metre. Only remark, as you say that there is no conceivable licence I shd. not be able to justify, that with all my licences, or rather laws, I am stricter than you and I might say than anybody I know. With the exception of the *Bremen* stanza,° which was, I think, the first written after 10 years' interval of silence, and before I had fixed my principles, my rhymes are rigidly good—to the ear—and such rhymes as *love* and *prove* I scout utterly. And my quantity is not like 'Fífty̆twō Bĕdfŏrd Squāre', where *fífty̆* might pass but *Bĕdfŏrd* I should never admit. Not only so but Swinburne's dactyls and anapaests are halting to my ear: I never allow e.g. *I* or *my* (that is diphthongs, for *I=a+i* and *my=ma+i*) in the short or weak syllables of those feet, excepting before vowels, semi-vowels, or *r*, and rarely then, or when the measure becomes (what is the word?) molossic°—thus: $\cup-\cup|\cup-\cup|\cup-\cup$, for then the first short is almost long. If you look again you will see. So that I may say my apparent licences are counterbalanced, and more, by my strictness. In fact all English verse, except Milton's, almost, offends me as 'licentious'. Remember this.

I do not of course claim to have invented *sprung rhythms* but only *sprung rhythm*; I mean that single lines and single instances of it are not uncommon in English and I have pointed them out in lecturing— e.g. 'why should this : desert be?'—which the editors have variously amended; 'There to meet : with Macbeth' or 'There to meet with Mac : beth'; Campbell has some throughout the *Battle of the Baltic*—'and their fleet along the deep : proudly shone'—and *Ye Mariners*—'as ye sweep : through the deep' etc; ... and, not to speak of *Pom pom*, in Nursery Rhymes, Weather Saws, and Refrains they are very common—but what I do in the *Deutschland* etc is to enfranchise them as a regular and permanent principle of scansion.

There are no outriding feet in the *Deutschland*. An outriding foot is, by a sort of contradiction, a recognized extra-metrical effect; it is and it is not part of the metre; not part of it, not being counted, but part of it by producing a calculated effect which tells in the general success. But the long, e.g. seven-syllabled, feet of the *Deutschland*, are strictly metrical. Outriding feet belong to counterpointed verse, which supposes a well-known and unmistakeable or unforgetable standard rhythm: the *Deutschland* is not counterpointed; counterpoint is excluded by sprung rhythm. But in some of my sonnets I have mingled the two systems°: this is the most delicate and difficult business of all.

The choruses in *Samson Agonistes* are intermediate between counter-pointed and sprung rhythm. In reality they are sprung, but Milton keeps up a fiction of counterpointing the heard rhythm (which is the same as the mounted rhythm) upon a standard rhythm which is never heard but only counted and therefore really does not exist. The want of a metrical notation and the fear of being thought to write mere rhythmic or (who knows what the critics might not have said?) even unrhythmic prose drove him to this. Such rhythm as French and Welsh poetry has is sprung, counterpointed upon a counted rhythm, but it differs from Milton's in being little calculated, not more perhaps than prose con-sciously written rhythmically, like orations for instance; it is in fact the *native rhythm* of the words used bodily imported into verse; whereas Milton's mounted rhythm is a real poetical rhythm, having its own laws and recurrence, but further embarassed by having to count.

Why do I employ sprung rhythm at all? Because it is the nearest to the rhythm of prose, that is the native and natural rhythm of speech, the least forced, the most rhetorical and emphatic of all possible rhythms, combining as it seems to me, opposite and, one wd. have thought, incompatible excellences, markedness of rhythm—that is rhythm's self—and naturalness of expression—for why, if it is forcible in prose to say 'lashed : rod', am I obliged to weaken this in verse, which ought to be stronger, not weaker, into 'láshed birch-ród' or something?

My verse is less to be read than heard, as I have told you before; it is oratorical, that is the rhythm is so. I think if you will study what I have here said you will be much more pleased with it and may I say? converted to it.

You ask may you call it 'presumptious jugglery'. No, but only for this reason, that *presumptious* is not English.

I cannot think of altering anything. Why shd. I? I do not write for the public. You are my public and I hope to convert you.

You say you wd. not for any money read my poem again. Nevertheless I beg you will. Besides money, you know, there is love. If it is obscure do not bother yourself with the meaning but pay attention to the best and most intelligible stanzas, as the two last of each part and the narrative of the wreck. If you had done this you wd. have liked it better and sent me some serviceable criticisms, but now your criticism is of no use, being only a protest memorialising me against my whole policy and proceedings.

I may add for your greater interest and edification that what refers to myself in the poem is all strictly and literally true and did all occur; nothing is added for poetical padding.

Believe me your affectionate friend

Gerard M. Hopkins S.J.

To Bridges

Mount St. Mary's College, Chesterfield. April 2 1878.

My dearest Bridges,—Your last letter was very kind indeed, but I should have lost all shame if under any circumstances I had allowed such a thing to be as for you to come hundreds of miles to cure me.°

I am overjoyed to hear of your and Mrs. Molesworth's° intercourse with Oak Hill.

It was pleasing and flattering to hear that Mr. Pater° remembers and takes an interest in me.

My muse turned utterly sullen in the Sheffield smoke-ridden air and I had not written a line till the foundering of the Eurydice the other day and that worked on me and I am making a poem—in my own rhythm but in a measure something like Tennyson's *Violet*° (bound with *Maud*); e.g.—

They say who saw one sea-corpse cold
How hé was of lovely manly mould,
    Every inch a tar,
Of the bést we bóast séamen áre.

Look, from forelock down to foot he,
Strung by duty is strained to beauty
    And russet-of-morning-skinned
With the sún, sált, and whírling wínd.

Oh! his nímble fínger, his gnárled gríp!
Léagues, léagues of séamanshíp
    Slumber in his forsaken
Bones and will not, will not waken.

I have consistently carried out my rhyming system, using the first letter of the next line to complete the rhyme in the line before it.

Well, write those things that 'will tickle me'.

The Deutschland would be more generally interesting if there were more wreck and less discourse, I know, but still it is an ode and not primarily a narrative. There is some narrative in Pindar but the principal business is lyrical. This poem on the Eurydice is hitherto almost all narrative however.

And what are you doing?

From notices in the *Athenaeum* it would appear that Gosse, Dobson, and Co. are still fumbling with triolets, villanelles, and what not.

Believe me your affectionate friend

Gerard M. Hopkins S.J.

April 3.

To Bridges

Stonyhurst College, Blackburn (or Whalley). May 13 1878.°

Dearest Bridges,—Remark the above address. After July I expect to be stationed in town—111 Mount Street, Grosvenor Square.

I hope your bad cold is gone.

I am very glad to hear the Rondeliers have come to see the beauty of your poetry. I have little acquaintance with their own. I have read a rondeau or rondel by Marzials° in the *Athenaeum* beginning and ending 'When I see you': it was very graceful and shewing an art and finish rare in English verse . . . I think that school is too artificial and exotic to take root and last, is it not?

I enclose you my Eurydice, which the *Month* refused. It is my only copy. Write no bilgewater about it: I will presently tell you what that is and till then excuse the term. I must tell you I am sorry you never read the Deutschland again.

Granted that it needs study and is obscure, for indeed I was not over-desirous that the meaning of all should be quite clear, at least unmistake-able, you might, without the effort that to make it all out would seem to have required, have nevertheless read it so that lines and stanzas should be left in the memory and superficial impressions deepened, and have liked some without exhausting all. I am sure I have read and enjoyed pages of poetry that way. Why, sometimes one enjoys and admires the very lines one cannot understand, as for instance 'If it were done when 'tis done' sqq.,° which is all obscure and disputed, though how fine it is everybody sees and nobody disputes. And so of many more passages in Shakspere and others. Besides you would have got more weathered to the style and its features—not really odd. Now they say that vessels sailing from the port of London will take (perhaps it should be / used once to take) Thames water for the voyage: it was foul and stunk at first as the ship worked but by degrees casting its filth was in a few days very pure and sweet and wholesomer and better than any water in the world. However that maybe, it is true to my purpose. When a new thing, such as my ventures in the Deutschland are, is presented us our first criticisms are not our truest, best, most homefelt, or most lasting but what come easiest on the instant. They are barbarous and like what the ignorant and the ruck say. This was so with you. The Deutschland on her first run worked very much and unsettled you, thickening and clouding your mind with vulgar mudbottom and common sewage (I see that I am going it with the image) and just then unhappily you *drew off* your criticisms all stinking (a necessity now of the image) and bilgy, whereas if you had let your thoughts cast themselves they would have been clearer in them-selves and more to my taste too. I did not heed them therefore, perceiving they were a first drawing-off. Same of the Eurydice—which being short and easy please read more than once.

Can you tell me who that critic in the *Athenaeum* is that writes very long reviews° on English and French poets, essayists, and so forth in a style like De Quincey's, very acute in his remarks, provoking, jaunty, and (I am sorry to say) would-be humorous? He always quotes Persian stories (unless he makes them up) and talks about Rabelæsian humour.

My brother's pictures, as you say, are careless and do not aim high, but

I don't think it would be much different if he were a batchelor. But, strange to say—and I shd. never even have suspected it if he had not quite simply told me—he has somehow in painting his pictures, though nothing that the pictures express, a high and quite religious aim; however I cannot be more explanatory.

Your bodysnatch story is ghastly, but so are all bodysnatch stories. My grandfather was a surgeon, a fellow-student of Keats', and once conveyed a body through Plymouth at the risk of his own.

Believe me your affectionate friend

Gerard M. Hopkins S.J.

May 21 1878

Please remember me very kindly to your mother.

To do the Eurydice any kind of justice you must not slovenly read it with the eyes but with your ears, as if the paper were declaiming it at you. For instance the line 'she had come from a cruise training seamen' read without stress and declaim is mere Lloyd's Shipping Intelligence; properly read it is quite a different thing. Stress is the life of it.

To Bridges

Stonyhurst, Blackburn. May 30 1878.

Dearest Bridges,—It gave me of course great comfort to read your words of praise. But however, praise or blame, never mingle with your criticisms monstrous and indecent spiritual compliments like something you have said there.

I want to remark on one or two things.

How are hearts of oak furled?° Well, in sand and sea water. The image comes out true under the circumstances, otherwise it could not hold together. You are to suppose a stroke or blast in a forest of 'hearts of oak' (=, ad propositum, sound oak-timber) which at one blow both lays them low and buries them in broken earth. *Furling* (*ferrule* is a blunder for *furl*, I think) is *proper* when said of sticks and staves.

So too of *bole*, I don't see your objection here at all. It is not only used by poets but seems technical and *proper* and in the mouth of timber merchants and so forth.

'This was that fell capsize' is read according to the above stresses—two cretics, so to say.

I don't see the difficulty about the 'lurch forward'? Is it in the scanning? which is imitative as usual—an anapaest, followed by a trochee, a dactyl, and a syllable, so that the rhythm is anacrustic or, as I should call it, 'encountering'.

'Cheer's death' = the death of cheer = the dying out of all comfort = despair.

'It is even seen'—You mistake the sense of this as I feared it would be mistaken. I believed Hare to be a brave and conscientious man: what I say is that 'even' those who seem unconscientious will act the right part at a great push.

About 'mortholes' I do wince a little but can not now change it. What I dislike much more however is the rhyme 'foot he' to *duty* and *beauty*. In fact I cannot stand it and I want the stanza corrected thus—

> Look, foot to forelock, how all things suit! he
> Is strung by duty, is strained to beauty,
>   And brown-as-dawning-skinned
> With brine and shine and whirling wind.

The difficulty about the Milky Way is perhaps because you do not know the allusion: it is that in Catholic times Walsingham Way was a name for the Milky Way, as being supposed a fingerpost to our Lady's shrine at Walsingham.

'O well wept' should be written asunder, not 'wellwept'. It means 'you do well to weep' and is framed like 'well caught' or 'well run' at a cricketmatch.

Obscurity I do and will try to avoid so far as is consistent with excellences higher than clearness at a first reading. This question of obscurity we will some time speak of but not now. As for affectation I do not believe I am guilty of it: you should point out instances, but as long as mere novelty and boldness strikes you as affectation your criticism strikes me as—as water of the Lower Isis.°

I see I have omitted one or two things. If the first stanza is too sudden it can be changed back to what it was at first—

> The Eurydice—it concerned thee, O Lord:
> 4   5   1   2   3
> O alas! three hundred hearts on board—

But then it will be necessary to change the third stanza as follows, which you will hardly approve—

> Did she pride her, freighted fully, on
> Bounden bales or a hoard of bullion?—

About 'grimstones' you are mistaken. It is not the remains of a rhyme to *brimstone*. I *could* run you some rhymes on it. You must know, we have a Father Grimstone in our province.

I shall never have leisure or desire to write much. There is one thing I should like to get done; an ode on the Vale of Clwyd begun therein. It would be a curious work if done. It contains metrical attempts other than any you have seen, something like Greek choruses, a peculiar eleven-footed line for instance.

What you have got of mine you may do as you like with about shewing to friends.

Is your own ode on Eurydice done? Will you send it, as well as other things; which shall be returned.

Believe me your affectionate friend

Gerard M. Hopkins S.J.

You are kind enough to want me to dine with you on coming up to town. I should have to go to our house at once. I shall have, no doubt, little time when in London but still we shall manage to meet.

May 31.

Postcard to Bridges

I forgot to answer about my metres (rhythms rather, I suppose). Do by all means° and you will honour them and me.

G.M.H.

June 9 1878

To Bridges

St. Giles's, Oxford. Feb. 15 '79.

Dearest Bridges,—I should have added in my last that the *Silver Jubilee*° had been published. It was printed at the end of a sermon, bearing the same title and due to the same occasion, of Fr. John Morris's of our Society. I have found it since I wrote and the copy I sent you from memory is not quite right. The third stanza should stand fourth and run—

> Not today we need lament
> Your lot of life is some way spent:
>   Toil has shed round your head
> Silver, but for Jubilee.

The thought is more pointed. Please correct it if you put it into your album.°

No, do not ask Gosse anything of the sort. (1) If I were going to publish, and that soon, such a mention would be 'the puff preliminary',° which it wd. be dishonourable of me to allow of. (2) If I did, a mention in one article of one review would do very little indeed, especially as publishing now is out of the question. (3) When I say that I do not mean to publish I speak the truth. I have taken and mean to take no step to do so beyond the attempt I made to print my two wrecks in the *Month*. If some one in authority knew of my having some poems printable and suggested my doing it I shd. not refuse, I should be partly, though not altogether, glad. But that is very unlikely. All therefore that I think of doing is to keep my verses together in one place—at present I have not even correct copies—, that, if anyone shd. like, they might be published after my death. And that again is unlikely, as well as remote. I could add other considerations, as that if I meant to publish at all it ought to be more or ought at least to be followed up, and how can that be? I cannot in conscience spend time on poetry, neither have I the inducements and inspirations that make others compose. Feeling, love in particular, is the great moving power and spring of verse and the only person that I am in love with seldom, especially now, stirs my hearts sensibly and when he does I cannot always 'make capital' of it, it would be a sacrilege to do so. Then again I have of myself made verse so laborious.

No doubt my poetry errs on the side of oddness. I hope in time to have a more balanced and Miltonic style. But as air, melody, is what strikes me most of all in music and design in painting, so design, pattern or what I am in the habit of calling 'inscape' is what I above all aim at in poetry. Now it is the virtue of design, pattern, or inscape to be distinctive and it is the vice of distinctiveness to become queer. This vice I cannot have escaped. However 'winding the eyes'° is queer only if looked at from the wrong point of view: looked at as a motion in and of the eyeballs it is what you say, but I mean that the eye winds/only in the sense that its focus or point of sight winds and that coincides with a point of the object and winds with that. For the object, a lantern passing further and further away and bearing now east, now west of one right line, is truly and properly described as winding. That is how it should be taken then.°

To Bridges

St. Giles's, Oxford. May 26 1879.

Dearest Bridges,—Your answerable letterage is three deep at least, but nevertheless work is work and of late Fr. Parkinson has sprung a leak (exema) in his leg and been laid up and I in consequence laid on all the harder: indeed he will never, I believe, be very active more, though now he does go about a little.

I shall be very glad to have your brother's book° when it appears, and to trace the prototype of you in it will be very interesting.

I have seen no more reviews of you.

The poem you send° is fine in thought, but I am not satisfied with the execution altogether: the pictures, except in the first stanza, are somewhat wanting in distinction (I do not of course mean distinctness), and I do not think the rhythm perfect, e.g. 'woodbine with' is a heavy dactyl. Since the syllables in sprung rhythm are not counted, time or equality in strength is of more importance than in common counted rhythm, and your times or strengths do not seem to me equal enough. The line you mark does resemble something in the Deutschland, now that you point it out, but there is no resemblance in the thought and it does not matter. I do not think the line very good; it is besides ambiguous. I understand, I believe everybody would understand, 'O if it were only for thee' to mean/If I had no guide (to nature's true meaning) but thee: the leading thought is that nature has two different, two opposite aspects, teaching opposite lessons of life—that one is between two stools with the two of them. Is it not? The whole mood and vein is remote; unknown to many temperaments; ineffective, I should think, with any; belonging to the world of imagination, but genuinely so. I believe you might have expressed it more pointedly though.

Of course I am very much pleased that you like my period–building (or whatever we are to call it) but do not see what is the matter with Patmore's.° It is his Unknown Eros you refer to, I suppose. The faults I see in him are bad rhymes; continued obscurity; and, the most serious, a certain frigidity when, as often, the feeling does not flush and fuse the language. But for insight he beats all our living poets, his insight is really profound, and he has an exquisiteness, farfetchedness, of imagery worthy of the best things of the Caroline age. However I cannot spend more time on his praises.

I agree with you that English terza rima is (so far as I have seen it) badly made and tedious and for the reason you give, but you are mistak[en] in thinking the triplet structure is unknown: Shelley's West Wind ode (if I mistake not) and some other ones are *printed* in detached 3-line stanzas. I wrote a little piece° so printed when at school and published it in *Once a Week*.

The sestet of the Purcell sonnet° is not so clearly worked out as I could wish. The thought is that as the seabird opening his wings with a whiff of wind in your face means the whirr of the motion, but also unaware gives you a whiff of knowledge about his plumage, the marking of which stamps his species, that he does not mean, so Purcell, seemingly intent only on the thought or feeling he is to express or call out, incidentally lets you remark the individualising marks of his own genius.

Sake is a word I find it convenient to use: I did not know when I did so first that it is common in German, in the form *sach*. It is the *sake* of 'for the sake of', *forsake, namesake, keepsake*. I mean by it the being a thing has outside itself, as a voice by its echo, a face by its reflection, a body by its shadow, a man by his name, fame, or memory, *and also* that in the thing by virtue of which especially it has this being abroad, and that is something distinctive, marked, specifically or individually speaking, as for a voice and echo clearness; for a reflected image light, brightness; for a shadow-casting body bulk; for a man genius, great achievements, amiability, and so on. In this case it is, as the sonnet says, distinctive quality in genius.

Wuthering is a Northcountry word for the noise and rush of wind: hence Emily Brontë's 'Wuthering Heights'.

By *moonmarks* I mean crescent shaped markings on the quillfeathers, either in the colouring of the feather or made by the overlapping of one on another.

My sister Kate is staying here with my aunt Mrs. Marsland Hopkins (who has now a house in Holywell).

Believe me your affectionate friend

Gerard M. Hopkins S.J.

May 31 1879.

To Bridges

St. Giles's, Oxford. Aug. 14 1879.

My dearest Bridges,—I must try and tersely scribble you something.

That German word is *sache*, not *sach*, except in compounds: you should have set me right.

Your Picnic verses° are very good, the rhymes capital, beyond the ingenuity I credited you with. Some lines however are faulty, as 'Anything more delicious'.

Muirhead,° who called here on Sunday, was on that party. I mean Muirhead, who was on that party, called etc.

I wish you would send me all the music you have,° to try. I wd. return it. I do not yet the present piece nor comment on it, as I have not had an opportunity of hearing it. I feel sure you have a genius in music—on the strength of the only piece I know 'O earlier': it is an inspiration of melody, but somewhat 'sicklied o'er', as indeed the words are.

To rejoin on some points of your criticisms. Though the analogy in the Candle sonnet° may seem forced, yet it is an 'autobiographical' fact that I was influenced and acted on the way there said.

I send a recast of the Handsome Heart.° Nevertheless the offence of the rhymes is repeated. I felt myself the objection you make and should only employ the device very sparingly, but you are to know that it has a particular effect, an effect of climax, and shd. so be read, with a rising inflection, after which the next line, beginning with the enclitic, grace-fully falls away. And in like manner with proclitics and so on: if a strong word and its epithet or other appendage are divided so that the appendage shall end one line and the supporting word begin the next, the last becomes emphasised by position and heads a fall-away or diminuendo. These little graces help the 'over-reaving' of the verse at which I so much aim, make it flow in one long strain to the end of the stanza and so forth.

I am somewhat surprised at your liking this sonnet so much. I thought it not very good. The story was that last Lent, when Fr. Parkinson was laid up in the country, two boys of our congregation gave me much help in the sacristy in Holy Week. I offered them money for their services, which the elder refused, but being pressed consented to take it laid out in a book. The younger followed suit; then when some days after I asked him what I shd. buy answered as in the sonnet. His father is Italian and therefore sells ices. I find within my professional experience now a good deal of matter to write on. I hope to enclose a little scene° that touched me at Mount St. Mary's. It is something in Wordsworth's manner; which is, I know, inimitable and unapproachable, still I shall be glad to know if you think it a success, for pathos has a point as precise as jest has and its happiness 'lies ever in the ear of him that hears, not in the mouth of him that makes'. I hope also soon to shew you a finer thing,° in a metre something like the Eurydice, not quite finished yet; also a little song° not unlike 'I have loved flowers that fade'. I have added some strokes to the Vale of Clwyd and have hopes of some day finishing it: it is more like your Hymn to Nature than anything else I can think of, the rhythm however widely unlike. Lastly I enclose a sonnet° on which I invite minute criticism. I endeavoured in it at a more Miltonic plainness and severity than I have anywhere else. I cannot say it has turned out severe, still less plain, but it seems almost free from quaintness and in aiming at one excellence I may have hit another.

I had quite forgotten the sonnet you have found, but can now recall almost all of it; not so the other piece, birthday lines° to me sister, I fancy.

Baliol is the old spelling and the one I prefer, but they have adopted Balliol and one must conform.

I was almost a great admirer of Barnes' Dorset (not Devon) poems. I agree with Gosse, not with you. A proof of their excellence is that you may translate them and they are nearly as good—I say nearly, because if the dialect plays any lawful part in the effect they ought to lose something in losing that. Now Burns loses prodigiously by translation. I have never however read them since my undergraduate days except the one quoted in Gosse's paper,° the beauty of which you must allow. I think the use of dialect a sort of unfair play, giving, as you say, 'a peculiar but shortlived charm', setting off for instance a Scotch or Lancashire joke which in standard English comes to nothing. But its lawful charm and use I take to be this, that it sort of guarantees the spontaneousness of the thought and puts you in the position to appraise it on its merits as coming from nature and not books and education. It heightens one's admiration for a phrase just as in architecture it heightens one's admiration of a design to know that it is old work, not new: in itself the design is the same but as taken together with the designer and his merit this circumstance makes a world of difference. Now the use of dialect to a man like Barnes is to tie him down to the things that he or another Dorset man has said or might say, which though it narrows his field heightens his effects. His poems use to charm me also by their Westcountry 'instress', a most peculiar product of England, which I associate with airs like Weeping Winefred, Polly Oliver, or Poor Mary Ann, with Herrick and Herbert, with the Worcestershire, Herefordshire, and Welsh landscape, and above all with the smell of oxeyes and applelofts: this instress is helped by particular rhythms and these Barnes employs; as, I remember, in 'Linden Ore'° and a thing with a refrain like 'Alive in the Spring' . . .

I should be very glad to see your prose of Michelangelo's sonnets° and also your verse, for though I do not like verse-renderings of verse (according to the saying *Traduttore traditore*), yet I think you could do them if anyone can. I have seen something of them, in particular a most striking one beginning—

Non ha l'ottimo artista alcun concetto.°

By the by, inversions—As you say, I do avoid them, because they weaken and because they destroy the earnestness or in-earnestness of the utterance. Nevertheless in prose I use them more than other people, because there they have great advantages of another sort. Now these advantages they should have in verse too, but they must not seem to be due to the verse; that is what is so enfeebling (for instance the finest of your sonnets to my mind has a line enfeebled by inversion plainly due to the verse, as I said once before "Tis joy the falling of her fold to view"°— but how it should be mended I do not see). As it is, I feel my way to their use. However in a nearly finished piece I have a very bold one indeed. So also I cut myself off from the use of *ere, o'er, wellnigh, what time, say not* (for *do not say*), because, though dignified, they neither belong to nor ever cd. arise from, or be the elevation of, ordinary modern speech. For it seems to me that the poetical language of an age shd. be the current language heightened, to any degree heightened and unlike itself, but not (I mean normally: passing freaks and graces are another thing) an obsolete one. This is Shakespeare's and Milton's practice and the want of it will be fatal to Tennyson's Idylls and plays, to Swinburne, and perhaps to Morris.

21 Trenchard Street, Bristol. Aug. 21. I am spending a few days here. I have roughly finished the little song and enclose it.

Remember me very kindly to Mrs. Molesworth and believe me your loving friend

Gerard M. Hopkins S.J.

To Bridges

Oct. 25 [1879] . . . I think then no one can admire beauty of the body more than I do, and it is of course a comfort to find beauty in a friend or a friend in beauty. But this kind of beauty is dangerous. Then comes the beauty of the mind, such as genius, and this is greater than the beauty of the body and not to call dangerous. And more beautiful than the beauty of the mind is beauty of character, the 'handsome heart'. Now every beauty is not a wit or genius nor has every wit or genius character. For though even bodily beauty, even the beauty of blooming health, is from the soul, in the sense, as we Aristotelian Catholics say, that the soul is the form of the body, yet the soul may have no other beauty, so to speak, than that which it expresses in the symmetry of the body—barring those blurs in the cast which wd. not be found in the die or the mould. This needs no illustration, as all know it. But what is more to be remarked is that in like manner the soul may have no further beauty than that which is seen in the mind, that there may be genius uninformed by character. I sometimes wonder at this in a man like Tennyson: his gift of utterance is truly golden, but go further home and you come to thoughts commonplace and wanting in nobility (it seems hard to say it but I think you know what I mean). In Burns there is generally recognized on the other hand a richness and beauty of manly character which lends worth to some of his smallest fragments, but there is a great want in his utterance; it is never really beautiful, he had no eye for pure beauty, he gets no nearer than the fresh picturesque expressed in fervent and flowing language (the most strictly beautiful lines of his that I remember are those in Tam o' Shanter: 'But pleasures are like poppies spread' sqq. and those are not). Between a fineness of nature which wd. put him in the first rank of writers and a poverty of language which puts him in the lowest rank of poets, he takes to my mind, when all is balanced and cast up, about a middle place. . . .

To R. W. Dixon (22 December 1880)

. . . insight is more sensitive, in fact is more perfect, earlier in life than later and especially towards elementary impressions: I remember that crimson and pure blues seemed to me spiritual and heavenly sights fit to draw tears once; now I can just see what I once saw, but can hardly dwell on it and should not care to do so. Another is—or it comes to one of the above—the greater demand for perfection in the work, the greater impatience with technical faults. In the particular case of Tennyson's Ode to Memory I find in my own case all these: it has a mysterious stress of feeling, especially in the refrain—I am to my loss less sensitive to that; it has no great meaning of any importance nor power of thought—I am to my advantage more alive to that; from great familiarity with the style I am deadened to its individuality and beauty, which is again my loss; and I perceive the shortcomings of the execution, which is my own advance in critical power. Absolutely speaking, I believe that if I were now reading Tennyson for the first time I should form the same judgment of him that I form as things are, but I should not feel, I should lose, I should never have gone through, that boyish stress of enchantment that this Ode and the *Lady of Shalott* and many other of his pieces once laid me under. Rose Hall, Lydiate (a country house where I sometimes spend a night as occasion requires and take the opportunity to write my letters). Jan. 11 1880. And here I must stop for tonight.

Jan. 14 8 Salisbury Street, Liverpool—The new prosody, Sprung Rhythm, is really quite a simple matter and as strict as the other rhythm. Bridges treats it in theory and practice as something informal and variable without any limit but ear and taste, but this is not how I look at it. We must however distinguish its $εἶναι$ and its $εὖ εἶναι$, the writing it somehow and the writing it as it should be written; for written anyhow it is a shambling business and a corruption, not an improvement. In strictness then and simple εἶναι it is a matter of accent only, like common rhythm, and not of quantity at all. Its principle is that all rhythm and all verse consists of feet and each foot must contain one stress or verse-accent: so far is common to it and Common Rhythm; to this it adds that the stress alone is essential to a foot and that therefore even one stressed syllable may make a foot and consequently two or more stresses may come running, which in common rhythm can, regularly speaking, never happen. But there may and mostly there does belong to a foot an unaccented portion or 'slack': now in common rhythm, in which less is made of stress, in which less stress is laid, the slack must be always one or else two syllables, never less than one and never more than two, and in most measures fixedly one or fixedly two, but in sprung rhythm, the stress being more *of* a stress, being more important, allows of greater variation in the slack and this latter may range from three syllables to none at all—*regularly*, so that paeons (three short syllables and one long or three slack and one stressy) are regular in sprung rhythm, but in common rhythm can occur only by licence; moreover may in the same measure have this range. Regularly then the feet in sprung rhythm consist of one, two, three, or four syllables and no more, and if for simplicity's sake we call feet by Greek names, taking accent for quantity, and also scan always as for rising rhythm (I call *rising rhythm* that in which the slack comes first, as in iambs and anapaests, *falling* that in which the stress comes first, as in trochees and dactyls), scanning thus, the feet in sprung rhythm will be monosyllables, iambs, anapaests, and fourth paeons, and no others. But for particular rhythmic effects it is allowed, and more freely than in common rhythm, to use any number of slack syllables, limited only by ear. And though it is the virtue of sprung rhythm that it allows of 'dochmiac' or 'antispastic' effects or cadences, when the verse suddenly changes from a rising to a falling movement, and this too is strongly felt by the ear, yet no account of it is taken in scanning and no irregularity caused, but the scansion always treated, conventionally and and for simplicity, as rising. Thus the line 'She had cóme from a crúise, tráining séamen' has a plain reversed rhythm, but the scanning is simply 'She had cóme | from a crúise | tráin | ing séa | men'—that is/rising throughout, having one monosyllabic foot and an overlapping syllable which is counted to the first foot of the next line. Bridges in the preface to his last issue says something to the effect that all sorts of feet may follow one another, an anapaest a dactyl for instance (which would make four slack syllables running): so they may, if we look at the real nature of the verse; but for simplicity it is much better to recognize, in scanning this new rhythm, only one movement, either the rising (which I choose as being commonest in English verse) or the falling (which is perhaps better in itself), and always keep to that.

In lyric verse I like sprung rhythm also to be *over-rove*, that is the scanning to run on from line to line to the end of the stanza. But for dramatic verse, which is looser in form, I should have the lines 'free-ended' and each scanned by itself.

Sprung rhythm does not properly require or allow of counterpoint. It does not require it, because its great variety amounts to a counterpointing, and it scarcely allows of it, because you have scarcely got in it that conventionally fixed form which you can mentally supply at the time when you are actually reading another one—I mean as when in reading 'Bý the wáters of life where'er they sat' you mentally supply 'By thé watérs', which is the normal rhythm. Nevertheless in dramatic verse I should sparingly allow it at the beginning of a line and after a strong caesura, and I see that Bridges does this freely in *London Snow* for instance. However by means of the 'outrides' or looped half-feet you will find in some of my sonnets and elsewhere I secure a strong effect of double rhythm, of a second movement in the verse besides the primary and essential one, and this comes to the same thing or serves the same purpose as counterpointing by reversed accents as in Milton.

But for the εὖ εἶναι of the new rhythm great attention to quantity is necessary. And since English quantity is very different from Greek or Latin a sort of prosody ought to be drawn up for it, which would be indeed of wider service than for sprung rhythm only. We must distinguish strength (or gravity) and length. About length there is little difficulty: plainly *bidst* is longer than *bids* and *bids* than *bid*. But it is not recognized by everybody that *bid*, with a flat dental, is graver or stronger than *bit*, with a sharp. The strongest and, other things being alike, the longest syllables are those with the circumflex, like *fire*. Any syllable ending in *ng*, though *ng* is only a single sound, may be made as long as you like by prolonging the nasal. So too *n* may be prolonged after a long vowel or before a consonant, as in *soon* or *and*. In this way a great number of observations might be made: I have put these down at random as samples. You will find that Milton pays much attention to consonant-quality or gravity of sound in his line endings. Indeed every good ear does it naturally more or less/ in composing. The French too say that their feminine ending is graver than the masculine and that pathetic or majestic lines are made in preference to end with it. One may even by a consideration of what the music of the verse requires restore sometimes the pronunciation of Shakspere's time where it has changed and shew for instance that *cherry* must have been *cher-ry* (like *her, stir, spur*) or that *heavy* was *heave-y* in the lines 'Now the heavy ploughman snores All with weary task foredone'. You speak of the word *over*. The *o* is long no doubt, but long *o* is the shortest of the long vowels and may easily be used in a weak place; I do not however find that Tennyson uses it so in the Ode to Memory: in the line 'Over the dewy dark [or 'dark dewy'] earth forlorn' it seems to be in a strong place.

I will inclose a little piece I composed last September in walking from Lydiate.° It is to have some plainsong music to it. I found myself quite unable to redeem my promise of copying you out the pieces you had not seen: time would not allow it. However I think you have seen them since in Bridges' book.° Liverpool is of all places the most museless. It is indeed a most unhappy and miserable spot. There is moreover no time for writing anything serious—I should say for composing it, for if it were made it might be written.

I do not despair of our coming to meet, for business might perhaps bring you here. Meanwhile believe me your affectionate friend

                                         Gerard M. Hopkins S.J.

Jan. 14 1881.
You will then send the poems, I hope, as soon as possible.
Jan. 16—I have added another piece, the *Brothers*.

                         To Bridges

. . . write more would weary you.

You asked me not long ago about my dramatic poem. It is a play:° I do not hold with dramatic poems. That is it will be if ever it gets done. But since I have been here it has made no way. At Hampstead I did some dozen lines or less. Every impulse and spring of art seems to have died in me, except for music, and that I pursue under almost an impossibility of getting on. Nevertheless I still put down my pieces, for the airs seem worth it; they seem to me to have something in them which other modern music has not got. I have now also one little piece harmonised: it is only two part counterpoint at present, but it sounds impressive and is a vast improvement on the naked air. If I could only finish the harmony to 'Thou didst delight mine eyes' I hope you would like it.

I could say plenty more, but this has been kept far too long already. Believe me your affectionate friend

                                         Gerard M. Hopkins S.J.

I expect not long hence to leave Liverpool. April 3 1881 . . .

To Bridges

St. Joseph's, North Woodside Road, Glasgow.
Sept. 16 1881.

Dearest Bridges,—How is it you do not know I am here? On Oct. 10 I am
to be at Roehampton (Manresa House, as of old) to begin, my 'tertianship', the third year (really ten months) of noviceship which we undergo
before taking our last vows. Till then I expect to be here mostly, but must
go to Liverpool to pack; for I came for a fortnight or so only and left my
things: indeed I am going to pieces as I stand.

I began a letter to you a short while ago, but tore it up. Meant to write I
have every day for long.

I am very glad you do improve. Still your recovery is very slow and I
cannot understand it. You did run well, like the Galatians:° how has your
good constitution been so unhappily bewitched? I hope nevertheless that
all your strength will return.

And the good Canon° too lies like a load on my heart. To him I am
every day meaning to write and last night it was I began, but it would not
do; however today I shall. Besides I have his poems—some here, some
left behind at Liverpool. I have some hopes of managing an interview on
my way south.

Things are pleasanter here than at Liverpool. Wretched place too
Glasgow is, like all our great towns; still I get on better here, though bad
is the best of my getting on. But now I feel that I need the noviceship very
much and shall be every way better off when I have been made more
spiritual minded.

There, I mean at Roehampton, I am pretty well resolved, I will
altogether give over composition for the ten months, that I may *vacare
Deo*° as in my noviceship proper. I therefore want to get some things done
first, but fear I never shall. One is a great ode on Edmund Campion S.J.°
For the 1st of December next is the 300th anniversary of his, Sherwin's
and Bryant's martyrdom, from which I expect of heaven some, I cannot
guess what, great conversion or other blessing to the Church in England.
Thinking over this matter my vein began to flow and I have by me a few
scattered stanzas, something between the *Deutschland* and *Alexander's
Feast*,° in sprung rhythm of irregular metre. But the vein urged by any
country sight or feeling of freedom or leisure (you cannot tell what a
slavery of mind or heart it is to live my life in a great town) soon dried and
I do not know if I can coax it to run again. One night, as I lay awake in a
fevered state, I had some glowing thoughts and lines, but I did not put
them down and I fear they may fade to little or nothing. I am sometimes

surprised at myself how slow and laborious a thing verse is to me when musical composition comes so easily, for I can make tunes almost at all times and places and could harmonise them as easily if only I could play or could read music at sight. Indeed if I could play the piano with ease I believe I could improvise on it. I have of late been finishing the air, but only the air, of *I have loved flowers that fade*.° I find now I can put a second part satisfactorily to myself, but about fuller chords I am timid and incapable as yet. It is besides very difficult to get at a piano. I have now also a certain power of counterpointing (I will not say harmonising) without an instrument: I do not, in my mind's ear, as a musician would do, hear the chords, but I have an instinct of what will do and verify by rule and reckoning the air which I do hear: this suggests itself and springs from the leading air which is to be accompanied.—I got a young lady this evening to play me over some of my pieces, but was not well pleased with them. What had sounded rich seemed thin. I had been trying several of them as canons, but this I found was unsatisfactory and unmeaning and the counterpoint drowns the air. If I could only get good harmonies to *I have loved flowers* it would be very sweet, I think; I shd. then send it you and should like Woolrych to see it too.

I am now writing to Canon Dixon.

I am promised before I leave Scotland two days to see something of the Highlands.

I can well understand that 'what there is unusual in expression in my verse is less pleasant when you are in that sort of weak state', for I find myself that when I am tired things of mine sound strange, forced, and without idiom which had pleased me well enough in the fresh heat of composition. But then the weaker state is the less competent and really critical. I always think however that your mind towards my verse is like mine towards Browning's: I greatly admire the touches and the details, but the general effect, the whole, offends me, I think it repulsive.

Your letters addressed to Liverpool were forwarded very leisurely.

Believe me your affectionate friend

Gerard M. Hopkins S.J.

Sept. 17 1881.

To R. W. Dixon (29 October 1881)

. . . On the Sonnet and its history a learned book or two learned books have been published of late and all is known about it—but not by me. The reason why the sonnet has never been so effective or successful in

England as in Italy I believe to be this: it is not so long as the Italian sonnet; it is not long enough, I will presently say how. Now in the form of any work of art the intrinsic measurements, the proportions, that is, of the parts to one another and to the whole, are no doubt the principal point, but still the extrinsic measurements, the absolute quantity or size[2] goes for something. Thus supposing in the Doric Order the Parthenon to be the standard of perfection, then if the columns of the Parthenon have so many semidiameters or modules to their height, the architrave so many, and so on these will be the typical proportions. But if a building is raised on a notably greater scale it will be found that these proportions for the columns and the rest are no longer satisfactory, so that one of two things—either the proportions must be changed or the Order abandoned. Now if the Italian sonnet is one of the most successful forms of composition known, as it is reckoned to be, its proportions, inward and outward, must be pretty near perfection. The English sonnet has the same inward proportions, 14 lines, 5 feet to the line, and the rhymes and so on may be made as in the strictest Italian type. Nevertheless it is notably shorter and would therefore appear likely to be unsuccessful, from want not of comparative but of absolute length. For take any lines from an Italian sonnet, as

> Non ha l'ottimo⌒artista⌒alcun concetto
> Che⌒un marmor solo⌒in se non circonscriva.°

Each line has two elisions and a heavy ending or 13 syllables, though only 10 or, if you like, 11 count in the scanning. An Italian heroic line then and consequently a sonnet will be longer than an English in the proportion 13:10, which is considerable. But this is not all: the syllables themselves are longer. We have seldom such a delay in the voice as is given to the syllable by doubled letters (as _ottimo_ and _concetto_) or even by two or more consonants (as _artista_ and _circonscriva_) of any sort, read as Italians read. Perhaps then the proportions are nearer 4:3 or 3:2. The English sonnet is then in comparison with the Italian short, light, tripping, and trifling.

This has been instinctively felt and the best sonnets shew various devices successfully employed to make up for the short-coming. It may be done by the mere gravity of the thought, which compels a longer dwelling on the words, as in Wordsworth (who otherwise is somewhat light in his versification), e.g.

> Earth has not anything to shew more fair—etc;

or by inversion and a periodic construction, which has something the same effect: there is a good deal of this in Bridges' sonnets; or by breaks and pauses, as

> Captain or colonel or knight-at-arms;

or by many monosyllables, as

> Both them I serve and of their train am I:

this is common with τοὺς περὶ° Swinburne; or by the weight of the syllables themselves, strong or circumflexed and so on, as may be remarked in Gray's sonnet, an exquisite piece of art, whatever Wordsworth may say,

> In vain to me the smiling mornings shine—

(this sonnet is remarkable for its falling or trochaic rhythm—

> In | vain to | me the | smiling | mornings | shine—

and not

> In vain | to me | the smil | ing morn | ings shine),

and it seems to me that for a mechanical difficulty the most mechanical remedy is the best: none, I think, meet it so well as these 'outriding' feet° I sometimes myself employ, for they more than equal the Italian elisions and make the whole sonnet rather longer, if anything, than the Italian is. Alexandrine lines (used throughout) have the same effect: this of course is a departure from the Italian, but French sonnets are usually in Alexandrines.

The above reasoning wd. shew that any metre (in the same rhythm) will be longer in Italian than in English and this is in fact, I believe, the case and is the reason perhaps why the *ottava rima* has never had the success in England it has had in Italy and why Spencer found it necessary to lengthen it in the ratio from 20 to 23 (= 80 to 92).

Surrey's sonnets are fine, but so far as I remember them they are strict in form. I look upon Surrey as a great writer and of the purest style. But he was an experimentalist, as you say, and all his experiments are not successful. I feel ashamed however to talk of English or any literature, of which I was always very ignorant and which I have ceased to read.

. . . This must be my last letter on literary matters while I stay here, for they are quite out of keeping with my present duties. I am very glad my criticisms should be of any service to you: they have involved a labour of love.

Nov. 2—My sister° is unwilling to send you the music, with which she is not satisfied, till I have seen it. It must therefore wait awhile.

I am ashamed at the expressions of high regard which your last letter and others have contained, kind and touching as they are, and do not know whether I ought to reply to them or not. This I say: my vocation puts before me a standard so high that a higher can be found nowhere else. The question then for me is not whether I am willing (if I may guess what is in your mind) to make a sacrifice of hopes of fame (let us suppose), but whether I am not to undergo a severe judgment from God for the lothness I have shewn in making it, for the reserves I may have in my heart made, for the backward glances I have° given with my hand upon the plough, for the waste of time° the very compositions you admire may have caused and their preoccupation of the mind which belonged to more sacred or more binding duties, for the disquiet and the thoughts of vainglory they have given rise to. A purpose may look smooth and perfect from without but be frayed and faltering from within. I have never wavered in my vocation, but I have not lived up to it. I destroyed the verse I had written when I entered the Society and meant to write no more; the *Deutschland* I began after a long interval at the chance suggestion of my superior, but that being done it is a question whether I did well to write anything else. However I shall, in my present mind, continue to compose, as occasion shall fairly allow, which I am afraid will be seldom and indeed for some years past has been scarcely ever, and let what I produce wait and take its chance; for a very spiritual man once told me that with things like composition the best sacrifice was not to destroy one's work but to leave it entirely to be disposed of by obedience. But I can scarcely fancy myself asking a superior to publish a volume of my verses and I own that humanly there is very little likelihood of that ever coming to pass. And to be sure if I chose to look at things on one side and not the other I could of course regret this bitterly. But there is more peace and it is the holier lot to be unknown than to be known.—In no case am I willing to write anything while in my present condition: the time is precious and will not return again and I know I shall not regret my forbearance. If I do get hereafter any opportunity of writing poetry I could find it in my heart to finish a tragedy° of which I have a few dozen lines written and the leading thoughts for the rest in my head on the subject of St. Winefred's martyrdom: as it happens, tomorrow is her feastday.

I hope you may have all happiness in your marriage. You have, I think, no children of your own, but Bridges told me he met your two step daughters at Hayton.

I am afraid our retreat will not begin tonight after all.
Believe me always your affectionate friend

                                   Gerard M. Hopkins S.J.

I should tell you that my letters now are opened.

To R. W. Dixon

Manresa House, Roehampton, S.W. Dec. 1 1881
(the very day 300 years ago of Father Campion's martyrdom).

My dear friend,—I am heartily glad you did not make away with, as you say you thought of doing, so warm and precious a letter as your last. It reached me on the first break or day of repose in our month's retreat; I began answering it on the second, but could not finish; and this is the third and last of them.

When a man has given himself to God's service, when he has denied himself and followed Christ, he has fitted himself to receive and does receive from God a special guidance, a more particular providence. This guidance is conveyed partly by the action of other men, as his appointed superiors, and partly by direct lights and inspirations. If I wait for such guidance, through whatever channel conveyed, about anything, about my poetry for instance, I do more wisely in every way than if I try to serve my own seeming interests in the matter. Now if you value what I write, if I do myself, much more does our Lord. And if he chooses to avail himself of what I leave at his disposal he can do so with a felicity and with a success which I could never command. And if he does not, then two things follow; one that the reward I shall nevertheless receive from him will be all the greater; the other that then I shall know how much a thing contrary to his will and even to my own best interests I should have done if I had taken things into my own hands and forced on publication. This is my principle and this in the main has been my practice: leading the sort of life I do here it seems easy, but when one mixes with the world and meets on every side its secret solicitations, to live by faith is harder, is very hard; nevertheless of God's help I shall always do so.

Our Society values, as you say, and has contributed to literature, to culture; but only as a means to an end. Its history and its experience shew that literature proper, as poetry, has seldom been found to be to that end a very serviceable means. We have had for three centuries often the flower of the youth of a country in numbers enter our body: among these how many poets, how many artists of all sorts, there must have been! But there have been very few Jesuit poets and, where they have been, I believe it would be found on examination that there was something exceptional in their circumstances or, so to say, counterbalancing in their career. For genius attracts fame and individual fame St. Ignatius looked on as the most dangerous and dazzling of all attractions . . . You see then what is against me, but since, as Solomon says, there is a time for everything, there is nothing that does not some day come to be, it may be that the time will come for my verses. I remember, by the by, once taking up a little book of the life of St. Stanislaus° told or commented on under emblems; it was much in the style of Herbert and his school and about that date; it was by some Polish Jesuit. I was astonished at their beauty and brilliancy, but the author is quite obscure. Brilliancy does not suit us. Bourdaloue° is reckoned our greatest orator: he is severe in style. Suarez° is our most famous theologian: he is a man of vast volume of mind, but without originality or brilliancy; he treats everything satisfactorily, but you never remember a phrase of his, the manner is nothing. Molina° is the man who *made* our theology: he was a genius and even in his driest dialectic I have remarked a certain fervour like a poet's. But in the great controversy on the Aids of Grace, the most dangerous crisis, as I suppose, which our Society ever went through till its suppression, though it was from his book that it had arisen, he took, I think, little part. The same sort of thing may be noticed in our saints. St. Ignatius himself was certainly, every one who reads his life will allow, one of the most extraordinary men that ever lived; but after the establishment of the Order he lived in Rome so ordinary, so hidden a life, that when after his death they began to move in the process of his canonisation one of the Cardinals, who had known him in his later life and in that way only, said that he had never remarked anything in him more than in any edifying priest . . . I quote these cases to prove that show and brilliancy do not suit us, that we cultivate the commonplace outwardly and wish the beauty of the king's daughter the soul to be from within.

I could say much more on all this, but it is enough and I must go on to other things. Our retreat ended on the 8th. The 'hoity toity' passage I have not seen; indeed I have never even had your book° in my hands except one day when waiting to see Bridges in his sickness I found it on the table and was just going to open it—but to the best of my remembrance I did not then open it either. I have for some years past had to put aside serious study. It is true if I had been where your book was easy of access I should have looked at it, perhaps read it all, but in Liverpool I never once entered the public library. However if, as I hope, the time for reading history should ever come I shall try to read this one. You said once you did not pretend not to have a side and that you must write as an Anglican: this is of course and you could not honestly be an Anglican and not write as one . . . My Liverpool and Glasgow experience laid upon my mind a conviction, a truly crushing conviction, of the misery of town life to the poor and more than to the poor, of the misery of the poor in general, of the degradation even of our race, of the hollowness of this century's civilisation: it made even life a burden to me to have daily thrust upon me the things I saw.

I have found to my dismay what I suspected before, that my sister° only sent you the music to two stanzas of your Song, whereas I made it for six. How she came to make so dreadful an oversight I cannot tell: the music changes and she had remarked on the change. But I must get her to send the rest and then you will be able to judge of the whole. I do not believe that my airs—if I can compare them with the work of an accomplished musician—would really be found to be like Mr. Metcalf's—to judge by the two pieces of his that you sent me.

I should tell you that I by no means objected to the couplet 'Rattled her keys', I admired it as a happy medley: I thought the fusion or rather the pieing was less happy in the opening of the poem.

About sonnet-writing I never meant to override your own judgment. I have put the objections to licentious forms and I believe they hold. But though many sonnets in English may in point of form be great departures from and degenerations of the type, put aside the reference to the type, and they may in themselves be fine poems of 14 lines. Still that fact, that the poet has tied himself within 14 lines and calls the piece a sonnet, lays him open to objection.

I must hold that you and Morris° belong to one school, and that though you should neither of you have read a line of the other's. I suppose the same models, the same masters, the same tastes, the same keepings, above all, make the school. It will always be possible to find differences, marked differences, between original minds; it will be necessarily so. So the species in nature are essentially distinct, nevertheless they are grouped into genera: they have one form in common, mounted on that they have a form that differences them. I used to call it the school of Rossetti: it is in literature the school of the Praeraphaelites. Of course that phase is in part past, neither do these things admit of hard and fast lines; still consider yourself, that you know Rossetti and Burne Jones, Rossetti through his sympathy for you and Burne Jones—was it the same or your sympathy for him? This modern medieval school is descended from the Romantic school (Romantic is a bad word) of Keats, Leigh

Hunt, Hood, indeed of Scott early in the century. That was one school; another was that of the Lake poets and also of Shelley and Landor; the third was the sentimental school, of Byron, Moore, Mrs. Hemans, and Haynes Bailey. Schools are very difficult to class: the best guide, I think, are keepings. Keats' school chooses medieval keepings, not pure nor drawn from the middle ages direct but as brought down through that Elizabethan tradition of Shakspere and his contemporaries which died out in such men as Herbert and Herrick. They were also great realists and observers of nature. The Lake poets and all that school represent, as it seems to me, the mean or standard of English style and diction, which culminated in Milton but was never very continuous or vigorously transmitted, and in fact none of these men unless perhaps Landor were great masters of style, though their diction is generally pure, lucid, and unarchaic. They were faithful but not rich observers of nature. Their keepings are their weak point, a sort of colourless classical keepings: when Wordsworth wants to describe a city or a cloudscape which reminds him of a city it is some ordinary rhetorical stage-effect of domes, palaces, and temples. Byron's school had a deep feeling but the most untrustworthy and barbarous eye, for nature; a diction markedly modern; and their keepings any gaud or a lot of Oriental rubbish. I suppose Crabbe to have been in form a descendant of the school of Pope with a strong and modern realistic eye; Rogers something between Pope's school and that of Wordsworth and Landor; and Campbell between this last and Byron's, with a good deal of Popery too, and a perfect master of style. Now since this time Tennyson and his school seem to me to have struck a mean or compromise between Keats and the medievalists on the one hand and Wordsworth and the Lake School on the other (Tennyson has some jarring notes of Byron in *Lady Clare Vere de Vere*, *Locksley Hall* and elsewhere). The Lake School expires in Keble and Faber and Cardinal Newman. The Brownings may be reckoned to the Romantics. Swinburne is a strange phenomenon: his poetry seems a powerful effort at establishing a new standard of poetical diction, of the rhetoric of poetry; but to waive every other objection it is essentially archaic, biblical a good deal, and so on: now that is a thing that can never last; a perfect style must be of its age. In virtue of this archaism and on other grounds he must rank with the medievalists.

This is a long ramble on literary matters, on which I did not want to enter.

At Torquay Bridges made at last a sudden and wonderful recovery: so I am told, for he has not written. He then went abroad with a common friend of ours, Muirhead, and is, I suppose, likely to be abroad for the winter. And I am afraid when he returns I shall not see him; for I may now be called away at any time.

Earnestly thanking you for your kindness and wishing you all that is best I remain your affectionate friend

Gerard M. Hopkins S.J.

Dec. 16 1881.

To Bridges

Stonyhurst College, Blackburn. Oct. 18 1882.

Dearest Bridges,—I have read of Whitman's (1) 'Pete'° in the library at Bedford Square (and perhaps something else; if so I forget), which you pointed out; (2) two pieces in the *Athenaeum* or *Academy*, one on the Man-of-War Bird, the other beginning 'Spirit that formed this scene'; (3) short extracts in a review by Saintsbury in the *Academy*:° this is all I remember. I cannot have read more than half a dozen pieces at most.

This, though very little, is quite enough to give a strong impression of his marked and original manner and way of thought and in particular of his rhythm. It might be even enough, I shall not deny, to originate or, much more, influence another's style: they say the French trace their whole modern school of landscape to a single piece of Constable's° exhibited at the Salon early this century.

The question then is only about the fact. But first I may as well say what I should not otherwise have said, that I always knew in my heart Walt Whitman's mind to be more like my own than any other man's living. As he is a very great scoundrel this is not a pleasant confession. And this also makes me the more desirous to read him and the more determined that I will not.

Nevertheless I believe that you are quite mistaken about this piece and that on second thoughts you will find the fancied resemblance diminish and the imitation disappear.

And first of the rhythm. Of course I saw that there was to the eye something in my long lines like his, that the one would remind people of the other. And both are in irregular rhythms. There the likeness ends. The pieces of his I read were mostly in an irregular rhythmic prose: that is what they are thought to be meant for and what they seemed to me to be. Here is a fragment of a line I remember: 'or a handkerchief designedly dropped'.° This is in a dactylic rhythm—or let us say anapaestic; for it is a great convenience in English to assume that the stress is always at the end of the foot; the consequence of which assumption is that in ordinary verse there are only two English feet possible, the iamb and the anapaest, and even in my regular sprung rhythm only one additional, the fourth paeon: for convenience' sake assuming this, then the above fragment is anapaestic—'or a $\overset{1}{\text{hánd}}$ $\overset{2}{}$ $\overset{3}{|}$ $\overset{1}{\text{kerchief}}$ $\overset{2}{\dots}$ $\overset{3}{|}$. $\overset{1}{\text{desígn}}$ $\overset{2}{|}$ edly $\overset{3}{\text{drópped}}$'—and there is a break down, a designed break of rhythm, after 'handkerchief', done no doubt that the line may not become downright verse, as it would be if he had said 'or a handkerchief purposely dropped'. Now you can of course say that he meant pure verse and that the foot is a paeon—'or a $\overset{1}{\text{hánd}}$ $\overset{2}{}$ $\overset{3}{|}$ $\overset{1}{\text{kerchief}}$ $\overset{2}{}$ $\overset{3}{\text{desígn}}$ $\overset{4}{|}$ $\overset{1}{\text{edly}}$ $\overset{2}{}$ $\overset{3}{\text{drópped}}$'; or that he means, without fuss, what I should achieve by looping the syllable *de* and calling that foot an outriding foot— for the result might be attained either way. Here then I must make the answer which will apply here and to all like cases and to the examples which may be found up and down the poets of the use of sprung rhythm—*if they could have done it they would*: sprung rhythm, once you hear it, is so eminently natural a thing and so effective a thing that if they had known of it they would have used it. Many people, as we say, have been 'burning', but they all missed it; they took it up and mislaid it again. So far as I know—I am enquiring and presently I shall be able to speak more decidedly—it existed in full force in Anglo saxon verse and in great beauty; in a degraded and doggrel shape in *Piers Ploughman* (I am reading that famous poem and am coming to the conclusion that it is not worth reading); Greene was the last who employed it at all consciously and he never continuously; then it disappeared—for one cadence in it here and there is not sprung rhythm and one swallow does not make a spring. (I put aside Milton's case, for it is altogether singular.) In a matter like this a thing does not exist, is not *done* unless it is wittingly and willingly done; to recognise the form you are employing and to mean it is everything. To apply this: there is (I suppose, but you will know) no sign that Whitman means to use paeons or outriding feet where these breaks in rhythm occur; it seems to me a mere extravagance to think he means people to understand of themselves what they are slow to understand even when marked or pointed out. If he does not mean it then he does not do it; or in short what he means to write—and writes—is rhythmic prose and that only. And after all, you probably grant this.

Good. Now prose rhythm in English is always one of two things (allowing my convention about scanning upwards or from slack to stress and not from stress to slack)—either iambic or anapaestic. You may make a third measure (let us call it) by intermixing them. One of these three simple measures then, all iambic or all anapaestic or mingled iambic and anapaestic, is what he in every case means to write. He dreams of no other and he *means* a rugged or, as he calls it in that very piece 'Spirit that formed this scene' (which is very instructive and should be read on this very subject), a 'savage' art and rhythm.

Extremes meet, and (I must for truth's sake say what sounds pride) this savagery of his art, this rhythm in its last ruggedness and decomposition into common prose, comes near the last elaboration of mine. For that piece of mine is very highly wrought. The long lines are not rhythm run to seed: everything is weighed and timed in them. Wait till they have taken hold of your ear and you will find it so. No, but what it *is* like is the rhythm of Greek tragic choruses or of Pindar: which is pure sprung rhythm. And that has the same changes of cadence from point to point as this piece. If you want to try it, read one till you have settled the true places of the stress, mark these, then read it aloud, and you will see. Without this these choruses are prose bewitched; with it they are sprung rhythm like that piece of mine.

Besides, why did you not say *Binsey Poplars* was like Whitman? The present piece is in the same kind and vein, but developed, an advance. The lines and the stanzas (of which there are two in each poem and having much the same relation to one another) are both longer, but the two pieces are greatly alike: just look. If so how is this a being untrue to myself? I am sure it is no such thing.

The above remarks are not meant to run down Whitman. His 'savage' style has advantages, and he has chosen it; he says so. But you cannot eat your cake and keep it: he eats his off-hand, I keep mine. It makes a very great difference. Neither do I deny all resemblance. In particular I noticed in 'Spirit that formed this scene' a preference for the alexandrine. I have the same preference: I came to it by degrees, I did not take it from him.

About diction the matter does not allow me so clearly to point out my independence as about rhythm. I cannot think that the present piece owes anything to him. I hope not, here especially, for it is not even spoken in my own person but in that of St. Winefred's maidens. It ought to sound like the thoughts of a good but lively girl and not at all like—not at all like Walt Whitman. But perhaps your mind may have changed by this.

I wish I had not spent so much time in defending the piece.

Believe me your affectionate friend

Gerard.

Oct. 19 1882. I am not sure I shall not ask C. D.° to let me see at least one packet of Mano. He should, every one should now, use one of these reproductive processes: it is next to printing and at least it secures one against irretrievable loss by the post. All our masters here use the gelatine process for flying sheets etc.

To Bridges

Stonyhurst College, Blackburn. Jan. 4 1883.

Dearest Bridges,—Since our holidays began I have been in a wretched state of weakness and weariness, I can't tell why, always drowsy and incapable of reading or thinking to any effect. And this must be why I was, before that, able to do so little on your *Prometheus*.°

I think the sonnet a fine work,° but should like the phrasing to be more exquisite in lines 2, 4, and perhaps elsewhere. Still it has to me an unspontaneous artificial air. I cannot consider the goblet and 'golden foil' a success. It is out of keeping with sons of toil and the unadornment of their brides. It is obscure too: it means, I suppose, that the goblet is of gold and that this gold sets off and is set off by the colour of the wine. This much resemblance there is, that as the goblet draws or swallows up and sort-of-drinks the liquid and the liquid at the same time swallows up and sort-of-drinks the material of the goblet so the body absorbs sleep and sleep the body. But the images of gold and crimson are out of keeping: brilliancy is only in the way. You were, you say, driven to it: I protest, and with indignation, at your saying I was driven to the same image. With more truth might it be said that my sonnet° might have been written expressly for the image's sake. But the image is not the same as yours and I do not mean by foil set-off at all; I mean foil in its sense of leaf or tinsel, and no other word whatever will give the effect I want. Shaken goldfoil gives off broad glares like sheet lightning and also, and this is true of nothing else, owing to its zigzag dints and creasings and network of small many cornered facets, a sort of fork lightning too. Moreover as it is the first rhyme, presumably it engendered the others and not they it. This reminds me that I hold you to be wrong about 'vulgar', that is obvious or necessary, rhymes. It follows from your principle that if a word has only one rhyme in the language it cannot be used in selfrespecting poetry at all. The truth seems to me that a problem is set to all, how to use that same pair (or triplet or any set) of rhymes, which are invariable, to the finest and most natural effect. It is nothing that the reader can say/ He had to say it, there *was* no other rhyme: you answer/shew me what better I could have said if there had been a million. Hereby, I may tell you, hangs a very profound question treated by Duns Scotus, who shews that freedom is compatible with necessity. And besides, common sense tells you that though if you say A_1 you cannot help saying A_2 yet you can help saying $A_1 + A_2$ at all; you could have said $B_1 + B_2$ or $C_1 + C_2$ etc. And is not music a sort of rhyming on seven rhymes and does that make it vulgar? The variety is more, but the principle the same. . .

Jan. 5—Hall Caine's 'Disquisition' on Rossetti's picture of Dante's Dream bought by the city of Liverpool reached me this morning, I suppose from the author. Noel Paton° is quoted as saying, with good-natured gush, that it may be ranked with the Madonna di San Sisto. Now, you know, it may *not*, and I am considering whether I shall tell Hall Caine so.

To return to your sonnet, could you not find another rhyme? there is *spoil, despoil, turmoil*, not to speak of *coil, boil, parboil*, and Hoyle on whist—the very sight of which dreary jugglery brings on yawns with me.

You speak of writing the sonnet in prose first. I read the other day that Virgil wrote the Aeneid in prose. Do you often do so? Is it a good plan? If it is I will try it; it may help on my flagging and almost spent powers. Years ago one of ours, a pupil of mine, was to write some English verses for me, to be recited: he had a real vein. He said he had no thoughts, but that if I would furnish some he would versify them. I did so and the effect was very surprising to me to find my own thoughts, with no variation to speak of, expressed in good verses quite unlike mine.

The sonnet on Purcell° means this: 1–4. I hope Purcell is not damned for being a Protestant, because I love his genius. 5–8. And that not so much for gifts he shares, even though it shd. be in higher measure, with other musicians as for his own individuality. 9–14. So that while he is aiming only at impressing me his hearer with the meaning in hand I am looking out meanwhile for his specific, his individual markings and mottlings, 'the sakes of him'. It is as when a bird thinking only of soaring spreads its wings: a beholder may happen then to have his attention drawn by the act to the plumage displayed.—In particular, the first lines mean: May Purcell, O may he have died a good death and that soul which I love so much and which breathes or stirs so unmistakeably in his works have parted from the body and passed away, centuries since though I frame the wish, in peace with God! so that the heavy condemnation under which he outwardly or nominally lay for being out of the true Church may in consequence of his good intentions have been reversed. 'Low lays him' is merely 'lays him low', that is/strikes him heavily, weighs upon him. (I daresay this will strike you as more professional than you had anticipated.) It is somewhat dismaying to find I am so unintelligible though, especially in one of my very best pieces. 'Listed', by the by, is 'enlisted'. 'Sakes' is hazardous: about that point I was more bent on saying my say than on being understood in it. The 'moonmarks' belong to the image only of course, not to the application; I mean not detailedly: I was thinking of a bird's quill feathers. One thing disquiets me: *I meant* 'fair fall' to mean *fair (fortune be) fall*; it has since struck me that perhaps 'fair' is an adjective proper and in the predicate and can only be used in cases like 'fair fall the day', that is, *may the day fall, turn out, fair*. My line will yield a sense that way indeed, but I never meant it so. Do you know any passage decisive on this?

Would that I had Purcell's music here.

Did you see Vernon Lee's° paper in the December *Contemp.*? I don't like it. She professes herself a disciple of a Mr. Edmund Gurney, who by way of reaction against the gush of programmes ('sturdy old tone-poet'—'inimitable drollery of the semi demiquavers in the dominant minor' and so on) says that we enjoy music because our apish ancestors serenaded their Juliet-apes of the period in rudimentary recitatives and our emotions are the survival—that sexual business will in short be found by roking the pot.° This is to swing from pap to poison. Would that I had my materials ready to talk sense.

Yours affectionately

Gerard Hopkins S.J.

Jan. 5 1883.

Is it not too much for two lines running to have the rhythm reversed in the 4th foot? as your 13 and 14. Perhaps not. Twelfth night.

To Bridges

University College, 85 & 86, Stephens Green, Dublin.°
March 7 1884

My dearest Bridges,—Remark the above address: it is a new departure or a new arrival and at all events a new abode. I dare say you know nothing of it, but the fact is that, though unworthy of and unfit for the post, I have been elected Fellow of the Royal University of Ireland in the department of classics. I have a salary of £400 a year, but when I first contemplated the six examinations I have yearly to conduct, five of them running, and to the Matriculation there came up last year 750 candidates, I thought that Stephen's Green (the biggest square in Europe) paved with gold would not pay for it. It is an honour and an opening and has many bright sides, but at present it has also some dark ones and this in particular that I am not at all strong, not strong enough for the requirements, and do not see at all how I am to become so. But to talk of weather or health and especially to complain of them is poor work.

The house we are in, the College, is a sort of ruin and for purposes of study very nearly naked. And I have more money to buy books than room to put them in.

I have been warmly welcomed and most kindly treated. But Dublin itself is a joyless place and I think in my heart as smoky as London is: I had fancied it quite different. The Phoenix Park is fine, but inconveniently far off. There are a few fine buildings.

It is only a few days since I sent the MS book to Mr. Patmore (and in packing I mislaid, I hope not lost, your copy of the poem 'Wild air, world-mothering air',° so that I had to send that unfinished): he acknowledged it this morning.

I enclose a poem of Tennyson's° which you may not have seen. It has something in it like your Spring Odes and also some expressions like my sonnet on Spring.

I shall also enclose, if I can find, two triolets° I wrote for the Stonyhurst Magazine; for the third was not good, and they spoilt what point it had by changing the title. These two under correction I like, but have fears that you will suspend them from a hooked nose: if you do, still I should maintain they were as good as yours beginning 'All women born'.

Believe me your affectionate friend

Gerard Hopkins S.J.

There was an Irish row over my election

To Bridges

University College, 85 & 86, Stephen's Green, Dublin.
Aug. 21 1884

Dearest Bridges,—I must let you have a line to acknowledge, with many thanks, the receipt of the MS book° and two or three very kind letters. I guessed whose was the elegant and legible hand on two of the addresses. As for the piece of a new garment, I came to the conclusion it was put in to bigout the enclosure. I also concluded that that new garment was a pair of wedding trousers. Circumstances may drive me to use my piece as a penwiper.

It is so near your wedding° that I do not know I ought to write of anything else. I could not ask to be present at it; and indeed, much as I desire to see you and your wife and her mother and Yattendon itself, perhaps that would not be so good a day for this after all as some other.

Only unhappily I do not see when that other is to be. However it is a fine buoyant saying, Non omnium rerum sol occidit.°

I had an interesting letter from Mr. Patmore all in praise of you.

Several things in your letters call for reply, but not now. If you do not like 'I yield, you do come sometimes,' (though I cannot myself feel the weakness you complain of it in it, and it has the advantage of being plain) will 'I yield, you foot me sometimes' do?° 'Own my heart' is merely 'my own heart', transposed for rhythm's sake and then *tamquam exquisitius*, as Hermann would say. 'Reave' is for rob, plunder, carry off.

I find that in correcting 'Margaret'° I wrote '*world* of wanwood' by mistake for 'worlds', as the sense requires.

Our society cannot be blamed for not valuing what it never knew of. The following are all the people I have let see my poems (not counting occasional pieces): some of them however, as you did, have shewn them to others. (1) The editor and sub-editor of our *Month* had the *Deutschland* and later the *Eurydice* offered them—(2) my father and mother and two sisters saw these, one or both of them, and I have sent them a few things besides in letters—(3) You—(4) Canon Dixon—(5) Mr. Patmore—(6) Something got out about the *Deutschland* and Fr. Cyprian Splaine, now of Stonyhurst, wrote to me to send it him and perhaps other poems of mine: I did so and he shewed it to others. They perhaps read it, but he afterwards acknowledged to me that in my handwriting he found it unreadable; I do not think he meant illegible—(7) On the other hand Fr. Francis Bacon, a fellownovice of mine, and an admirer of my sermons saw all and expressed a strong admiration for them which was certainly sincere. They are therefore, one may say, unknown. It always seems to me that poetry is unprofessional, but that is what I have said to myself, not others to me. No doubt if I kept producing I should have to ask myself what I meant to do with it all; but I have long been at a standstill, and so the things lie. It would be less tedious talking than writing: now at all events I must stop.

I must tell you a humorous touch of Irish Malvolio or Bully Bottom, so distinctively Irish that I cannot rank it: it amuses me in bed. A Tipperary lad, one of our people, lately from his noviceship, was at the wicket and another bowling to him. He thought there was no one within hearing, but from behind the wicket he was overheard after a good stroke to cry out 'Arrah, sweet myself!'°

I must write once more against the 3rd.

Believe me always your affectionate friend

Gerard M. Hopkins S.J.

Aug. 24 1884.

To his sister Kate°

University College, Stephen's Green, Dublin. Dec. 9 1884

Me dear Miss Hopkins,—Im intoirely ashamed o meself. Sure its a wonder I could lave your iligant corspondance so long onanswered. But now Im just afther conthroiving a jewl of a convaniance be way of a standhen desk and tis a moighty incurgement towards the writin of letters intoirelee. Tis whoy ye hear from me this evenin.

It bates me where to commince, the way Id say anything yed be interistud to hear of. More be token yell be plased tintimate to me mother Im intirely obleeged to her for her genteel offers. But as titchin warm clothen tis undher a misapprehinsion shes labourin. Sure twas not the inclimunsee of the saysons I was complainin of at all at all. Twas the povertee of books and such like educational convaniences.

And now, Miss Hopkins darlin, yell chartably exkees me writin more in the rale Irish be raison I was never rared to ut and thats why I do be so slow with my pinmanship, bad luck to ut (savin your respects), but for ivery word I delineate I disremember two, and thats how ut is with me.

(The above very fair.)

The weather is wild and yet mild.

I have a kind of charge of a greenhouse.

I am hoping to hear Dvorak's Stabat Mater at Trinity College tomorrow. I think you heard it.

I have an invitation for Xmas to Lord Emly's.

A dear old French Father, very clever and learned and a great photographer, who at first wanted me to take to photography with him, which indeed in summer would be pleasant enough, finding that once I used to draw, got me to bring him the few remains I still have, cows and horses in chalk done in Wales too long ago to think of, and admired them to that degree that he is urgent with me to go on drawing at all hazards; but I do not see how that could be now, so late: if anybody had said the same 10 years ago it might have been different.

You spoke in your last of meeting Baillie.° He was always the kindest and best of friends and I always look upon myself in the light of a blackguard when I think of my behaviour to him and of his to me. In this case however he has not written since you met him and I hope to be beforehand with him.

Tis a quare thing I didn't finish this letter yet. Ill shlip me kyard in betune the sheets the way yell know Im not desaivin ye. Believe me your loving brother

Gerard.

Dec. 13 1884

To Bridges

University College, St. Stephen's Green, Dublin. Sept. 1 1885.

Dearest Bridges,—I have just returned from an absurd adventure, which when I resigned myself to it I could not help enjoying. A hairbrained fellow took me down to Kingstown and on board his yacht and, whereas I meant to return to town by six that evening, would not let me go either that night or this morning till past midday. I was afraid it would be compromising,° but it was fun while it lasted.

I have been in England. I was with my people first at Hampstead, then at Midhurst in Sussex in a lovely landscape: they are there yet. And from there I went to Hastings to Mr. Patmore's for a few days. I managed to see several old friends and to make new ones, amongst which Mr. W. H. Cummings the tenor singer and composer, who wrote the Life of Purcell: he shewed me some of his Purcell treasures and others and is going to send me several things. I liked him very much but the time of my being with him was cut short. I did not attempt to see you: I did not know that visitors wd. at that time be very welcome and it wd. have been difficult to me in any case to come. I am very sorry to hear of Mrs. Bridges' disappointment:° somehow I had feared that would happen.

I shall shortly have some sonnets to send you, five or more.° Four° of these came like inspirations unbidden and against my will. And in the life I lead now, which is one of a continually jaded and harassed mind, if in any leisure I try to do anything I make no way—nor with my work, alas! but so it must be.

Mr. Patmore lent me Barnes' poems°—3 volumes, not all, for indeed he is prolific. I hold your contemptuous opinion an unhappy mistake: he is a perfect artist and of a most spontaneous inspiration; it is as if Dorset life and Dorset landscape had taken flesh and tongue in the man. I feel the defect or limitation or whatever we are to call it that offended you: he lacks fire; but who is perfect all round? If one defect is fatal what writer could we read?

An old question of yours I have hitherto neglected to answer, am I thinking of writing on metre? I suppose thinking too much and doing too little. I do greatly desire to treat that subject; might perhaps get something together this year; but I can scarcely believe that on that or on anything else anything of mine will ever see the light—of publicity nor even of day. For it is widely true, the fine pleasure is not to do a thing but to feel that you could and the mortification that goes to the heart is to feel it is the power that fails you: *qui occidere nolunt Posse volunt*; it is the refusal of a thing that we like to have. So with me, if I could but get on, if I could but produce work I should not mind its being buried, silenced, and going no further; but it kills me to be time's eunuch and never to beget. After all I do not despair, things might change, anything may be; only there is no great appearance of it. Now because I have had a holiday though not strong I have some buoyancy; soon I am afraid I shall be ground down to a state like this last spring's and summer's, when my spirits were so crushed that madness seemed to be making approaches—and nobody was to blame, except myself partly for not managing myself better and contriving a change.

Believe me, with kind wishes to Mrs. Bridges, your affectionate friend

Gerard M. Hopkins S.J.

Sept. 8 '85.

To Bridges (13 October 1886)

... By the bye, I say it deliberately and before God, I would have you and Canon Dixon and all true poets remember that fame, the being known, though in itself one of the most dangerous things to man, is nevertheless the true° and appointed air, element, and setting of genius and its works. What are works of art for? to educate, to be standards. Education is meant for the many, standards are for public use. To produce then is of little use unless what we produce is known, if known widely known, the wider known the better, for it is by being known it works, it influences, it does its duty, it does good. We must then try to be known, aim at it, take means to it. And this without puffing in the process or pride in the success. But still. Besides, we are Englishmen. A great work by an Englishman is like a great battle won by England. It is an unfading bay tree. It will even be admired by and praised by and do good to those who hate England (as England is most perilously hated), who do not wish even to be benefited by her. It is then even a patriotic duty τῇ ποιήσει ἐνεργεῖν° and to secure the fame and permanence of the work. Art and its fame do not really matter, spiritually they are nothing, virtue is the only good; but it is only by bringing in the infinite that to a just judgment they can be made to look infinitesimal or small or less than vastly great; and in this ordinary view of them I apply to them, and it is the true rule for dealing with them, what Christ our Lord said of virtue, Let your light shine before men that they may see your good works (say, of art) and glorify yr. Father in heaven° (that is, acknowledge that they have an absolute excellence in them and are steps in a scale of infinite and inexhaustible excellence) ...

To Bridges (28 October 1886)

... One blot is no great matter, I mean not a damning matter. One blot may be found in the works of very learned clerks indeed. *Measure for Measure* is a lovely piece of work, but it was a blot, as Swinburne raving was overheard for hours to say, to make Isabella marry the old Duke. *Volpone* is one of the richest and most powerful plays ever written, but a writer in a late *Academy*° points out a fault of construction (want of motive, I think, for Bonario's being at Volpone's house when Celia was brought there): it will stand that one fault. True you say that in Stevenson's book there are many such: but I do not altogether believe there are.

This sour severity blinds you to his great genius. *Jekyll and Hyde* I have read. You speak of 'the gross absurdity' of the interchange.° Enough that it is impossible and might perhaps have been a little better masked: it must be connived at, and it gives rise to a fine situation. It is not more impossible than fairies, giants, heathen gods, and lots of things that literature teems with—and none more than yours.° You are certainly wrong about Hyde being overdrawn: my Hyde is worse. The trampling scene is perhaps a convention: he was thinking of something unsuitable for fiction.

I can by no means grant that the characters are not characterised, though how deep the springs of their surface action are I am not yet clear. But the superficial touches of character are admirable: how can you be so blind as not to see them? e.g. Utterson frowning, biting the end of his finger, and saying to the butler 'This is a strange tale you tell me, my man, a very strange tale'. And Dr. Lanyon: 'I used to like it, sir [life]; yes, sir, I liked it. Sometimes I think if we knew all' etc. These are worthy of Shakespeare. Have you read the *Pavilion on the Links* in the volume of *Arabian Nights* (not one of them)? The absconding banker is admirably characterised, the horror is nature itself, and the whole piece is genius from beginning to end.

In my judgment the amount of gift and genius which goes into novels in the English literature of this generation is perhaps not much inferior to what made the Elizabethan drama, and unhappily it is in great part wasted. How admirable are Blackmore and Hardy! Their merits are much eclipsed by the overdone reputation of the Evans—Eliot—Lewis—Cross woman (poor creature! one ought not to speak slightingly, I know), half real power, half imposition. Do you know the bonfire scenes in the *Return of the Native* and still better the sword-exercise scene in the *Madding Crowd*, breathing epic? or the wife-sale in the *Mayor of*

Casterbridge (read by chance)? But these writers only rise to their great strokes; they do not write continuously well: now Stevenson is master of a consummate style and each phrase is finished as in poetry. It will not do at all, your treatment of him. . .

To Bridges

University College, St. Stephen's Green, Dublin. Feb. 17 '87

Dearest Bridges,—I am joyed to see your hand again and delighted to hear your praise of the Canon's book.° I too have thought there is in him a vein of truly matchless beauty: it is not always the whole texture but a thread in it and sometimes the whole web is of that. But till you spoke I had almost despaired of my judgment° and quite of publishing it. The pathetic imagination of *Sky that rollest ever* seems to me to have nothing like it but some of Coleridge in our literature.

Mrs. Waterhouse has not written: it never indeed entered my head that she would, the piece being a fragment° too. But I wanted to pay her a compliment and conceived she would like this particular poem. It is in a commoner and smoother style than I mostly write in, but that is no harm: I am sure I have gone far enough in oddities and running rhymes (as even in some late sonnets you have not seen) into the next line. I sent a later and longer version to C. D., who much admired and urged me to write lots of it. It should run to about twice the present length and when complete I daresay you will like it. I am amused and pleased at Maurice W. expounding it: it is not at all what I wanted to happen, but after all if I send verse to Mrs. Waterhouse I cannot suppose no one else in the house will see it.

Tomorrow morning I shall have been three years in Ireland, three hard wearying wasting wasted years. (I met the blooming Miss Tynan° again this afternoon. She told me that when she first saw me she took me for 20 and some friend of hers for 15; but it won't do: they should see my heart and vitals, all shaggy with the whitest hair.) In those I have done God's will (in the main) and many many examination papers. I am in a position which makes it befitting and almost a duty to write anything (bearing on classical study) which I may feel that I could treat well and advance learning by: there is such a subject; I do try to write at it; but I see that I cannot get on, that I shall be even less able hereafter than now. And of course if I cannot do what even my appliances make best and easiest, far less can I anything else. Still I could throw myself cheerfully into my day's work? I cannot, I am in a prostration. Wales set me up for a while, but the effect is now past. But out of Ireland I shd. be no better, rather worse probably. I only need one thing—a working health, a working strength: with that, any employment is tolerable or pleasant, enough for human nature; without it, things are liable to go very hardly with it.

Now come on Mrs. Gaskell. What ails poor Mrs. Gaskell? One book of hers I have read through, *Wives and Daughters*: if that is not a good book I do not know what a good book is. Perhaps you are so barbarous as not to admire Thomas Hardy—as you do not Stevenson; both, I must maintain, men of pure and direct genius.

Have you followed the course of late Homeric criticism? The pendulum is swinging heavily towards the old view of a whole original Iliad. In the track of the recent dialectic investigations I have made out, I think, a small but (as a style-test) important point; but my induction is not yet complete.

I will bear in mind to send for the *Feast of Bacchus*° at an early opportunity, if (but that is not certain) one should occur.

I am almost afraid I have offended, not offended but not pleased, Mr. Patmore by a late letter: I hope it is not so bad. I hope you will enjoy yourselves there: let me see, do you know Mrs. Patmore? If you do you cannot help liking her. With best love to Mrs. Bridges I am your affectionate friend

Gerard

Yesterday Archbishop Walsh° had a letter in the *Freeman* enclosing a subscription to the defence of Dillon and the other traversers on trial for preaching the Plan of Campaign and saying that the jury was packed and a fair trial impossible. The latter was his contribution to the cause of concord and civil order. Today Archbp. Croke° has one proposing to pay no taxes. One archbishop backs robbery, the other rebellion; the people in good faith believe and will follow them. You will see, it is the beginning of the end: Home Rule or separation is near. Let them come: anything is better than the attempt to rule a people who own no principle of civil obedience at all, not only to the existing government but to none at all. I shd. be glad to see Ireland happy, even though it involved the fall of England, if that could come about without shame and guilt. But Ireland will not be happy: a people without a principle of allegiance cannot be; moreover this movement has throughout been promoted by crime. Something like what happened in the last century between '82 and 1800 will happen in this: now as then one class has passed off its class-interests as the interests of the nation and so got itself upheld by the support of the nation; now as then it will legislate in its own interest and the rest will languish; distress will bring on some fresh convulsion; beyond that I cannot guess.

The ship I am sailing in may perhaps go down in the approaching gale: if so I shall probably be cast up on the English coast.

After all I have written above my trouble is not the not being able to write a book; it is the not being fit for my work and the struggling vainly to make myself fitter.

Feb. 18 1887.

To Bridges

University College, St. Stephen's Green, Dublin. July 30 1887

. . . The drama ought to grow up with its audience; but now the audience is, so to say, jaded and senile and an excellence it knows of already cannot move it. Where a real novelty is presented to it, like Gilbert's and Sullivan's operas, which are a genuine creation of a type, it responds. However I cannot write more now and I have not the proper knowledge of the subject.

I have been reading the Choephoroi carefully and believe I have restored the text and sense almost completely in the corrupted choral odes. Much has been done in this way by dint of successive effort; the recovery, from the 'pie' of the MSS, of for instance the last antistrophe of the last ode is a beautiful thing to see and almost certain; but both in this and the others much mere pie remains and it seems to me I have recovered nearly all. Perhaps I might get a paper on it into the *Classical Review* or *Hermathena*:° otherwise they must wait to be put into a book; but when will that book or any book of mine be? Though I have written a good deal of my book on Metre. But it is a great pity for Aeschylus' choruses to remain misunderstood, for it is his own interpretation of the play and his own moral to the story.

What a noble genius Aeschylus had! Besides the swell and pomp of words for which he is famous there is in him a touching consideration and manly tenderness; also an earnestness of spirit and would-be piety by which the man makes himself felt through the playwright. This is not so with Sophocles, who is only the learned and sympathetic dramatist; and much less Euripides.

On Irish politics I had something to say, but there is little time. 'It only needs the will,' you say: it is an unwise word. It is true, it (that is, to govern Ireland) does 'only need the will'; but Douglas Jerrold's joke is in place, about Wordsworth (or whoever it was) that could write plays as good as

Shakespeare's 'if he had the mind', and 'only needed the mind'. It is a just reproach to any man not to do what lies in his own power and he could do if he would: to such a man you may well say that the task in question only needs the will. But where a decision does *not* depend on us and we cannot even influence it, then it is only wisdom to recognize the facts—the will or want of will in those, not us, who have control of the question; and that is the case now. The will of the nation is divided and distracted. Its judgment is uninformed and misinformed, divided and distracted, and its action must be corresponding to its knowledge. It has always been the fault of the mass of Englishman to know and care nothing about Ireland, to let be what would there (which, as it happened, was persecution, avarice, and oppression): and now, as fast as these people wake up and hear what wrong England has done (and has long ceased doing) to Ireland, they, like that woman in Mark Twain, 'burst into tears and rushing upstairs send a pink silk parasol and a box of hairpins to the seat of war'. If you in your limited but appreciable sphere of influence can bring people to a just mind and a proper resolution about Ireland (as you did, you told me, take part in your local elections) do so: you will then be contributing to that will which 'only is wanting'; but do not reproach me, who on this matter have perhaps both more knowledge and more will than most men. If however you think you could do but little and are unwilling even to do that (for I suppose while you are writing plays you cannot be canvassing electors), then recognise with me that with an unwavering will, or at least a flood of passion, on one, the Irish, side and a wavering one or indifference on the other, the English, and the Grand Old Mischiefmaker loose,° like the Devil, for a little while and meddling and marring all the fiercer for his hurry, Home Rule is in fact likely to come and even, in spite of the crime, slander, and folly with which its advance is attended, may perhaps in itself be a measure of a sort of equity and, considering that worse might be, of a kind of prudence.

I am not a judge of the best way to publish. Though double columns are generally and with reason objected to yet I thought *Nero*° looked and read well with them. (I am convinced it is one of the finest plays ever written.)...

I know scarcely anything of American literature and if I knew much I could not now write about it.

I hope soon to write to Canon Dixon. Give him my best love I am happy to think of your being together.

I daresay I shall be at Haslemere within the week. Court's Hill Lodge is the name of the house and is probably now not necessary.

Monsignor Persico° is going about. His coming will certainly do good.

I should like to talk to him, perhaps may. I have met him at a great dinner.
  Your affectionate friend

                                                           Gerard

Aug. 1 1887. 'Getting old'—you should never say it. But I was fortythree
on the 28th of last month and already half a week has gone.

                        To Bridges (12 January 1888)

. . . At Monasterevan° I tried to get some outstanding and accumulated
sonnets ready for hanging on the line, that is in my book of MS, the one
you wrote most of, and so for sending to you. All however are not ready
yet, but they will soon be. I could send one tonight if time served, but if
possible I should like to despatch this letter. It is now years that I have
had no inspiration of longer jet than makes a sonnet, except only in that
fortnight in Wales: it is what, far more than direct want of time, I find
most against poetry and production in the life I lead. Unhappily I cannot
produce anything at all, not only the luxuries like poetry, but the duties
almost of my position, its natural outcome—like scientific works. I am
now writing a quasi-philosophical paper on the Greek Negatives: but
when shall I finish it? or if finished will it pass the censors? or if it does will
the *Classical Review* or any magazine take it? All impulse fails me: I can
give myself no sufficient reason for going on. Nothing comes: I am a
eunuch—but it is for the kingdom of heaven's sake.
  Did you see Wooldridge° in town? No doubt. And how is he getting
on? painting, music, and all. I am sure he is right in the advice he gave me,
to be very contrapuntal, to learn that well. I want to do so if I can; it is the
only way. I have fooled at it too much. I have found a thing that, if I had
my counterpoint well at my fingers' ends, wd. be most valuable: it is that
the tunes I make are very apt to fall into fugues and canons, the second
strain being easy counterpoint to the first or to its fugal answer. E.g. my
Crocus, which you once expressed an admiration for, makes a canon
with itself at the octave two bars off and, as far as I have found, at one bar
off too. This is a splendid opening for choral treatment. And I have a fine
fugue on hand to 'Orpheus with his lute';° but I shall not hurry with it,
but keep the counterpoint correct. There seems to be, I may remark, no
book that bridges the gap between double counterpoint and fugue. For
instance, I have Ouseley° on both and Higgs° on Fugue and neither
breathes a word on so simple a point as this, that the answer in Bach and
Handel enters, that is that the counterpoint begins, freely on an
unprepared discord. But this is contrary to the elements of counterpoint

proper. What I ought to do, or somebody else rather to have done, is to tabulate Bach's practice and principles.

We are suffering from the region-fog, as it seems to be. I have been a little ill and am still a little pulled down; however I am in good spirits. Term has begun.

There was more to say, I forget what. It seems this will not go tonight . . .

There, I have copied one—*Tom's Garland*. It has many resemblances to *Harry Ploughman*, a fault in me the sonnetteer, but not a fault that can be traced home to either of the sonnets. They were conceived at the same time: that is how it is. But I have too much tendency to do it, I find. 'There is authority for it'—not the lady of the strachey,° but Aeschylus: he is always forgetting he said a thing before. Indeed he never did, but tried to say it two or three times—something rich and profound but not by him distinctly apprehended; so he goes at it again and again like a canary trying to learn the Bluebells of Scotland. To bed, to bed: my eyes are almost bleeding.

With best wishes to Mrs. Bridges, I am your affectionate friend

Gerard M. Hopkins.

By saying you are going to register your little daughter as an Elizabeth I take you to signify that you reserve her for Mr. Beeching to christen at Yattendon.

To Bridges

Univ. Coll., Stephen's Green, Dublin. April 29 1889.

Dearest Bridges,—I am ill to-day, but no matter for that as my spirits are good. And I want you too to 'buck up', as we used to say at school, about those jokes over which you write in so dudgeonous a spirit. I have it now down in my tablets that a man may joke and joke and be offensive; I have had several warnings lately leading me to make the entry, tho' goodness knows the joke that gave most offence was harmless enough and even kind. You I treated to the same sort of irony as I do myself; but it is true it makes all the world of difference whose hand administers. About Daniel° I see I was mistaken: if he pays you more than and sells you as much as other publishers (which however is saddening to think of: how many copies is it? five and twenty?) my objections do not apply. Then you ought to remember that I did try to make you known in Dublin and had some little success. (Dowden° I will never forgive; could you not kill Mrs. Bridges? then he might take an interest in you). Nay I had great success and placed you on the pinnacle of fame; for it is the pinnacle of fame to become educational and be set for translation into Gk. iambics, as you are at Trinity: this is to be a classic; 'this', as Lord Beaconsfield said to a friend who told him he found his young daughter reading *Lothair*, 'O this is fame indeed'. And Horace and Juvenal say the same thing. And here I stop, for fear of it ripening into some kind of joke.

I believe I enclose a new sonnet.° But we greatly differ in feeling about copying one's verses out: I find it repulsive, and let them lie months and years in rough copy untransferred to my book. Still I hope soon to send you my accumulation. This one is addressed to you.

Swinburne has a new volume out,° which is reviewed in its own style: 'The rush and the rampage, the pause and the pull-up of these lustrous and lumpophorous lines'. It is all now a 'self-drawing web'; a perpetual functioning of genius without truth, feeling, or any adequate matter to be at function on. There is some heavydom, in long waterlogged lines (he has no real understanding of rhythm, and though he sometimes hits brilliantly at other times he misses badly) about the *Armada*, that pitfall of the patriotic muse; and *rot* about babies, a blethery bathos into which Hugo and he from opposite coasts have long driven Channel-tunnels. I am afraid I am going too far with the poor fellow. Enough now, but his babies make a Herodian of me.

My song° will be a very highly wrought work and I do hope a fine one. Do you think canon wd. spoil the tune? I hope not, but the contrary. But if the worst came to the worst, I could, since a solo voice holds its own against instruments, give the canon-following to a violin. I shall hear what Sir Robert Stewart says about it. This is how it now stands. I tried at first to make the air such that it shd. be rigidly the same in every note and rhythm (always excepting the alterations to save the tritone) in all its shifts; but I found that impracticable and that I had reached the point where art calls for loosing, not for lacing. I now make the canon strict in each verse, but allow a change, which indeed is besides called for by the change of words, from verse to verse. Indeed the air becomes a generic form which is specified newly in each verse, with excellent effect. It is like a new art this. I allow no modulation: the result is that the tune is shifted into modes, viz. those of La, Mi, and Sol (this is the only way I can speak of them, and they have a character of their own which is neither that of modern major and minor music nor yet of the plain chant modes, so far as I can make out). The first shift is into the mode of La: this shd. be minor, but the effect is not exactly that; rather the feeling is that Do is still the keynote, but has shifted its place in the scale. This impression is helped by the harmony, for as the Third is not flattened the chords appear major. The chord at the beginning of every bar is the common chord or first inversion; the $\frac{6}{4}$ may appear in course of the bar and discords are in passing or prepared. Perhaps the harmony may be heavy, but I work according to the only rules I know. I can only get on slowly with it and must hope to be rewarded in the end. Now I must lie down.

Who is Miss Cassidy? She is an elderly lady who by often asking me down to Monasterevan and by the change and holiday her kind hospitality provides is become one of the props and struts of my existence. St. Ernin founded the monastery: a singular story is told of him. Henry VIII confiscated it and it became the property of Lord Drogheda. The usual curse on abbey lands attends it and it never passes down in the direct line. The present Lord and Lady Drogheda have no issue. Outside Moore Abbey, which is a beautiful park, the country is flat, bogs and river and canals. The river is the Barrow, which the old Irish poets call the dumb Barrow. I call it the burling Barrow Brown. Both descriptions are true. The country has nevertheless a charm. The two beautiful young people° live within an easy drive.

With kind love to Mrs. Bridges and Mrs. Molesworth, I am your affectionate friend

Gerard.

SERMONS AND DEVOTIONAL WRITINGS

From SERMON FOR OCT. 5 1879, GIVEN AT BEDFORD LEIGH°

... (2) How best to say the Rosary. One way to dwell on the persons of the Mystery: examples. Another by giving a special meaning to the words of the Hail Mary, as *Full of grace*, according to the mystery in hand: examples

Though a devotion in honour of B. V., all the Mysteries but two to do with our Lord and the sorrowful Mysteries exclusively so. It is dwelling on Christ's life in Mary his mother's company, calling her to witness how good he is, how merciful, how afflicted, how glorified by God, and so forth

(3) See text: number of *classes* of men recommended to the B. V.'s protection. cf. titles of her litany. She is in fact *the universal mother*; however unlike her children loves them all. No wonder she can, for in her met things that are thought to be and even are opposite and incompatible, viz. maidenhood and motherhood; then courage and meekness, height and lowliness, wisdom and silence, retirement and renown

St. Bernard's saying, All grace given through Mary: this a mystery. Like blue sky, which for its richness of colour does not stain the sunlight, though smoke and red clouds do, so God's graces come to us unchanged but all through her. Moreover she gladdens the Catholic's heaven and when she is brightest so is the sun her son: he that sees no blue sees no sun either, so with Protestants.

God is holiness, loves only holiness, cares only for it, created the world for it (which, without man, if churned or pressed would yield God none). The B. V. like God in this, loves and cares for this only. Now holiness God promotes by giving grace; the grace he gives not direct but as if stooping and drawing it from her vessel, taking it down from her storehouse and cupboard. It is in some way laid up in her. She sympathises for all of us as though she and not we were in the circumstance, a sinner doing penance, though so innocent; a soldier in battle, though herself a woman. But when all is said heart cannot think her greatness, tongue cannot tell her praise.

SERMON FOR SUNDAY EVENING NOV. 23 1879 AT BEDFORD LEIGH—Luke ii 33. *Et erat pater ejus et mater mirantes super his quae dicebantur de illo*° (text taken at random)

... Our Lord Jesus Christ, my brethren, is our hero,° a hero all the
world wants. You know how books of tales are written, that put one man
before the reader and shew him off handsome for the most part and
brave and call him My Hero or Our Hero. Often mothers make a hero of
a son; girls of a sweetheart and good wives of a husband. Soldiers make a
hero of a great general, a party of its leader, a nation of any great man that
brings it glory, whether king, warrior, statesman, thinker, poet, or
whatever it shall be. But Christ, he is the hero. He too is the hero of a
book or books, of the divine Gospels. He is a warrior and a conqueror; of
whom it is written he went forth conquering and to conquer. He is a king,
Jesus of Nazareth king of the Jews, though when he came to his own
kingdom his own did not receive him, and now, his people having cast
him off, we Gentiles are his inheritance. He is a statesman, that drew up
the New Testament in his blood and founded the Roman Catholic
Church that cannot fail. He is a thinker, that taught us divine mysteries.
He is an orator and poet, as in his eloquent words and parables appears.
He is all the world's hero, the desire of nations. But besides he is the hero
of single souls; his mother's hero, not out of motherly foolish fondness
but because he was, as the angel told her, great and the son of the Most
High and all that he did and said and was done and said about him she
laid up in her heart. He is the truelove and the bridegroom of men's
souls: the virgins follow him whithersoever he goes; the martyrs follow
him through a sea of blood, through great tribulation; all his servants take
up their cross and follow him. And those even that do not follow him, yet
they look wistfully after him, own him a hero, and wish they dared answer
to his call. Children as soon as they can understand ought to be told
about him, that they may make him the hero of their young hearts. But
there are Catholic parents that shamefully neglect their duty: the grown
children of Catholics are found that scarcely know or do not know his
name. Will such parents say they left instruction to the priest or the
schoolmaster? Why, if they sent them very early to the school they might
make that excuse, but when they do not what will they say then? It is at the
father's or the mother's mouth first the little one should learn. But the
parents may be gossipping or drinking and the children have not heard of
their lord and saviour. Those of you, my brethren, who are young and yet
unmarried resolve that when you marry, if God should bless you with
children, this shall not be but that you will have more pity, will have pity
upon your own.

There met in Jesus Christ all things that can make man lovely and
loveable. In his body he was most beautiful. This is known first by the
tradition in the Church that it was so and by holy writers agreeing to suit

those words to him / Thou art beautiful in mould above the sons of men:°
we have even accounts of him written in early times. They tell us that he
was moderately tall, well built and slender in frame, his features straight _Jesus_
and beautiful, his hair inclining to auburn, parted in the midst, curling
and clustering about the ears and neck as the leaves of a filbert, so they
speak, upon the nut. He wore also a forked beard and this as well as the
locks upon his head were never touched by razor or shears; neither, his
health being perfect, could a hair ever fall to the ground. The account I
have been quoting (it is from memory, for I cannot now lay my hand upon
it) we do not indeed for certain know to be correct, but it has been
current in the Church and many generations have drawn our Lord
accordingly either in their own minds or in his images.

. . . I come to his mind. He was the greatest genius that ever lived. You
know what genius is, brethren—beauty and perfection in the mind. For
perfection in the bodily frame distinguishes a man among other men his
fellows: so may the mind be distinguished for its beauty above other
minds and that is genius. Then when this genius is duly taught and
trained, that is wisdom; for without training genius is imperfect and again
wisdom is imperfect without genius. But Christ, we read, advanced in
wisdom and in favour with God and men: now this wisdom, in which he
excelled all men, had to be founded on an unrivalled genius. Christ then
was the greatest genius that ever lived. You must not say, Christ needed
no such thing as genius; his wisdom came from heaven, for he was God.
To say so is to speak like the heretic Apollinaris, who said that Christ had
indeed a human body but no soul, he needed no mind and soul, for his
godhead, the Word of God, thát stood for mind and soul in him. No, but
Christ was perfect man and must have mind as well as body and that
mind was, no question, of the rarest excellence and beauty; it was genius.
As Christ lived and breathed and moved in a true and not a phantom
human body and in that laboured, suffered, was crucified, died, and was
buried; as he merited by acts of his human will; so he reasoned and
planned and invented by acts of his own human genius, genius made
perfect by wisdom of its own, not the divine wisdom only.

A witness to his genius we have in those men who being sent to arrest
him came back empty handed, spellbound by his eloquence, saying /
Never man spoke like this man.

A better proof we have in his own words, his sermon on the mount, his
parables, and all his sayings recorded in the Gospel. My brethren, we are
so accustomed to them that they do not strike us as they do a stranger that
hears them first, else we too should say / Never man etc. No stories or
parables are like Christ's, so bright, so pithy, so touching; no proverbs or

sayings are such jewellery: they stand off from other men's thoughts like stars, like lilies in the sun; nowhere in literature is there anything to match the Sermon on the Mount: if there is let men bring it forward . . .

And if you wish for another still greater proof of his genius and wisdom look at this Catholic Church that he founded, its ranks and constitution, its rites and sacraments.

Now in the third place, far higher than beauty of the body, higher than genius and wisdom the beauty of the mind, comes the beauty of his character, his character as man. For the most part his very enemies, those that do not believe in him, allow that a character so noble was never seen in human mould. Plato the heathen, the greatest of the Greek philosophers, foretold of him:° he drew by his wisdom a picture of the just man in his justice crucified and it was fulfilled in Christ. Poor was his station, laborious his life, bitter his ending: through poverty, through labour, through crucifixion his majesty of nature more shines. No heart as his was ever so tender, but tenderness was not all: this heart so tender was as brave, it could be stern. He found the thought of his Passion past bearing, yet he went through with it. He was feared when he chose: he took a whip and singlehanded cleared the temple. The thought of his gentleness towards children, towards the afflicted, towards sinners, is often dwelt on; that of his courage less. But for my part I like to feel that I should have feared him. We hear also of his love, as for John and Lazarus; and even love at first sight, as of the young man that had kept all the commandments from his childhood. But he warned or rebuked his best friends when need was, as Peter, Martha, and even his mother. For, as St. John says, he was full both of grace and of truth.

But, brethren, from all that might be said of his character I single out one point and beg you to notice that. He loved to praise, he loved to reward. He knew what was in man, he best knew men's faults and yet he was the warmest in their praise. When he worked a miracle he would grace it with / Thy faith hath saved thee, that it might almost seem the receiver's work, not his . . .

SERMON FOR MONDAY EVENING OCT. 25 1880 *on Divine Providence and the Guardian Angels*, THE 24TH BEING THE FEAST OF ST. RAPHAEL (I have preached on Monday the 4th and the 11th also, but could put down no notes)

Notes—God knows infinite things, all things, and heeds them all in particular. We cannot 'do two things at once', that is cannot give our full heed and attention to two things at once. God heeds all things at once.

He takes more interest in a merchant's business than the merchant, in a vessel's steering than the pilot, in a lover's sweetheart than the

[In consequence of this word *sweetheart* I was in a manner suspended and at all events was forbidden (it was some time after) to preach without having my sermon revised. However when I was going to take the next sermon I had to give after this regulation came into force to Fr. Clare for revision he poohpoohed the matter and would not look at it]

lover, in a sick man's pain than the sufferer, in our salvation than we ourselves. The hairs of our heads are numbered before him. He heeds all things and cares about all things, but not alike; he does not care for nor love nor provide for all alike, not for little things so much as great, brutes as men, the bad as the good, the reprobate who will not come to him as the elect who will. It was his law that the ox should not be muzzled that trod out the corn, but this provision was made for an example to men, not for the sake of the beast; for: Does God care for oxen? asks St. Paul; that is to say, compared with his care for men he does not care for them. Yet he does care for them and for every bird and beast and finds them their food. Not a sparrow,° our Lord says, falls to the ground without your Father, that is / without his noticing and allowing and meaning it. But we men, he added, are worth many, that is / any number of, sparrows. So then God heeds all things and cares and provides for all things but for us men he cares most and provides best.

Therefore all the things we see are made and provided for us, the sun, moon, and other heavenly bodies to light us, warm us, and be measures to us of time; coal and rockoil for artificial light and heat; animals and vegetables for our food and clothing; rain, wind, and snow again to make these bear and yield their tribute to us; water and the juices of plants for our drink; air for our breathing; stone and timber for our lodging; metals for our tools and traffic; the songs of birds, flowers and their smells and colours, fruits and their taste for our enjoyment. And so on: search the whole world and you will find it a million-million fold contrivance of providence planned for our use and patterned for our admiration.

But yet this providence is imperfect, plainly imperfect. The sun shines too long and withers the harvest, the rain is too heavy and rots it or in floods spreading washes it away; the air and water carry in their currents the poison of disease ... everything is full of fault, flaw, imperfection, shortcoming; as many marks as there are of God's wisdom in providing for us so many marks there may be set against them of more being needed still, of something having made of this very providence a shattered frame and a broken web.

Let us not now enquire, brethren, why this should be; we most sadly feel and know that so it is. But there is good in it; for if we were not forced from time to time to feel our need of God and our dependance on him, we should most of us cease to pray to him and to thank him. If he did everything we should treat him as though he did nothing, whereas now that he does not do all we are brought to remember how much he does and to ask for more. And God desires nothing so much as that his creatures should have recourse to him.

But there is one great means he has provided for every one of us to make up for the shortcomings of his general and common providence. This great and special providence is the giving each of us his guardian angel° ...

... He counts all our steps, he knows every hair of our heads, he is witness of all our good deeds and all our evil; he sees all and remembers all. Even our hearts he searches, for he sees them in the light of God's knowledge and God reveals to him all that can be of service to him in his charge and duty of leading the human being entrusted to him to the kingdom of heaven. But though he knows and remembers all the harm we have done he will not be our accuser; where he cannot help us he will be silent; he will speak but of our right deeds and plead in our defence all the good he has observed in us. His whole duty is to help us to be saved, to help us both in body and soul. We shall do well therefore to be ashamed of ourselves before our guardian angel, but not to have no other feelings than shame and dread towards him; for he is our good faithful and charitable friend, who never did and never could sleep one moment at his post, neglect the least thing that could be of service to us, or leave a stone unturned to help us all the days that we have been in his keeping. We should deeply trust him, we should reverence and love him, and often ask his aid.

Here, brethren, I must meet an objection which may be working on your minds. If everyone has so watchful and so strong a keeper at his side why is there such a thing as sudden death, as catching fever, as taking poison by mistake, as being shot or any way injured, even as a stumble or a fall, a scald or a sprain? What are the guardian angels doing that they let such things be?—To begin with, many mischiefs that might befall us our guardian angels do ward off from us: that is the first answer to be made. Next their power over us depends in part on the power we give them and by willingly putting ourselves into their hands, by expressly asking them to help us, we enable them to do so; for always God's special providences are for his special servants. They are not to save us from all the consequences of our own wickedness or folly or even from the wickedness and folly of other men; for we are our own masters, are free to act and then must take the consequences; moreover man is his brother's keeper and may be well or ill kept, as Abel was by wicked Cain not kept but killed. But the fullest answer is this—that in appointing us guardian angels God never meant they should make us proof against all the ills that flesh is heir to, that would have been to put us in some sort back into the state of Paradise which we have lost; but he meant them, accompanying us through this world of evil and mischance, sometimes warding off its blows and buffets, sometimes leaving them to fall, always to be leading us to a better; which better world, my brethren, when you have reached and with your own eyes opened look back on this you will see a work of wonderful wisdom in the guidance of your guardian angel. In the meantime God's providence is dark and we cannot hope to know the why and wherefore of all that is allowed to befall us. . .

THE SPIRITUAL EXERCISES: (*PRINCIPIUM SIVE FUNDAMENTUM*)

EXERCISE:

Man was created to praise, reverence and serve God Our Lord, and by so doing to save his soul. And the other things on the face of the earth were created for man's sake and to help him in the carrying out of the end for which he was created. Hence it follows that man should make use of creatures so far as they help him to attain his end and withdraw from them so far as they hinder him from so doing. For that, it is necessary to make ourselves indifferent in regard to all created things in so far as it is left to the choice of our free will and there is no prohibition; in such sort that we do not on our part seek for health rather than sickness, for riches rather than poverty, for honour rather than dishonour, for a long life rather than a short one; and so in all other things, desiring and choosing only those which may better lead us to the end for which we were created.

G. M. H.'S NOTES:

'Homo creatus est'°—Aug. 20 1880: during this retreat, which I am making at Liverpool, I have been thinking about creation and this thought has led the way naturally through the exercises hitherto. I put down some thoughts.—We may learn that all things are created by consideration of the world without or of ourselves the world within. The former is the consideration commonly dwelt on, but the latter takes on the mind more hold. I find myself both as man and as myself something most determined and distinctive, at pitch, more distinctive and higher pitched than anything else I see; I find myself with my pleasures and pains, my powers and my experiences, my deserts and guilt, my shame and sense of beauty, my dangers, hopes, fears, and all my fate, more important to myself than anything I see. And when I ask where does all this throng and stack of being, so rich, so distinctive, so important, come from / nothing I see can answer me. And this whether I speak of human nature or of my individuality, my selfbeing. For human nature, being more highly pitched, selved, and distinctive than anything in the world, can have been developed, evolved, condensed, from the vastness of the world not anyhow or by the working of common powers but only by one of finer or higher pitch and determination than itself and certainly than any that elsewhere we see, for this power had to force forward the starting or stubborn elements to the one pitch required. And this is much more true when we consider the mind; when I consider my selfbeing, my consciousness and feeling of myself, that taste of myself, of *I* and *me* above and in all things, which is more distinctive than the taste of ale or alum, more distinctive than the smell of walnutleaf or camphor, and is incommunicable by any means to another man (as when I was a child I used to ask myself: What must it be to be someone else?). Nothing else in nature comes near this unspeakable stress of pitch, distinctiveness, and selving, this selfbeing of my own. Nothing explains it or resembles it, except so far as this, that other men to themselves have the same feeling. But this only multiplies the phenomena to be explained so far as the cases are like and do resemble. But to me there is no resemblance: searching nature I taste *self* but at one tankard, that of my own being. The development, refinement, condensation of nothing shews any sign of being able to match this to me or give me another taste of it, a taste even resembling it. . .

Aug. 7 1882°—God's utterance of himself in himself is God the Word, outside himself is this world. This world then is word, expression, news of God. Therefore its end, its purpose, its purport, its meaning, is God and its life or work to name and praise him. Therefore praise put before reverence and service . . . the world, man, should after its own manner give God being in return for the being he has given it or should give him back that being he has given. This is done by the great sacrifice. To contribute then to that sacrifice is the end for which man was made (ibidem)

NOTES ON SUAREZ, *DE MYSTERIIS VITAE CHRISTI*°

... there is a scale or range of pitch which is also infinite and terminates upwards in the directness or uprightness of the 'stem' of the godhead and the procession of the divine persons. God then can shift the self that lies in one to a higher, that is / better, pitch of itself; that is / to a pitch or determination of itself on the side of good. But here arises a darker difficulty still; for how can we tell that each self has, in particular, any such better self, any such range from bad to good? In the abstract there is such a range of pitch and conceivably a self to be found, actually or possibly, at each pitch in it, but how can *each* self have all these pitches? for this seems contrary to its freedom; the more so as if we look at the exhibition of moral freedom in life, at men's lives and history, we find not only that in the same circumstances and seemingly with the same graces they behave differently, not only they do not range as fast from bad to good or good to bad one as another, but, even what is most intrinsic to a man, the influence of his own past and of the preexisting disposition of will with which he comes to action seems irregular and now he does well, now he sins, bids fair to be a sinner and becomes a saint or bids fair to be a saint and falls away, and indeed goes through vicissitudes of all sorts and changes times without number.

This matter is profound; but so far as I see this is the truth. First, though self, as personality, is prior to nature it is not prior to pitch. If there were something prior even to pitch, of which that pitch would be itself the pitch, then we could suppose that that, like everything else, was subject to God's will and could be pitched, could be determined, this way or that. But this is really saying that a thing is and is not itself, is and is not A, is and is not. For self before nature is no thing as yet but only possible; with the accession of a nature it becomes properly a self, for instance a person: only so far as it is prior to nature, that is to say / so far as it is a definite self, the possibility of a definite self (and not merely the possibility of a number or fetch of nature) it is identified with pitch, moral pitch, determination of right and wrong. And so far, it has its possibility, as it will have its existence, from God, but not so that God makes pitch no pitch, determination no determination, and difference indifference. The indifference, the absence of pitch, is in the nature to be superadded. And when nature is superadded, then it cannot be believed, as the Thomists think, that in every circumstance of free choice the person is of himself indifferent towards the alternatives and that God determines which he shall, though freely, choose. The difficulty does not lie so much in his being determined by God and yet choosing freely, for on one side that may and must happen, but in his being supposed equally disposed or pitched towards both at once. This is impossible and destroys the notion of freedom and of pitch.

Nevertheless in every circumstance it is within God's power to determine the creature to choose, and freely choose, according to his will; but not without a change or access of circumstance, over and above the bare act of determination on his part. This access is either of grace, which is 'supernature', to nature or of more grace to grace already given, and it takes the form of instressing the affective will, of affecting the will towards the good which he proposes. So far this is a necessary and constrained affection on the creature's part, to which the *arbitrium* of the creature may give its avowal and consent. Ordinarily when grace is given we feel first the necessary or constrained act and after that the free act on our own part, of consent or refusal as the case may be. This consent or refusal is given to an act either hereafter or now to be done, but in the nature of things such an act must always be future, even if immediately future or of those futures which arise in acts and phrases like 'I must ask you' to do so-and-so, 'I wish to apologise', 'I beg to say', and so on. And ordinarily the motives for refusal are still present though the motive for consent has been strengthened by the motion, just over or even in some way still working, of grace. And therefore in ordinary cases refusal is possible not only physically but also morally and often takes place. But refusal remaining physically possible becomes morally (and strictly) impossible in the following way.

... Therefore in that 'cleave' of being which each of his creatures shews to God's eyes alone (or in its 'burl' of being / uncloven) God can choose countless points in the strain (or countless cleaves of the 'burl') where the creature has consented, does consent, to God's will in the way above shewn. But these may be away, may be very far away, from the actual pitch at any given moment existing. It is into that possible world that God for the moment moves his creature out of this one or it is from that possible world that he brings his creature into this, shewing it to itself gracious and consenting; nay more, clothing its old self for the moment with a gracious and consenting self. This shift is grace. For grace is any action, activity, on God's part by which, in creating or after creating, he carries the creature to or towards the end of its being, which is its selfsacrifice to God and its salvation. It is, I say, any such activity on God's part; so that so far as this action or activity is God's it is divine stress, holy spirit, and, as all is done through Christ, Christ's spirit; so far as it is action, correspondence, on the creature's it is *actio salutaris*; so far as it is looked at *in esse quieto* it is Christ in his member on the one side, his member in Christ on the other. It is as if a man said: That is Christ playing at me and me playing at Christ, only that it is no play but truth; That is Christ <u>being me</u> and me being Christ.

It is plain then how true it is what is said at p. 53 about correspondence being a grace and even the grace of graces. For the momentary and constrained correspondence, being a momentary shift from a worse, ungracious / to a better, a gracious self, is a grace, a favour, and it is grace in the strict sense of that word; it is grace bestowed for the moment and offered for a continuance. But the continued and unconstrained correspondence is a greater blessing and therefore still more a grace and as it was only possible through the first or constrained one (the 'forestall'), the second multiplies the first and is a grace upon a grace or, as St. John says, χάρις ἀντὶ χάριτος, freer on man's part and also doubly free, unconstrained, gratis, grace / on God's.

What then does Marie Lataste° (or Christ in speaking to her) mean by its being granted in answer to prayer?—First that salvation and every necessary grace is so granted. But taking the matter more in particular, I understand at bottom to be meant the simple act of *arbitrium*. For prayer is the expression of a wish to God and, since God searches the heart, the conceiving even of the wish is prayer in God's eyes (see Rom. viii 26, 27.). For there must be something which shall be truly the creature's in the work of corresponding with grace: this is the *arbitrium*, the verdict on God's side, the saying Yes, the 'doing-agree' (to speak barbarously), and looked at in itself, such a nothing is the creature before its creator, it is found to be no more than the mere wish, discernible by God's eyes, that it might do as he wishes, might correspond, might say Yes to him; correspondence itself is on man's side not so much corresponding as the wish to correspond, and this least sigh of desire, this one aspiration, is the life and spirit of man. For beyond this, all the work of actual correspondence, all whatever that has any score or 'afterleave' to any eye but God's, in the corresponding with grace needs fresh grace or the continuance of the offered grace, grace concomitant.

And remark that prayer understood in this sense, this sigh or aspiration or stirring of the spirit towards God, is a *forestall* of the thing to be done, as on the other side grace prevenient is God's forestall of the same, and it is here that one creature, one man, differs so much from another: in one God finds only the constrained correspondence with his forestall (as explained near the top of p. 220); in another he finds after this an act of choice properly so called. And by this infinitesimal act the creature does what in it lies to bridge the gulf fixed between its present actual and worser pitch of will and its future better one ...

Now though it is true that God can raise anyone from any pitch of will and being however low to one in which he shall be gracious and consenting to God and in St. Paul's case wonderfully did so and though we cannot know why he does so to one and not to another, not to all, yet it is easy to see certain things which help to explain such cases as those of St. Paul or St. Matthew or of the penitent thief. St. Paul, if he was in sin and vehemently resisting the grace he then had at the time he was struck down, yet when he had been raised to his grace of conversion, even if that had been, which indeed it was not, a violence done to his will for that moment, corresponded most earnestly then and ever after. But if the will is always and even morally and practically free while in pilgrimage, *in via*, he need not have done so; he might still have come to be a cast away: he speaks so himself ...

God's forestalling of man's action by prevenient grace, which carries with it a consenting of man's will, seems to stand to the action of free choice which follows and to which, by its continued strain and breathing on and man's responding aspiration or drawing in of breath, it leads / as the creation of man and angels in sanctifying grace stands to the act by which they entered with God into the covenant and commonwealth of original justice; further / as the infused virtues of baptism stand to the acts of faith etc which long after follow. This agrees well with the light I once had upon the nature of faith, that it is God / in man / knowing his own truth.° It is like the child of a great nobleman taught by its father and mother a compliment of welcome to pay to the nobleman's father on his visit to them: the child does not understand the words it says by rote, does not know their meaning, yet what they mean it means. The parents understand what they do not say, the child says what it does not understand, but both child and parents mean the welcome.

The will is surrounded° by the objects of desire as the needle by the points of the compass. It has play then in two dimensions. This is to say / it is drawn by affection towards any one, A, and this freely, and it can change its direction towards any other, as free, B, which implies the moving through an arc. It has in fact, more or less, in its affections a tendency or magnetism towards every object and the *arbitrium*, the elective will, decides which: this is the needle proper. But in fallen man all this motion, both these dimensions, κεῖται ἐν τῷ πονηρῷ;° so that the uplifting action of supernatural grace takes place as if in a third dimension, motion, in which man is totally incapable. And here remark what now clearly appears, that the action of such assisting grace is twofold, to help man determine the will towards the right object in one field and at the same time from that field or plane to lift it to a parallel and higher one; besides all the while stimulating its action, in the right plane and in the right direction, towards the right object; so that in fact it is threefold, not twofold—(1) quickening,° stimulating, towards the object, towards good: this is especially in the affective will, might be a natural grace, and in a high degree seems to be the grace of novices; (2) corrective, turning the will from one direction or pitting into another, like the needle through an arc, determining its choice (I mean / stimulating that determination, which it still leaves free): this touches the elective will or the power of election and is especially the grace of the mature mind; (3) elevating,° which lifts the receiver from one cleave of being to another and to a vital act in Christ: this is truly God's finger touching the very vein of personality, which nothing else can reach and man can respond to by no play whatever, by bare acknowledgment only, the counter stress which God alone can feel ('subito probas eum'°), the aspiration in answer to his inspiration. Of this I have written above and somewhere else long ago.

When man was created in grace, that is / in the elevated, the supernatural / state, and his will addressed towards God, the work of actual grace was all of the first sort. This may be called creative grace, the grace which destined the victim for the sacrifice, and which belongs to God the Father. After the Fall there came too 'medicinal', corrective, redeeming / grace, by the restrictions of the Law, by the exhortations of the Prophets, and by Christ himself. And all Christ's words, it seems to me, are either words of cure, as 'Veniam et curabo eum',° 'Volo, mundare',° or corrections of some error or fault; their function is always *ramener à la route*.° This then is especially Christ's grace, it is a purifying and a mortifying grace, bringing the victim to the altar and sacrificing it. And as creative grace became insufficient by the Fall: so this grace of Christ's did not avail when he was no longer present to keep bestowing it or when its first force was spent. At Pentecost the elevating grace was given which fastened men in good. This is especially the grace of the Holy Ghost and is the acceptance and assumption of the victim of the sacrifice.

CREATION AND REDEMPTION
THE GREAT SACRIFICE

NOV. 8 1881 (*Long Retreat*)

Time has 3 dimensions and one positive pitch or direction. It is therefore not so much like any river or any sea as like the Sea of Galilee, which has the Jordan running through it and giving a current to the whole.

Though this one direction of time if prolonged for ever might be considered to be parallel to or included in the duration of God, the same might be said of any other direction in time artificially taken. But it is truer to say that there is no relation between any duration of time and the duration of God. And in no case is it to be supposed that God creates time and the things of time, that is to say / this world, in that duration of himself which is parallel with the duration of time and was before time. But rather as the light falls from heaven upon the Sea of Galilee not only from the north, from which quarter the Jordan comes, but from everywhere / so God from every point, so to say, of his being creates all things. But in so far as the creation of one thing depends on that of another, as suppose trees were created *for* man and *before* man, so far does God create in time or in the direction or duration of time.

There is therefore in the works of creation an order of time, as the order of the Six Days, and another order, the order of intention, and that not only intention in understanding and intention in will but also intention or forepitch of execution, of power or activity. In the order of intention 'other things on the face of the earth' are created after man; the more perfect first, the less after. From this it follows that the more perfect is created in its perfection, that is to say / if perfectible and capable of greater and less perfection, it is created at its greatest. And thus it is said 'Ipsius enim sumus factura, *creati in Christo Jesu* in operibus bonis, quae praeparavit Deus ut in illis ambulemus'° (Eph. ii 10., and he had already dwelt on these Gentiles and on himself too as having been children of wrath, dead in sins, and so on). And further it follows that man himself was created for Christ as Christ's created nature for God (cf. 'omnia enim vestra sunt, vos autem Christi, Christus autem Dei'° 1 Cor. iii 22, 23.). And in this way Christ is the firstborn among creatures. The elect then were created in Christ some before his birth, as Abraham, some before their own, as St. Ignatius; that so their correspondence with grace and seconding of God's designs is like a taking part in their own creation, the creation of their best selves. And again the wicked and the lost are like halfcreations and have but a halfbeing.

The first intention then of God outside himself or, as they say, *ad extra*, outwards, the first outstress of God's power, was Christ; and we must believe that the next was the Blessed Virgin. Why did the Son of God go thus forth from the Father not only in the eternal and intrinsic procession of the Trinity but also by an extrinsic and less than eternal, let us say aeonian one?—To give God glory and that by sacrifice, sacrifice offered in the barren wilderness outside of God, as the children of Israel were led into the wilderness to offer sacrifice. This sacrifice and this outward procession is a consequence and shadow of the procession of the Trinity, from which mystery sacrifice takes its rise; but of this I do not mean to write here. It is as if the blissful agony or stress of selving in God had forced out drops of sweat or blood, which drops were the world, or as if the lights lit at the festival of the 'peaceful Trinity' through some little cranny striking out lit up into being one 'cleave' out of the world of possible creatures. The sacrifice would be the Eucharist, and that the victim might be truly victim like, like motionless, helpless, or lifeless, it must be in matter. Then the Blessed Virgin was intended or predestined to minister that matter. And here then was that mystery of the woman clothed with the sun which appeared in heaven. She followed Christ the nearest, following the sacrificial lamb 'whithersoever he went'.

In going forth to do sacrifice Christ went not alone but created angels to be his company, lambs to follow him the Lamb, the flower of the flock, 'whithersoever he went', that is to say, first to the hill of sacrifice, then after that back to God, to beatitude. They were to take part in the sacrifice and he was to redeem them all, that is to say / for the sake of the Lamb of God who was God himself God would accept the whole flock and for the sake of one ear or grape the whole sheaf or cluster; for redeem may be said not only of the recovering from sin to grace or perdition to salvation but also of the raising from worthlessness before God (and all creation is unworthy of God) to worthiness of him, the meriting of God himself, or, so to say, godworthiness. In this sense the Blessed Virgin was beyond all others redeemed, because it was her more than all other creatures that Christ meant to win from nothingness and it was her that he meant to raise the highest.

. . . A coil or spiral is then a type of the Devil, who is called the old (or original) serpent, and this I suppose because of its 'swale' or subtle and imperceptible drawing in towards its head or centre, and it is a type of death, of motion lessening and at last ceasing. *Invidia autem diaboli mors intravit in mundum*: God gave things a forward and perpetual motion; the Devil, that is / thrower of things off the track, upsetter, mischiefmaker, clashing one with another brought in the law of decay and consumption in inanimate nature, death in the vegetable and animal world, moral death and original sin in the world of man . . .

The snake or serpent a symbol of the Devil.° So also the Dragon. A dragon is or is taken to be a reptile. And first a dragon is a serpent with any addition you make, as of feet or of wings or something less. I found some Greek proverb 'The serpent till he has devoured a serpent does not become a dragon' and the snakes found in China and preserved in temples for adoration are called dragons in virtue of some supposed incarnation which has taken place in them, but they are and look ordinary snakes. So that if the Devil is symbolised as a snake he must be an archsnake and a dragon. Mostly dragons are represented as much more than serpents, but always as in some way reptiles. Now among the vertebrates the reptiles go near to combine the qualities of the other classes in themselves and are, I think, taken by the Evolutionists as nearest the original vertebrate stem and as the point of departure for the rest. In this way clearly dragons are represented as gathering up the attributes of many creatures: they are reptiles always, but besides sometimes have bat's wings; four legs, sometimes those of the mammal quadrupeds, sometimes birds' feet and talons; jaws sometimes of crocodiles, but sometimes of eagles; armouring like crocodiles again, but also sturgeons and other fish, or lobsters and other crustacea; or like insects; colours like the dragonfly and other insects; sometimes horns; and so on. And therefore I suppose the dragon as a type of the Devil to express the universality of his powers, both the gifts he has by nature and the attributes and sway he grasps, and the horror which the whole inspires. We must of course remember how the Cherubim are in Scripture represented as composite beings, combinations of eagles, lions, oxen, and men, and that the religions of heathendom have sphinxes, fauns and satyrs, 'eyas-gods', 'the dog-Anubis', and so on. The dragon then symbolises one who aiming at every perfection ends by being a monster, a 'fright'.

THE PRINCIPLE OR FOUNDATION

Homo creatus est°—CREATION THE MAKING OUT OF NOTHING, bringing from nothing into being: once there was nothing, then lo, this huge world was there. How great a work of power!

. . . WHY DID GOD CREATE?—Not for sport, not for nothing. Every sensible man has a purpose in all he does, every workman has a use for every object he makes. Much more has God a purpose, an end, a meaning in his work. He meant the world to give him praise, reverence and service: *to give him glory.* . . .

The creation does praise God, does reflect honour on him, is of service to him, and yet the praises fall short; the honour is like none, less than a buttercup to a king; the service is of no service to him. In other words *he does not need it.* He has infinite glory without it and what is infinite can be made no bigger. Nevertheless he takes it: he wishes it, asks it, he commands it, he enforces it, he gets it.

The sun and the stars shining glorify God. They stand where he placed them, they move where he bid them. 'The heavens declare the glory of God'. They glorify God, *but they do not know it*. The birds sing to him, the thunder speaks of his terror, the lion is like his strength, the sea is like his greatness, the honey like his sweetness; they are something like him, they make him known, they tell of him, they give him glory, but they do not know they do, they do not know him, they never can, they are brute things that only think of food or think of nothing. This then is poor praise, faint reverence, slight service, dull glory. Nevertheless what they can *they always do*.

But AMIDST THEM ALL IS MAN, man and the angels: we will speak of man. Man was created. Like the rest then to praise, reverence, and serve God; to give him glory. He does so, even by his being, beyond all visible creatures: 'What a piece of work is man!' (Expand by 'Domine, Dominus, quam admirabile etc ... Quid est homo ... Minuisti eum paulo minus ab angelis'.°) But man can know God, *can mean to give him glory*. This then was why he was made, to give God glory and to mean to give it; to praise God fréely, willingly to reverence him, gládly to serve him. Man was made to give, and mean to give, God glory.

I WAS MADE FOR THIS, each one of us was made for this.

Does man then do it? Never mind others now nor the race of man: DO I DO IT?—If I sin I do not: how can I dishonour God and honour him? wilfully dishonour him and yet be meaning to honour him? choose to disobey him and mean to serve him? ...

... we can repent our sins and BEGIN TO GIVE GOD GLORY. The moment we do this we reach the end of our being, we do and are what we were made for, we make it worth God's while to have created us. This is a comforting thought: we need not wait in fear till death; any day, any minute we bless God for our being or for anything, for food, for sunlight, we do and are what we were meant for, made for—things that give and mean to give God glory. This is a thing to live for. Then make haste so to live.

For IF YOU ARE IN SIN YOU ARE GOD'S ENEMY, you cannot love or praise him. You may say you are far from hating God; but if you live in sin you are among God's enemies, you are under Satan's standard and enlisted there; you may not like it, no wonder; you may wish to be elsewhere; but there you are, an enemy to God. It is indeed better to praise him than blaspheme, but the praise is not a hearty praise; it cannot be. You cannot mean your praise if while praise is on the lips there is no reverence in the mind; there can be no reverence in the mind if there is no obedience, no submission, no service. And there can be no obeying God while you disobey him, no service while you sin. Turn then, brethren, now and give God glory. You do say grace at meals and thank and praise God for your daily bread, so far so good, but thank and praise him now for everything. When a man is in God's grace and free from mortal sin, then everything that he does, so long as there is no sin in it, gives God glory and what does not give him glory has some, however little, sin in it. It is not only prayer that gives God glory but work. Smiting on an anvil, sawing a beam, whitewashing a wall, driving horses, sweeping, scouring, everything gives God some glory if being in his grace you do it as your duty. To go to communion worthily gives God great glory, but to take food in thankfulness and temperance gives him glory too. To lift up the hands in prayer gives God glory, but a man with a dungfork in his hand, a woman with a sloppail, give him glory too. He is so great that all things give him glory if you mean they should. So then, my brethren, live.

A. M. D. G.

MEDITATION ON HELL

Preparatory prayer.

1st prelude—to see with the eyes of the imagination the length, breadth, and depth of Hell. Not known where Hell is, many say beneath our feet. It is at all events *a place of imprisonment*, a prison; *a place of darkness*; and *a place of torment* and that by fire (case of Dives). The devils wander, but bear their torment with them. There is such a place: shutting our eyes will not do away with a steeple or a sign-post nor will not thinking of or not believing in Hell put it out of being.

2nd prelude—To ask for what we want, which here is such an inward feeling of the pain the damned suffer that if we ever come to forget the love of the eternal Lord, through our faults (our venial sins, lukewarmness, worldliness, negligence), the fear of hell-pains at least may help us then and keep us from falling into mortal sin. The great evil of hell is the loss of God, but we think little enough of this: let us think then of what we dread even here, the pain of fire and others, that we understand.

1st point—*with the eyes of the imagination to see* those huge flames and the souls of the lost as if in bodies of fire. The lost now lying in hell are Devils without bodies and disembodied souls, they suffer nevertheless torment as of bodily fire. Though burning and other pains afflict us through our bodies yet it is the soul that they afflict, the mind: if the mind can be deadened, as by chloroform, no pain is felt at all: God can then if he chooses bodily afflict the mind that is out of or never had a body to suffer in. No one in the body can suffer fire for very long, the frame is destroyed and the pain comes to an end; not so, unhappily, the pain that afflicts the indestructible mind, nor after the Judgment day the incorruptible body. Christ speaks of the lost as being salted, that is preserved, with fire and some things, like asbestos-cloth or fireclay, burn but are unchanged. This fire afflicts the lost only so far as they have sinned: therefore the rich glutton asks for a drop of water for *his tongue*, for by the tongue he sinned on earth. Let all consider this: we are our own tormentors, for every sin we then shall have remorse and with remorse torment and the torment fire. The murderer suffers one way, the drunkard another, but all can say they are tormented in that flame. The glutton had there no tongue to torment and yet he was tormented in that flame, for God punishing him through his own guilty thoughts made him seem to suffer in the part that had offended. So of all. In that flame then see them now. They have no bodies there, flame is the body that they wear. You have seen a glassblower breathe on a flame; at once it darts out into a jet taper as a lancehead and as piercing too. The breath of God's anger first kindled the fire of hell; (Is. xxx 33.) it strikes with a distinct indignation still on each distinct unforgiven sin; the wretched soul starts into a flame that has some frightful and fantastic likeness to its sin; so sinners are themselves the flames of hell. O hideous and ungainly sight! which will cease only when at the day of Judgment the body and the soul are at one again and the sinful members themselves and in themselves receive their punishment, a punishment which lasts for ever. *Their worm*, our Lord says, *does not die and their fire is never quenched.*

2nd point—Hear with the ears the wailings (of despair), howls (of pain), cries (of self reproach), blasphemies against Christ our Lord and all his saints because they are in heaven and *they* lost in hell. They do not all blaspheme, there is a sullen dreadful silence of despair; but blaspheme or not, they know and are convinced their chastisement is just. They know that too well; their consciences, their own minds judge them and tell them so. They appeared before Christ at death, their mind's eye was opened, they saw themselves, condemned themselves, despaired, asked for no mercy, and turned from that sight to bury themselves anywhere, even in hell; as a frightened or shamed child buries its head in the pillow they bury theirs in the pit. Neither do they cry with throat and tongue, they have none, but their wailing is an utterance that passes in their woeful thoughts. Nevertheless spirits as they are, they hear and understand each other and add to each other's woe. And if it fears you, brethren, to think of it, but to imagine it, when your ears are open to other and to cheerful living sounds, believe that it is worse to them that have nothing else to do but wail or listen to but wailing.

3rd point—Smell with an imaginary smell the smoke, the brimstone, the dregs and bilgewater of that pit, all that is foul and loathsome. The same must be said of here as has been said of eyes and ears. It is sin that makes them fuel to that fire: the blinding stifling tear-drawing remembrance of a crowd of sins is to them like smoke; stinging remorse is like the biting brimstone; their impurity comes up before them, they loved it once and breathed it, now it revolts them, it is to them like vomit and like dung, and they cannot quit themselves of it: why not? because they are guilty of it; it is *their own sin*; they wallowed in it willingly, now against their will they must for ever wallow.

Sight does not shock like hearing, sounds cannot so disgust as smell, smell is not so bitter as proper bitterness, which is in taste; therefore for the

Fourth point—taste as with taste of tongue all that is bitter there, the tears ceaselessly and fruitlessly flowing; the grief over their hopeless loss; the worm of conscience, which is the mind gnawing and feeding on its own most miserable self. It is still the same story: *they*, their sins are the bitterness, tasted sweet once, now taste most bitter; no worm but themselves gnaws them and gnaws no one but themselves—none from within, but devils have some power upon them from without; for Satan that they served, they are his slaves and he has them in prison, his own prison . . .

And still bitterness of taste is not so cruel as the pain that can be touched and felt. Seeing is believing but touch is the truth, the saying goes. Therefore in

The fifth point—touch and feel how those flames touch and are felt like burning by the souls. How keen and searching is the pain of fire! Worst of all to feel it in the naked soul or in a body that it cannot with its heat consume.

Can these things be? It is terrible to me to have to speak of them, but Christ spoke of them: they must then be true. Are they just?—Yes, because God is just. But you can yourselves see they are just: if you tell your child: Let your sister alone, do not beat her, or I will beat *you* / are you unjust to threaten him? And if he disobeys you and torments her still are you unjust to carry out your threat? Are *we* not warned by God? Is it unjust to forbid us murder, adultery, theft? Is it unjust to threaten punishment if we disobey? or if we have disobeyed to carry out the threat?

Why, we were warned, I say: it is we who carry out the threat; we walk over hell's brink. Or is it that the punishment is greater than the fault? But, brethren, God is wise and just, he must be the judge: if he is so earnest to be obeyed that he forbids sin even under pain of hell, then surely great must that offence be which needs a penalty so great to fright us back from it. All will be punished according to their guilt, according to their knowledge and their power; there are light pains and heavy pains, *many stripes* and *few*; the heathen unwarned by faith, warned indeed but dimly warned by reason, may suffer little in that flame, but we that know what sin is and what hell is and, knowing, do the sin and brave the punishment, we shall suffer much. Alas! it is enough to say we shall suffer everlastingly. But no, my brethren, because by God's mercy we shall never see that place at all.

If you are terrified (I wish you, brethren, for once this much evil, I wish you to feel terror) *turn* where?—*to Christ our Lord* and to his heart that feels and understands what pain and fear and desolation are—all in a word that you can ever feel on earth . . .

ON DEATH

Preparatory prayer.

Introductory remark—*Death is certain* and uncertain, certain to come, uncertain when and where . . .

. . . *1st prelude* to the meditation—*We shall die in these bodies*. I see you living before me, with the mind's eye, brethren, I see your corpses: those same bodies that sit there before me are rows of corpses that will be. And I that speak to you, you hear and see me, you see me breathe and move: this breathing body is my corpse and I am living in my tomb. This is one thing certain of your place of death; you are there now, you sit within your corpses; look no farther: there where you are you will die.

2nd prel.—What we want is so deep a sense of the certainty and uncertainty of death, to have death so before us, that we may dread to sin now and when we die die well.

1st point—*The terrors of death*—(1) It is the greatest of earthly evils. It robs us of our all. Do you love sunshine, starlight, fresh air, flowers, fieldsports?—Despair then: you will see them no more; they will be above ground, you below; you will lose them all. Do you love townlife, homelife, the cheerful hearth, the sparkling fire, company, the social glass, laughter, frolic among friends?—Despair then: you will have no more of them for ever, the churchyards are full of such men as you are now, that feasted once and now worms feast on; ... This is the first terror of death: it is the worst of earthly evils and robs us of our all, and it is the only evil certain to come.

(2) The next terror of death are *the pains of death*. Death mostly is the end of fatal sickness and when is sickness, fatal sickness, without pain? And its pain is not as other pains, which either we surmount and get the better of or at least we can keep up with; but fatal sickness and its pains are for the dying man a losing battle, he bears them and is worse, he may have patience and they do not spare, bad they may be but they will be worse, things will come to the worse and then not mend, making the proverb a lie; the pains of fatal sickness are the pains of death; a woman is in pangs and she brings forth a child, she is at peace and from her pangs has come new life, but we shall be in pangs and bring forth death. I do not mean that the pain of dying is always great; I know well and have seen that often it is not so, so far as from outside we can judge; but often it is; and great pain or little or none, it is terror enough that life is ebbing away. And even for those who seem to die peacefully, if they have their senses to the last, one cannot without shrinking think of that very last moment when flesh and spirit are rent asunder and the soul goes out into the cold, leaving the body its companion dear a corpse behind. This will be to every one of you; I see your corpses here before me.

But there are worse pangs of death than those of the body. There is the sweat of fear, *there is the dread of what is to come after*. Saints have feared: St. Hilarion, St. Jerome, that all their lives did penance, trembled when they came to pass away. The Devil rages then, for he knows that his time is short: one thought of mortal sin at the last gasp is enough, it will do his work for ever, and he watches well, he knows when death is near. And what, my brethren, if you should find yourselves dying and in mortal sin? hurriedly you will send for the priest, counting the minutes till he comes; and what if he should be away, some miscarriage happen, as there always may, even where all seems best provided for? I could tell you even from my own experience tales to make you tremble and more from that of others. The last sacraments are a tower of strength if you can get them, but who can tell you that you will? God does not promise that. Will you trust a priest? May not even a wise and zealous man for once be careless or misjudge and think there is no danger when there is? Few are the parish priests that never, as the saying is, let one slip through their hands ... If things like this should happen to any of you make an act of sorrow for your sins, earnestly asking God to give you the grace, which he will: do this and you will be saved; but yet I say it is a wretched wretched thing to die in fear, without the sacraments.

(3) The third terror of death is its uncertainty, that *it may be sudden* and find us unprepared. This is worse than the rest. Few people die sudden deaths, but a few do and what of those few? Here you are many, most of you then will not die suddenly, but some few of you will; some few, some one or two that hear me, will die suddenly. There are dangers by land and sea, wrecks, railway accidents, lightning, mischances with machinery, fires, falls; there are murders—I have talked with the widows and the orphaned children of suddenly murdered men; there are heart complaints and other sudden strokes of death. Now a sudden death *need* not be a bad death; holy men have sometimes prayed to die suddenly and their prayer been heard; it has its advantages, for if we then are in the grace of God we have no time to fall away; but yet consider, brethren, how many men live in unbroken mortal sin, who if they died suddenly must therefore die in mortal sin; consider how many more go indeed to confession and are forgiven, but sin again shortly and so spend most of their days in mortal sin and if they were cut off suddenly would (not certainly indeed, but) *most likely*, O dreadful likelihood! be then in mortal sin. Others would be as likely as not, others not so likely, but even one chance, how terrible it is! And then follows the judgment of damnation. This is a terror then that far exceeds the rest.

Second point—The comforts of death—God our maker, who knows the clay we are made of; God our father, who never can forget his child; God our Redeemer, who died to save us, has a special providence over death. He knows our utter need, he is most helpful when our need is sorest. Therefore he has provided these three things, the last sacraments, the grace of contrition, and holy hope.

(1) *The Last Sacraments*—These are the sacrament of Penance, the sacrament of Extreme Unction, and the Holy Communion. The first is the most necessary, but by all means have them all. Of *Penance* I need say nothing here except that whether it is for yourself or for others, you should send for the priest in time while the sick man has his senses, that he may make his confession duly; nevertheless if it is too late for that the priest's absolution without confession is enough to save the soul, supposing—always supposing this—there was any sorrow for sin there. We may always suppose there was some sorrow for sin if the sick man wished himself to confess; that is a sign of itself almost certain and where that has been the case and absolution has been given the soul will be saved. But where there is no willingness to confess the priest's absolution is of no use; it cannot forgive, for there is no sorrow. This is the first of God's mercies over death, the first comfort of the deathbed, that you may be forgiven all the sins of a lifetime by one short confession; nay, if you

wish to confess and cannot you may still be forgiven if the priest absolves you.

The *Holy Communion* can after confession be given. It cannot be given when the senses have failed, therefore again send in time. When the priest gives communion with the words: *Accipe, frater (or soror), viaticum* etc / Receive, brother (or sister), the provision of the Body of our Lord J. C., and may he preserve thee from the malicious enemy and bring thee safe to life everlasting. Amen instead of the common ones: *Corpus Dni N. J. C.* etc he supposes you to be near death or at least in danger of death, against which he provides you; for *viaticum* means money, provision, for a journey, that is / the journey to the other world. And so given the sacrament may be received after taking food and at any time of day and even if one had been to communion that very morning. Otherwise there is no difference and whether we are to live or die the communion always does the soul good. Alas! it is often the only communion that the dying man has made or the first for many, many years; and a weary sick and dying man is in poor dispositions for this most holy and most heavenly sacrament. Nevertheless let him have it; nay, he is bound, he must have it. But if he is ready to vomit he cannot be allowed to receive it, the risk to the Lord's Body is too great; therefore, brethren, pray, as our Lord said *Pray that your flight may not be in the winter*, so I say here, Pray now that you may not suffer from that evil in your last sickness, but may be able to receive the Lord's Body.

This is the second of God's deathbed mercies, the second comforting sacrament of death, the Holy Communion. And indeed what a thing it is to think that Christ should be willing so to be carried to a dying sinner. Is not this the good shepherd going to seek the lost sheep! Going to seek it, and where? Alas, brethen, to what filthy places have I not myself carried the Lord of Glory! and worse than filthy places, dens of shame. Pray that when your last hour comes you may make him a welcome in a place becoming and respectable and, much more, in a humble penitent desirous heart.

The third of the last sacraments is the *Holy Oil*. Now this is called one of the last sacraments, but it is not meant for death only; rather it is meant for life, to raise up from sickness and to save life, and this I have seen it do and suddenly and so that the physicians were amazed at it. Do not then be such fools as to neglect it, for that neglect may cost your life; God may have meant to raise you up by it if you had asked for it in time, but after a while may mean that it should not help your body; nay, beware that it may not come too late to help your soul. For now we are speaking of death, when, as I am supposing, this sacrament is to be a *last* sacrament in truth.

In that case it is meant to strengthen you for your death agony, for the last struggle with your spiritual enemy: that is its effect in this life. Moreover it quits you of either all or some of the remains of sin; of the debt of penalty left, as you know, after the guilt of sin has been forgiven; it is a sacrament of indulgence, better than any indulgence you could gain or others for you after you are gone: that is its effect in the other life. Moreover this holy oil works with virtue on the man that lies stark and insensible as well as on the one that has his senses. Have it then, send for it; it may save your soul, it may save you from the fiery bath of purgatory. This is the third comforter among the sacraments of death.

(2) But now suppose you cannot get the last sacraments or that they delay to come and you are in danger and in fear and trembling, knowing or at least fearing you are in a state of sin: God has provided still another way. Make an act of sorrow for your sins, *an act of contrition: repent.* True contrition, that is / perfect sorrow for sin, is when we are sorry, and sorry for God's sake. Now people think that it is hard to have so pure a sorrow; that to be sorry because we are in danger to be damned may be easy, only that is not enough without the sacraments; but to be sorry for God's sake and not our own—which is contrition—is a thing for saints or the devout and not for ordinary men or loth and lingering sinners. But easy or hard, remember, brethren, this and root and rivet it into your hearts, that sorrow for sin is a gift of God and that if you ask for it it will be given. Do not enquire if you have got it: ask for it, beg and pray for it, with tears, with inward tears at least, and strong cries of the heart beseech God to give it you. Summon your last strength for that. You will have it, you will get it in time enough. God knows our need. God knows our need. I repeat it a third time, God knows our need. Luke xi 11–13. 'which of you asks his father for bread? and will he give him a stone? or a fish? will he for the fish give him a serpent? or if he ask for an egg will he reach him a scorpion? If then you that are bad know how to give good gifts to your children how much more will your heavenly father give the good spirit', as the spirit of contrition, of sorrow for sin—'to those that ask him!' He will give it, he will not see you perish. Pray for it too to the Sacred Heart of Christ. Get a crucifix hung, if you can, before your bed: looking towards that pray your just dying saviour to look on you now at your death. You will never be refused. Say 'Agnus Dei' etc over and over again. And pray to the Blessed Virgin: 'pray for us sinners now *and at the hour of our death*'. Get others too to pray round you. Make an act of sorrow for your sins, then fall again to praying God to give you sorrow. And for your comfort, know that as soon as the wish to be forgiven is truly

in your heart you either are forgiven there and then or that before you pass away infallibly you will be.

(3) This leads me to the third thing that sheds comfort over death, *holy hope*. Though you should be unable either to get the sacraments or to make (or at least feel sure that you have made) a true act of sorrow for your sins and that you are forgiven, nevertheless go on hoping *and praying* (as far as your state allows) that God will save you, though it should be, as it might be, only at the very instant of the soul leaving the body. Hope is an anchor cast in heaven: as long as you do not let it go, hold it must and lost you cannot be. . .

. . . You should know that *there is a special providence over death*. God provides for everything but most for man: *not a sparrow falls to the ground*, Christ says, *without your Father* and *You*, he says, *are worth more than many*, that is / any number of, *sparrows*. He provides for all that happens to us, but most for the most important; therefore most for death: *the hairs of your head*, Christ says, *are numbered*; if when the hair is parted God counts and knows how many hairs fall to the right side and how many to the left how much more does he take account of the parting of soul and body! It is seen again and again, I have seen it myself and speak of what I know, that people get the last sacraments just in time, that some happy chance or other falls out in their favour. And when we do not see the providence it may still be there and working in some secret way. Hope for it then and pray for it and yet fear and tremble, *work out your salvation in fear and trembling*. For I must end as I began. One of God's providences is by warnings—the deaths of others, sermons, dangers, sicknesses, a sudden thought: beware, beware of neglecting a warning. This very discourse of mine, this meditation, is a warning. A warning leaves a man better or worse, does him good or harm; never leaves him as it finds him. Some are cut off suddenly in mortal sin without time for either the sacraments or an act of contrition or even of hope. If the holy oil is administered before death while they lie insensible, yet without *any* sorrow for sin it cannot save them; and many are for long spaces of time together habitually sinning and without sorrow, hope, or the wish to repent. Does God then cut them off purposely in sin?—No, for his providence here is *always in our favour*; but he may with a provoking sinner use no special providence and let things take their ordinary course: now in the ordinary course of things a certain number must die suddenly by accident, malice, or disease. This is a terrible possibility. But, I repeat, there is a special providence over death.

Here recall the two preludes and end by exhorting them to hope, to repent, to begin now their preparation for death, and always to pray for final perseverance or final penitence. See also the meditation before for a colloquy to be made, if suitable, on their behalf.

RETREAT AT BEAUMONT
Sept. 3–10 incl. 1883

[Sept. 8.] . . . During this retreat I have much and earnestly prayed that God will lift me above myself to a higher state of grace, in which I may have more union with him, be more zealous to do his will, and freer from sin. Yesterday night it was 15 years exactly since I came to the Society. In this evening's meditation on the Temptation I was with our Lord in the wilderness in spirit and again begged this, acknowledging it was a great grace even to have the desire. For indeed it is a pure one and it is long since I have had so strong and spiritual a one and so persistent. I recommended to him our novices newly come today or last night and those who should this morning have taken their vows. And I had other good thoughts I do not put down.

Also in some med. today I·earnestly asked our Lord to watch over my compositions, not to preserve them from being lost or coming to nothing, for that I am very willing they should, but they might not do me harm through the enmity or imprudence of any man or my own; that he would have them as his own and employ or not employ them as he would see fit. And this I believe is heard

Sept. 9. In meditating on the Crucifixion I saw how my asking to be raised to a higher degree of grace was asking also to be lifted on a higher cross. Then I took it that our Lord recommended me to our Lady and her to me

Sept. 10. The walk to Emmaus. This morning in thanksgiving after mass much bitter thought but also insight in things. And the above meditation was made in a desolate frame of mind; but towards the end I was able to rejoice in the comfort our Lord gave those two men, taking that for a sample of his comfort and them for representatives of all men comforted, and that it was meant to be of universal comfort to men and therefore to me and that this was all I really needed; also that it was better for me to be accompanying our Lord in his comfort of them than to want him to come my way to comfort me

RETREAT NOTES

Jan 1. 1889.° St. Stanislaus' College, Tullabeg.

Principium seu Fundamentum: 'Homo creatus est ut laudet' etc—All moral good, all man's being good, lies in two things—in being right, being in the right, and in doing right; in being on the right side, on the side of good, and on that side of doing good. Neither of these will do by itself. Doing good but on the wrong side, promoting a bad cause, is rather doing wrong. Doing good but in no good cause is no merit: of whom or what does the doer deserve well? Not at any rate of God. Nor plainly is it enough to be on the right side and not promote it.

But men are variously constituted to make much of one of these things and neglect the other. The Irish think it enough to be Catholics or on the right side and that it is no matter what they say and do to advance it; practically so, but what they think is that all they and their leaders do to advance the right side is and must be right. The English think, as Pope says for them, he can't be wrong whose life is in the right. Marcus Aurelius seems in his Meditations to be leading the purest and most unselfish life of virtue; he thinks, though with hesitation, that Reason governs the Universe and that by this life he ranks himself on the side of that Reason; and indeed, if this was all he had the means of doing, it was enough; but he does not know of any particular standard the rallying to which is the appointed signal of taking God the sovereign Reason's, God the Word's, side; and yet that standard was then raised in the world and the Word and sovereign Reason was then made flesh and he persecuted it. And in any case his principles are principles of despair and, again, philosophy is not religion.

But how is it with me? I was a Christian from birth or baptism, later I was converted to the Catholic faith, and am enlisted 20 years in the Society of Jesus. I am now 44. I do not waver in my allegiance, I never have since my conversion to the Church. The question is how I advance the side I serve on. This may be inwardly or outwardly. Outwardly I often think I am employed to do what is of little or no use. Something else which I can conceive myself doing might indeed be more useful, but still it is an advantage for there to be a course of higher studies for Catholics in Ireland and that that should be partly in Jesuit hands; and my work and my salary keep that up. Meantime the Catholic Church in Ireland and the Irish Province in it and our College in that are greatly given over to a partly unlawful cause, promoted by partly unlawful means, and against my will my pains, laborious and distasteful, like prisoners made to serve the enemies' gunners, go to help on this cause. I do not feel then that outwardly I do much good, much that I care to do or can much wish to prosper; and this is a mournful life to lead. In thought I can of course divide the good from the evil and live for the one, not the other: this justifies me but it does not alter the facts. Yet it seems to me that I could lead this life well enough if I had bodily energy and cheerful spirits. However these God will not give me. The other part, the more important, remains, my inward service.

I was continuing this train of thought this evening when I began to enter on that course of loathing and hopelessness which I have so often felt before, which made me fear madness and led me to give up the practice of meditation except, as now, in retreat, and here it is again. I could therefore do no more than repeat *Justus es, Domine, et rectum judicium tuum* and the like, and then being tired I nodded and woke with a start. What is my wretched life? Five wasted years almost have passed in Ireland. I am ashamed of the little I have done, of my waste of time, although my helplessness and weakness is such that I could scarcely do otherwise. And yet the Wise Man warns us against excusing ourselves in that fashion. I cannot then be excused; but what is life without aim, without spur, without help? All my undertakings miscarry: I am like a straining eunuch. I wish then for death: yet if I died now I should die imperfect, no master of myself, and that is the worst failure of all. O my God, look down on me

Jan. 2—This morning I made the meditation on the Three Sins, with nothing to enter but loathing of my life and a barren submission to God's will. The body cannot rest when it is in pain nor the mind be at peace as long as something bitter distills in it and it aches. This may be at any time and is at many: how then can it be pretended there is for those who feel this anything worth calling happiness in this world? There is a happiness, hope, the anticipation of happiness hereafter: it is better than happiness, but it is not happiness now. It is as if one were dazzled by a spark or star in the dark, seeing it but not seeing by it: we want a light shed on our way and a happiness spread over our life

Afternoon: on the same—more loathing and only this thought, that I can do my spiritual and other duties better with God's help. In particular I think it may be well to resolve to make the examen every day at 1.15 and then say vespers and compline if not said before. I will consider what next

Jan. 3—Repetition of 1st and 2nd exercise—Helpless loathing. Then I went out and I said the Te Deum and yet I thought what was needed was not praise of God but amendment of life

Jan. 5th.—Repetition of meditations on Incarnation and Nativity—All that happens in Christendom and so in the whole world affected, marked, as a great seal, and like any other historical event, and in fact more than any other event, by the Incarnation; at any rate by Christ's life and death, whom we by faith hold to be God made man. Our lives are affected by the events of Roman history, by Caesar's victory and murder for instance. Yet one might perhaps maintain that at this distance of time individuals could not find a difference in their lives, except in what was set down in books of history and works of art, if Pompey instead of Caesar had founded the Empire or Caesar had lived 20 years longer.

But our lives and in particular those of religious, as mine, are in their whole direction, not only inwardly but most visibly and outwardly, shaped by Christ's. Without that even outwardly the world could be so different that we cannot even guess it. And my life is determined by the Incarnation down to most of the details of the day. Now this being so that I cannot even stop it, why should I not make the cause that determines my life, both as a whole and in much detail, determine it in greater detail still and to the greater efficiency of what I in any case should do, and to my greater happiness in doing it?

It is for this that St Ignatius speaks of the angel *discharging his mission*, it being question of action leading up to, as now my action leads from, the Incarnation. The Incarnation was for my salvation and that of the world: the work goes on in a great system and machinery which even drags me on with the collar round my neck though I could and do neglect my duty in it. But I say to myself that I am only too willing to do God's work and help on the knowledge of the Incarnation. But this is not really true: I am not willing enough for the piece of work assigned me, the only work I am given to do, though I could do others if they were given. This is my work at Stephen's Green. And I thought that the Royal University was to me what Augustus's enrolment was to St Joseph: *exiit sermo a Caesare Augusto° etc.*; so resolution of the senate of the R. U. came to me, inconvenient and painful, but the journey to Bethlehem was inconvenient and painful; and then I am bound in justice, and paid. I hope to bear this in mind

Jan. 6th. Epiphany.—Yesterday I had ever so much light on the mystery of the feast and the historical interpretation of the gospel and last night on the Baptism and today on that and the calling of Nathanael and so on, more than I can easily put down. However I had better have at least some notes . . .

. . . John was the Baptist and must baptise them. For this probably he used *affusion*, throwing water on them, and for this some shell or scoop, as he is represented. And he seems to allude to this in contrasting himself

with Christ: *ego quidem aqua baptizo . . . cuius ventilabrum in manu eius*°
Luke iii. 16, 17.—*he* baptises with breath and fire, as wheat is winnowed
in the wind and sun, and uses no shell like this which only washes once
but a fan that thoroughly and for ever parts the wheat from the chaff. For
the fan is a sort of scoop, a shallow basket with a low back, sides sloping
down from the back forwards, and no rim in front, like our dustpans, it is
said. The grain is either scooped into this or thrown in by another, then
tossed out against the wind, and this vehement action St John compares
to his own repeated 'dousing' or affusion. The separation it makes is very
visible too: the grain lies heaped on one side, the chaff blows away the
other, between them the winnower stands; after that nothing is more
combustible than the chaff, and yet the fire he calls unquenchable. It will
do its work at once and yet last, as this river runs forever, but has to do its
work over again. Everything about himself is weak and ineffective, he and
his instruments; everything about Christ strong.

NOTES

Abbreviations used in the Notes are listed on p. xiii.

References to the Bible are to the Authorized Version except where otherwise stated.

For explanations of such terms as standard rhythm, counterpoint, sprung rhythm, over-rove, see pp. 107–8, 238 and 243.

The metrical marks shown are all Hopkins's own, taken generally from the manuscript chosen as text (see Note on the Text, p. xli). Hopkins's stresses are incorporated in the text while the rest of his metrical symbols are included in the notes. A list of most of these, with his explanations, follows:

~   quiver or circumflexion, making one syllable nearly two.

⎯⎯   over three or more syllables gives them the time of one half foot.

:   great colons signify a stress on the syllables either side of the colon; a sprung opening, i.e. a great colon at the beginning of a line indicates a stress on the initial syllable (C. L. P.).

∽ ∾   counterpoint signs (see p. 108).

⁀   between syllables slurs them into one.

⎯   outride; under one or more syllables makes them extrametrical.

″   heavy stress.

˝   stresses of sense; independent of the natural stress of the verse.

⌐⎯⎯   over two neighbouring syllables means that, though one has and the other has not the metrical stress, in recitation-stress they are to be about equal.

Other symbols are explained in notes to poems in which they occur.

G. M. H. also used a number of musical terms such as rallentando (slowing down), staccato (▼; short, detached sounds), sforzando (suddenly loud), and the pause (⌒ ; a dwelling on a syllable which need not have the metrical stress).

---

1 *The Escorial.* Text from *C*. Dated 'Easter 1860'. Hopkins won the Poetry Prize at Highgate School with this poem on a set subject. The entry had to be anonymous and Hopkins chose as identifying motto Theocritus (Idyll VII, l. 41), 'I compete like a frog against the cicadas'. The text of the poem and its accompanying notes (listed below) were most probably written out by his mother, Mrs Manley Hopkins. Hopkins's letters and diaries show the same keen interest in architecture and art evident in the poem. For many of the details he seems to have used William H. Prescott's *Reign of Philip the Second* (London, 1859), which the family owned: see vol. iii, pp. 372–90. The footnotes, and the symbols used for cues, are G. M. H.'s.

Motto: MS βάτραχός.

A line above l. 22, and stanza 9, were probably omitted to comply with the stipulated length.

4 l. 98. Apostrophe in 'age's' is editorial.

5 l. 106. MS altered to 'continually' in pencil evidently not the reading submitted and perhaps not changed by G. M. H.

6 *Promêtheus Desmotês*. Text from autograph in *B* II (school notebook also used at Oxford). In a letter of 3 September 1862 Hopkins told E. H. Coleridge that he had been reading *Prometheus Bound* and considered it 'immensely superior to anything else of Aeschylus'' he had read, 'really full of splendid poetry'. (*L* III, p. 6). He enclosed ll. 20–36 of his translation in the letter.

7 *Il Mystico*. MS Texas. Printed in *L* III, pp. 9–13 (3 September 1862). Hopkins sent 'extracts' from it to E. H. Coleridge, saying that it was the best thing that he had done recently and was in imitation of 'Il Penseroso'. It remained unfinished.

8 ll. 48–64. See Ezekiel 1, 8:12, 36.

9 l. 65. *lark*: skylark. See 'The Sea and the Skylark'.

   ll. 75–6. 'I had formerly instead of the lines resembling them which I have put in the enclosed copy, "And when the silent heights were won, Alone in air to face the sun". Now is that or is it not a plagiarism from Tennyson's *Eagle*, "Close to the sun in lonely lands," (see the poem)? I am in that state that I want an unprejudiced decision' (*L* III, p. 14).

10 l. 119. See Rev. 21: 19–20.

11 *A windy day in summer*. MS Texas. Printed *L* III, p. 13. Enclosed in a letter to E. H. Coleridge of 3 September 1862.

    A fragment of anything you like. Printed *L* III, p. 13, written below 'A windy day' in the letter mentioned above.

    A Vision of the Mermaids. Text from autograph in *C*. Transcription in *A*. Limited facsimile edition in 1929. The autograph is written below one of Hopkins's circular drawings. It shows groups of mermaids standing waist-deep in the sea admiring a spectacular sunset (reproduced *J*, pl. 3). See Hopkins's description of his family's seaside holiday in July 1863 (*L* III, p. 201): 'My brothers and cousin catch us shrimps, prawns and lobsters, and keep aquariums . . . large red scar.'

15 *Winter with the Gulf Stream*. Published in *Once a Week* (14 February 1863) and reproduced *L* III, p. 437. Text from a transcription by Fr. F. E. Bacon, SJ of a revised version of August 1871, the original of which is now lost.

16 l. 27. *Pactolus*. The river whose golden sands are said to have come from Midas, who bathed in it to lose his unwanted gift of turning everything he touched into gold.

    Spring and Death. Text from undated autograph in *C*. 1863? Transcription by Bridges in *A*. The fanciful tone of the poem recalls the illustrated letter-headings Hopkins drew to entertain his younger brothers and sisters (See *J*, pls. 5, 7, and 8).

Fragments: (i) 'The wind that passes by so fleet'. *C* I (printed *J*, p. 9). MS cancelled in pencil. Late 1863.

(ii) 'Whose braggart 'scutcheon' and 'The villain shepherds'. *C* I (printed *J*, p. 18) from an outline of a possible death-speech for Ajax based on Sophocles' play, *Ajax* (*J*, p. 306 n. 17. 8). February? 1864. The journals contain more single lines with poetic flavour than are printed in this edition.

(iii) 'The sparky air'. *C* I (printed *J*, p. 20). 19 March 1864.

(iv) '—and on their brittle green quils'. *C* I (printed *J*, p. 22). Between 10 and 14 April 1864.

18 *Pilate*. *C* I and *C* II. June 1864 (noted *J*, p. 25); the final two and a half stanzas are from a fair copy of October (noted *J*, p. 49). The sequence is uncertain, stanzas 2, 3, 5–7 are numbered; the other stanzas are printed in the order in which they occur in the manuscript. The layout is from a fair copy in *C* II.

G. M. H. may have been influenced by Tennyson's 'St. Simeon Stylites', which he greatly admired (*L* III, p. 8), both in choosing to write the poem as a dramatic monologue and for the descriptions of Pilate's sufferings. The imagery suggests that he also had *Prometheus Bound* in mind.

ll. 1–4. Hopkins began a revision but did not carry through the changes necessary to incorporate it:

> Betwixt the morsels of the snow,
> Under the mastering blue black heat,
> When the winds blow, when strong rains twist and flow
> Along my face and hands and feet. (N. H. M.)

20 l. 94. Two pages were cut out of *C* II at this point. On one occurs the first half of a stanza, printed *J*, p. 49:

> But if this overlast the day
> Undone, and I must wait the year,
> Yet no delay can serve to grate away
> A purpose desperately dear

21 '*She schools the flighty pupils*'. *C* I (printed *J*, p. 26), June 1864.

Richard. *C* I (printed *J*, p. 27), first version May–June 1864. Written in Spenserian stanza (N. H. M.). See p. 49 for further drafts.

Uncancelled alternatives:
l. 4. The listening downs and breezes seemèd he.
l. 8. 'drinking' later alternative to 'one drunk'.
l. 9. 'True' later alternative to 'In'.
l. 10. 'forehead' later alternative to 'frontal'.

A Soliloquy of One of the Spies left in the Wilderness. *C* I, July 1864 (part of stanza 2 is from *C* II). See Exodus 12, 16, and 17 and Numbers 10, 11, 13–14. There are many variants (see *J*, pp. 28–9). N. H. M. has reordered the stanzas to follow the manuscript more closely. Final variants have been chosen for stanzas 4 and 9.

23 l. 45. *to-day*. Hyphen editorial.

The Lover's Stars. *C* I, between drawings dated 16 July and 18 July (printed *J*, pp. 29–30). Hopkins called it 'a trifle in something like Coventry Patmore's style' (*L* III, p. 213).

ll. 5–8 were revised to

> The other leaves the West behind
> Or it may be the prodigal South,
> Passes the seas and comes to find
> Acceptance round his mistress' mouth.

Since the consequent revisions required to stanzas 4 and 5 were not made, the earlier version has been retained.

24 *'During the eastering'*. *C* I (printed *J*, p. 30), between drawings dated 18 July and 22 July 1864.

'—*Hill / Heaven*'. *C* I (printed *J*, p. 31), follows previous poem and precedes drawing of 22 July 1864.

25 *'Distance / Dappled'*. *C* I (printed *J*, p. 31) between 22 and 25 July 1864.

The peacock's eye. *C* I, the second of a number of fragments written 22–5 July 1864 (see *J*, p. 31). Hopkins commented below, 'Overloaded apparently'. The first draft ran:

> The peacock's eye
> Winks away its azure sheen
> Barter'd for a ring of green.
> The bean-shaped pupil of moist jet
> Is the silkiest violet.

See *J*, pp. 209–10 (17 May 1871) for a later description of the peacock's opened train: 'I have thought it looks like a tray or green basket or fresh-cut willow hurdle set all over with Paradise fruits cut through—first through a beard of golden fibre and then through wet flesh greener than greengages or purpler than grapes . . . and then within all a sluggish corner drop of black or purple oil.'

l. 5. *jet*: deep, glossy black.

'Love preparing to fly'. *C* I immediately after 'The peacock's eye'.

l. 2. *glassy*. See *J*, p. 231 (11 May) where 'glassy' means 'a notable glare the eye may abstract and sever from the . . . colour of light'.

Barnfloor and Winepress. Printed in *The Union Review*, vol. iii (1865) p. 579. Rough draft in *C* I (noted *J*, p. 32); anonymous transcription, dated 1865 (Bodl.), and another in *A*. Text as printed except for the following where the manuscript readings seemed preferable: l. 3 first fruits; l. 6 thrashing floor; l. 7 this head; l. 11 'Thou' for 'Those'; l. 13 'fenced' for 'pierced'.

26 l. 16. *Gethsemane*. See Matt. 26: 36–57.

l. 21. *Joseph's garden*. See Matt. 27: 57–60.

l. 28. *Libanus*. Mount Lebanon, famous for its forests and fruit.

Cf. George Herbert's 'The Bunch of Grapes'.

New Readings. For an early draft see *J*, p. 32; text is from a transcription of a later version made by V. S. S. Coles (MS *A*). R. B. pointed out the similarity with George Herbert's poem, 'The Sacrifice'.

l. 2. See Luke 6: 44.　　　　　　　　ll. 6–10. See Matt. 13: 4–7.

27　ll. 11–12. See Luke 9: 12–17.　　　ll. 14–15. See Matt. 26: 53–4.

'He hath abolished the old drouth'. *C* I, July 1864 (noted *J*, p. 32). It may have been included under the title 'New Readings'. References to the Psalms are to the Prayer Book version.

ll. 1–3. See Psalm 107: 35.　　　　l. 4. Psalm 40: 3.

ll. 6–7. Psalm 118: 17.　　　　　　ll. 8, 14–15. See Matt. 13: 30.

ll. 16–17. Psalm 65: 14.

Heaven–Haven. Originally titled 'Rest' in *C* I, July 1864 (see *J*, p. 33). A similar version of the first two drafted stanzas occurs on the same sheet as 'For a Picture of Saint Dorothea' (MS *C*) but is titled 'Fair Havens—The Nunnery' and replaces the semi-colons of the second lines with commas and l. 7 'green' with 'great'. The version in the Dolben Family Papers (Northants Record Society, Lamport Hall) is the same as the latter except for l. 7 where it reads 'green' not 'great'. It is titled 'Fair Havens; or The Convent'. Text is from autograph in *A*. For the title see George Herbert's poem, 'The Size' (last line), 'These seas are tears, and heav'n the haven'. Cf. Tennyson's 'Morte d'Arthur' ll. 260–4 for his description of Avilion:

> Where falls not hail, or rain or any snow,
> Nor ever wind blows loudly; but it lies
> Deep-meadowed, happy, fair with orchard-lawns
> And bowery hollows crowned with summer sea.
> (*J*, p. 356 n. 140. 7).

28　*'I must hunt down the prize'*. *C* I (noted *J*, p. 33), between 25 July and 1 August 1864, when Hopkins left for a reading holiday in Wales with Edward Bond and A. E. Hardy (see *L* III, pp. 211–14).

'Why should their foolish bands'. *C* I (noted *J*, p. 34), between 1 August and 14 August 1864. A youthful criticism of mourners not comforted by Christian doctrine. Stop (l. 5) and comma (l. 7) editorial.

l. 8. Uncancelled alternative: 'Far from its head an angel's hoverings'.

'Why if it be so'. *C* I, where it follows the previous poem. Noted *J*, p. 34. Hopkins wrote to Baillie (*L* III, pp. 212–13) 'I was lost in storms of rain on the mountains between Bala and Ffestiniog. It really happened what is related in novels and allegories, "the dry beds of the morning were now turned into the channels of swollen torrents", etc. At last a river ran across the road and cut me off entirely. I took refuge in a shepherd's hut and slept amongst the Corinthians. They, I mean the shepherd and family, gorged me with eggs and bacon and oaten cake and curds and whey. Thus I did what old

gentlemen tell you with a sort of selfish satisfaction that you must learn to do,

ROUGHED IT;

I believe it means irritating the skin on sharp-textured blankets. These old gentlemen have always had to do it when they were your age.' Written at Bala.

29 *'Or else their cooings'*. *C* I (printed *J*, p. 34), early August 1864, Maentwrog.

'It was a hard thing'. *C* I (noted *J*, p. 34), shortly before a drawing dated 14 August 1864. Written at Maentwrog.

'Glimmer'd along the square-cut steep'. *C* I (printed *J*, p. 35), probably written at Maentwrog on 14 August 1864 or shortly afterwards.

30 *'Late I fell in the ecstacy'*. *C* I (printed *J*, p. 35), mid August 1864.

l. 2. *the flood*. Noah's flood, see Genesis 6: 5.

'Think of an opening page'. *C* I (printed *J*, p. 35), mid August 1864. Written at Maentwrog.

'Miss Story's character'. *C* I (noted *J*, p. 35), August–September 1864. Hopkins wrote to Baillie (*L* III, p. 213) 'We have four Miss Storys staying in the house, girls from Reading. This is a great advantage—but not to reading.' He seems to have drafted some of the verses 'in the van between Ffestiniog and Bala' and others in the 'train from Chalk Farm to Croydon' (*J*, p. 317), where he stayed with his grandfather (*L* III, p. 222).

ll. 19–20. Earlier variant: '. . . but less than female tact, / Sees the right thing to do and does not act'.

31 *'Her prime of life'*. *C* I (printed *J*, p. 35), August 1864. H. H. suggests that it is a parody by G. M. H.

'Did Helen steal'. *C* I (printed *J*, p. 36), August 1864.

32 *(Woods in Spring)*. *C* I (printed *J*, p. 36), August 1864.

'Like shuttles fleet the clouds'. *C* I (printed *J*, p. 36), August 1864.

(Seven Epigrams). *C* I (noted *J*, p. 37), August 1864. (i) written 'in the van between Ffestiniog and Bala'.

33 *By Mrs. Hopley*. *C* I (printed *J*, p. 37), August 1864. A first version, including ll. 1–2 of the text, was cancelled. The poem seems to have been written in reaction to articles in *The Times* (14, 15, 16, and 18 July 1864) about a suit for legal separation. A Mr Thomas Hopley had been jailed for beating a pupil to death. On his release his wife sued for separation on the grounds of cruelty to herself and their children. Mr Hopley produced letters from his wife, written to him while he was in prison, suggesting that the family were all very fond of him. The jury had difficulty in reaching a decision and eventually refused the suit because, although Mrs Hopley's charges were believed, she appeared to have condoned her husband's behaviour, negating grounds for separation. Mrs Hopley was identified by Mrs E. E. Duncan-Jones (*TLS*, 10 October 1968, p. 1159).

l. 1. *theory*. *The Times* mentions that Mr Hopley held 'absurd theories' with regard to the management of women in pregnancy and young children, amounting to maltreatment. He also had strong theories about education and intended to establish a model school (15 and 18 July).

34 *Io.* C I and C II (see *J*, pp. 38 and 48), September 1864 and later. Hopkins wrote to Baillie on 10 September 1864 that besides beginning to turn 'Floris in Italy' into a play he had 'done very little' since last writing (14 August) 'except three verses, a fragment, being a description of Io (transformed into a heifer.) It sounds odd' (*L* III, p. 221). The story of Io occurs in Aeschylus' *Prometheus Bound* (see pp. 308 and 268 above). According to Aeschylus, Zeus fell in love with Io and changed her into a white heifer to hide her from his wife, Hera. The story is also told by Ovid in *Heroides* (xiv. 85–108) and *Metamorphoses* (i. 588 ff.). Ovid says that Hera set Argus, a huntsman with many eyes, to guard Io and prevent Zeus from seducing her. To the same end, and from anger, Hera sent a gadfly to torment Io and this drove her, maddened, through many lands.

l. 6. Variant:
She rests half-meshing from the too-bright sky.

ll. 11–12. Variants:
> Her hue's a honied brown and creamy lakes,
> Like a cupp'd chestnut damaskèd ⎰ with ⎱ breaks.
> As – – – 's – ⎱ in ⎰ –

. (See *J*, p. 48.)

l. 14. Variant: 'The knot of feathery locks'.

ll. 17–18. Variants:
> Day brings not back his basilisking stare,
> Nor night beholds a single flame-ring flare

and

> Night is not blown with flame-rings everywhere,
> Nor day new-basilisks his tireless stare

ll. 16–18 refer to Argus who kept two of his eyes, his 'vigil-organ', open at all times until Hermes, sent by Zeus, lulled him to sleep with stories and songs and cut off his head. Argus' many eyes have sometimes been seen as a symbol of the stars. This may explain 'flame-rings', imagery that Hopkins often applied to stars.

(*Fragments*). C I, August–September 1864
(i) printed *J*, p. 38. *heifer's*. Apostrophe editorial.
(ii) printed *J*, p. 39. *wester'd*: lying to the west.
(iii) printed *J*, p. 39.
35 (iv) printed *J*, p. 39. Possibly part of 'Io'.
(v) printed *J*, p. 39.

The rainbow. C I (printed *J*, p. 39), early September 1864.

l. 2. 'his'. MS reads 'its', a remnant of an earlier version when 'He' read 'It'.

'*—Yes for a time*'. C I (printed *J*, pp. 39–40), 7 or 8 September 1864.

36 *Fragments of 'Floris in Italy'.* C I and C II (see *J*, pp. 36, 40–3, 45, 47, 65).
Probably begun as a poem in June or July 1864 (see *L* III, p. 213), written
perhaps on some of the twenty-two pages torn out of *C* I. In August or
September 1864 Hopkins decided to rewrite it as a play (see *L* III, p. 221).
The extant, dramatic fragments are disordered but seem to come from three
different scenes:

  (i) a night scene in which Giulia steals away from the sleeping Floris
        (printed *J*, pp. 40–2);
 (ii) a soliloquy about Guinevere's adultery (printed *J*, pp. 45–6), which may
        not have been intended for this play;
(iii) a speech in which Floris, having discovered that Giulia loves him,
        argues against reciprocating her love (variants printed *J*, pp. 65,
        70).

There are as well two prose scenes (see pp. 186–9):

  (i) A Fool dictates a letter to Giulia (printed *J*, pp. 42–3, written 9
        September);
 (ii) Set outside the cave of a hermit. Gabriel, half-mad because of
        Guinevere's adultery, meets a shepherd and his wife sheltering
        from a storm (printed *J*, pp. 44–5, written between 10 and 14
        September 1864). This may not have been intended for *Floris in
        Italy*.

Both prose scenes are much influenced by Shakespeare.

Uncancelled variants:

	(i) l. 8.	Methinks my laughter is more perilous
37	l. 45.	'Tis so conceivèd in his lineament.
	l. 61.	Sexing and ranking with our ruder files
		Enroll'd and sexing — – – —
38	l. 65.	Most like the tuft of plighted silver round
		— – – plighted tuft of — —
	(ii) l. 10.	In blazon, gilt and images of bronze,
		– gilt and blazon and bronze statuary,
	l. 13.	Such heathenish misadventure dogs one sin.

(iii) Commas in scene description editorial.

	ll. 1–2.	Beauty it may be is the meet of lines,
		Or careful-spacèd sequences of sound,
		(stop in text editorial).
	l. 8.	To spread the compass on the all-starr'd sky:
		(colon in text from this variant).
	ll. 9–12.	Or try with eyesight to divide
		One star out from the daylight air.
		And find it will not be described
		Because its place is charted there.

   or

But only try with eyesight to divide
One star by daylight from the strong blue air.

ll. 13–20.   My love in lists of loves I would not find,
Much less all love in one conscribed spot.
Though true love is by narrowest bands confined,
New love is free love, or true love 'tis not. (*J*, p. 70)

39 ll. 17–18.   Is to give regimen to the imperfect wind,
And slender element to piece and plot.
The —— elements – — – –

These lines (17–18) are later variants, but Hopkins did not make the changes necessary to incorporate them.

ll. 21–5.   Thus he ties spider's web across his sight,
And gives for tropes his judgment all away,
Gilds with some sparky fancies his black night
And stumbling swears he walks by light of day.
Blindness! A learned fool and well-bred churl

(*Fragments*): *C* II, September 1864.

(*a*) (*Stars*) (i) noted *J*, p. 44. (ii) printed *J*, p. 44. (iii) printed *J*, p. 44. (iv) printed *J*, p. 46. (v) printed *J*, p. 47.

40 (*b*) (i) printed *J*, p. 43. 'live' and 'vive' are suggested alternatives to 'red' (l. 4). (ii) printed *J*, p. 46.

'*I am like a slip of comet*'. *C* II (noted *J*, p. 46), just before a drawing dated 14 September 1864. Perhaps intended to be part of 'Floris in Italy'. A number of important comets had been sighted in the late 1850s and 1860s. Hopkins incorporates details mentioned in contemporary reports and astronomical books.

l. 11. *Gideon's fleece*. See Judges 6: 37–8.

41 '*No, they are come*'. *C* II (noted *J*, p. 46) follows a drawing dated 14 September 1864.

l. 9. Variant: 'armed' instead of 'unsteady'.

l. 10. Variants: Heave their unsteady columns;
or
Heave through their flushing columns.

'*Now I am minded*'. *C* II immediately following previous poem.

l. 3. *hursts*: woods.   l. 16. MS indented.

l. 17. *hunters' moon*. The next full moon after harvest moon; approximately the third week in October.

42 *A Voice from the World*. Fragments scattered through *C* I and *C* II, June 1864–January 1865. See *J*, p. 52 for G. M. H.'s suggested arrangement of these. The scheme was cancelled and most of the fragments alluded to are missing. 'But what indeed' (see p. 67) may have been intended as an addition to the poem. *L* III, p. 36 (January 1867) suggests that G. M. H. may later

have completed the poem, retitled it 'Beyond the Cloister', and submitted it to *Macmillan's*.

See Appendix A for Christina Rossetti's poem, 'The Convent Threshold'. G. M. H. rendered parts of this in Latin (see p. 89).

l. 13. *bluebell sheaves*. See *J*, p. 22 'sheaves of bluebells', and *J*, p. 209 (prose pp. 207–8) for a later, detailed description.

l. 15. *to thrill*. MS uses abbreviation for 'the' in error.

l. 24. *spent*. Comma editorial.

44 l. 85. Two pages of the MS at this place are badly smudged: words in brackets are conjectural (H. H.).

ll. 94–5. See Matt. 25: 31–3.

l. 102. This line is at the bottom of the page, and quite illegible (H. H.).

45 l. 116. Apparently not completed in MS, though space is left for the missing words (H. H.). The sense is revealed by cancelled variants (N. H. M.).

l. 126. *Esau's cry*. See Genesis 27: 34.   ll. 139–41. See 2 Kings 5: 18.

46 l. 166. *foods*. Comma editorial.

'The cold whip-adder'. *C* II (noted *J*, p. 48), October 1864.

47 *(Fragments)*: *C* II, October–November 1864.

(i), (iii) printed *J*, p. 48.   (vi) printed *J*, p. 50.

48 *For a Picture of Saint Dorothea*. *C* II, November? 1864. Two almost identical autographs in *A* and *C*. Text from *A*. This was the first poem G. M. H. showed to R. B. (in 1866?). Hopkins later wrote an expanded dramatic version with three characters (see p. 90). See 'Lines for a Picture of St. Dorothea', p. 84 and p. 90.

Because Dorothea was Christian, she was tortured by the Governor of Caesarea, Sapricius. She declared that God is everywhere and that he invites us to his Paradise 'where the woods are ever adorned with fruit, and lilies ever bloom white, and roses ever flower; where the fields are green, the mountains wave with fresh grass, and the springs bubble up eternally' (S. Baring-Gould, *Lives of the Saints*, February, p. 176). A lawyer called Theophilus who was present derisively suggested that Dorothea send him some of the heavenly apples and roses. Later, just before she was executed, Dorothea prayed that his request might be fulfilled and an angel, disguised as a beautiful youth, appeared carrying three apples and three red roses. When Theophilus received the gifts he too became a Christian and was likewise executed. (Dorothea's feast-day is 6 February.) MS title in block capitals.

'Proved Etherege'. *C* II (printed *J*, p. 50), November? 1864.

49 *Fragments of 'Richard'*. *C* II (noted *J*, p. 51), (i) and (ii) October–November 1864; (iii) and (iv) 16–24 July 1865. Hopkins wrote to Baillie that he had begun a story to be called 'Richard' (*L* III, pp. 213–14, 20 July–14 August 1864). For the first drafts (*C* I, May–June 1864) see p. 21. Hopkins applied

the language and conventions of pastoral to university students and devoted the bulk of the lines extant to his detailed observations of nature.

(ii) l. 6. MS 'of the round'.   (iii) l. 2. *much-dreaded*. Hyphen editorial.

50 l. 9. *Cumnor Hill*. West of Oxford.

51 *'All as that moth'*. *C* II (printed *J*, p. 51), November–December 1864.

l. 3. *underplighted*: underwoven.

The Queen's Crowning. *C* II (noted *J*, p. 51), December 1864.

53 l. 52. MS reads 'θ' = G. M. H.'s abbreviation for 'the' in mistake for 'to'.

l. 53. Comma editorial.   l. 57 *happ'd*: from 'hap', to cover.

56 *'Tomorrow meet you?'* *C* II following previous poem (noted *J*, p. 51).

Fragment of 'Stephen and Barberie'. *C* II (printed *J*, p. 52), December? 1864 or January 1865. Cf. Tennyson's poem, 'Mariana'.

l. 5. *descend*. Conjectural: the MS is badly smudged here, but *de . . .* is visible (4th edn.).

l. 7. *size*: OED 'To assume size; to increase in size'.

57 ll. 13–14. *Willow*. A traditional, Elizabethan song of unrequited love. See *Othello*, IV. iii.

'Boughs being pruned'. *C* II (noted *J*, p. 52), December 1864–January 1865.

l. 2. *squinch*. 'Masonry built across the top corners of a tower to support the oblique sides of a spire' (Francis Bond, *An Introduction to English Church Architecture* (London, 1913), vol. I, p. xxvii).

l. 3. *chamfer'd*. 'An edge of wood or stone sliced off in a sloping direction', ibid., p. xxi.

l. 8. *lessen'd mill*. With edges worn.

'A silver scarce-call-silver'. *C* II (printed *J*, p. 52), December 1864–January 1865.

'I hear a noise of waters'. *C* II (printed *J*. p. 54), February 1865. A number of descriptions of bluebells and primroses at this point in the diary show Hopkins delighting in the early spring. The following, four entries after the fragment, may also belong to the poem:

> and then as thick as fast
> The crystal-ended hyacinths blow.
>
> The dented primrose and bead-budded may.

58 *(Dawn)*. *C* II, January 1865. (i) and (ii) printed *J*, p. 55; (iii) printed *J*, pp. 55–6.

l. 6. *'sperses*: disperses.

'When eyes that cast'. *C* II (printed *J*, p. 56), February 1865.

59 *The Summer Malison*. *C* II (noted *J*, p. 56), February or March 1865.

malison: curse.

ll. 2 and 6 recall Pharaoh's dream of 'kine' and 'seven ears of corn', Genesis 41.

l. 6. *lodgèd*: laid flat by the weather.

St. Thecla. G. M. H.'s autograph of this and 'In Theclam Virginem' (p. 123) (the Latin version facing the English text) was found in 1952 by Fr. R. Burke Savage, SJ, among the papers of the late Fr. Connolly, SJ, editor of *Studies*. The MS is now in the archives of the Jesuit community at 35 Lower Leeson St., Dublin (4th edn.). The poem was probably originally written in 1864–5 (N. H. M.). The extant fair copy would seem, from the handwriting, to belong to the mid 1870s.

St Thecla is considered to be the first female martyr. Mrs Jameson notes that 'such was the veneration paid to this saint in the East, and in the early ages of Christianity, that it was considered the greatest praise that could be given a woman to compare her to St. Thecla' (*Sacred and Legendary Art*, 2nd edn. (1850), p. 329).

Thecla, who lived at Iconium, was engaged to be married, but overhearing Paul preaching, was converted and abandoned her former plans to follow him. She was subsequently sentenced as a Christian to be burned at the stake but was rescued. Later, when she was recaptured, she was sentenced to be thrown to wild beasts but was again miraculously saved. She was known as the Virgin Thecla and was famous for her purity and her ability to heal the sick.

l. 8. *Pegasus*. The winged horse sent by the gods to help Bellerophon fight the Chimaera. After killing the monster, Bellerophon is said to have tried to fly to heaven to join the gods and for this presumptuous act Jove sent a gadfly that tormented Pegasus into unseating Bellerophon, who was lamed by the fall.

l. 9. *were none*: were mythological, not real.

l. 10. Paul was from Tarsus (Acts 22: 3).

60 *Easter Communion*. C II (printed *J*, p. 57), Lent 1865. First draft written 2–12 March: text from copy of 26 June.

Cancelled variants: l. 3. Come, scored in secret . . .

ll. 5–6. You whom the pursuant cold so wastes and nips,

l. 12.   Give fragrant-threaded
        change of fragrant-threaded gold raiment.
    Give gladden* fold of fragrant-threaded ease
        fragrant-threaded folds of ease,
   *[?] written over 'golden'? (H. H.)

61 *'From any hedgerow'*. C II (printed *J*, p. 57), March 1865.

'O Death, Death'. C II (printed *J*, p. 58), March 1865). See the Apostles' Creed: 'He descended into Hell. The third day he rose again . . . He ascended into Heaven . . .'

ll. 8–10. See Psalm 24: 9.

'*A basket broad*'. *C* II (printed *J*, p. 58), March 1865. It may have been intended for one of the poems on St Dorothea.

(*Fragments*): *C* II, March–April 1865.

(i) printed *J*, p. 58; (ii) printed *J*, p. 59.

62 '*Love me as I love thee*'. *C* II (noted *J*, p. 58), March 1865. The original is:

Εἴ με φιλοῦντα φιλεῖς, δισσὴ χάρις· εἰ δέ με μισεῖς,
τόσσον μὴ μισῆς, ὅσσον ἐγώ σε φιλῶ.

(Anonymous)

To Oxford. C II (noted *J*, p. 63), dated 'Low Sunday and Monday, 1865' (23 and 24 April). Hopkins sent transcriptions, now lost, to Addis. Text is from rough drafts in *C* II. G. M. H. wrote above them, they 'are not quite right'.

l. 1. *terms*: academic terms.   l. 6. *pleasaunce*. Archaic. See Dict. 'pleasance'.

l. 10. *towers musical*. College bell towers.

l. 11. *window-circles*: rose windows.

(*continued*) l. 5. *visual compulsion*: optical illusion.

63 l. 12. 'to' editorial (N. H. M.).

'*Where art thou friend*'. *C* II (noted *J*, p. 60), 25–7 April 1865. The friend addressed may well be the reader (see Rudy Bremer, 'Where art thou friend . . .', *Hopkins Quarterly*, vii, no. 1 (spring 1980), pp. 9–14).

Variants: l. 3. Either unknown to me in the age that is

l. 14. No, no, no, but for Christ who knew and loved thee

'*Bellisle*'. *C* II (printed *J*, p. 60), shortly after 24 April 1865. See 'To Oxford', pp. 62, 68.

l. 2. *echoing the sound*. Balliol?

'*Confirmed beauty*'. *C* II (printed *J*, pp. 60–1), April 1865; follows 'Bellisle'.

l. 5. *Tantalean*. In Greek mythology Tantalus was punished by the gods either for stealing their sacred nectar or killing his son as a sacrifice. He was placed in a pool up to his chin but the water receded whenever he tried to quench his thirst and ripe fruit dangling from a bough above his head evaded his reach.

64 l. 6. *Pharaoh's ears*. See Genesis 41: 6.

The Beginning of the End. MS *A* is subtitled 'a neglected lover's address to his mistress'. Draft in *C* II (noted *J*, p. 62), 6–8 May 1865. The first sonnet is followed by:

> Some men may hate their rivals and desire
> Secretive moats, knives, smothering-cloths, drugs, flame;
> But I am so consumèd with my shame
> I dare feel envy scarcely, never ire.
> O worshipful the man that she sets higher

Text of the first and third sonnets is from R. B.'s transcription in *A* of later versions. R. B. noted that they were 'in Italian form and Shakespearian mood (refused by *Cornhill Magazine*) . . .'. Layout, stop (i, l. 1), l.c. 'winter' (l. 5), italics (ll. 12–13), l. 14 'unpassion'd' from *C* II.

65   *The Alchemist in the City*. *C* II (noted *J*, p. 62), 15 May 1865.

67   *'But what indeed'*. *C* II (printed *J*, p. 62), between 15 May and 24 June 1865. Perhaps intended as part of 'The Voice from the World' (see p. 41).

'Myself unholy'. *C* II (noted *J*, p. 63), June 1865. MS has semicolon after 'brother' (l. 9) and comma after 'other' (l. 13).

Cancelled variants:

l. 4.   White clouds to furnace-eaten regions coaly:

l. 7.   And so my trust confusedly is shook.
                   confidence is struck and shook.

ll. 9–14.   He has a fault of mine, he its near brother,
                   And part I like and part I hate the fall;
                         is sweet to me and part is gall;
                   In him this fault I found, in him another:
                   And though they each have one and I have all,
                   This time it serves not. I can seek no other
                   Than Christ: to Christ I look, on Christ I call.

68   *To Oxford*. *C* II (printed *J*, pp. 63–4), June 1865. Hopkins noted above the poem that he had given a copy to V. S. S. Coles, and below it 'the last two lines I have forgotten and must get'. See 'Bellisle', p. 63.

l. 6. *Belleisle*. Balliol (Mrs E. E. Duncan-Jones).

l. 9. *'As when . . . blest'*. Tennyson, 'A Dream of Fair Women', l. 281.

'See how Spring opens'. *C* II (noted *J*, p. 63), dated 26 June 1865. The poem may well be about G. M. H.'s recent conviction of the truth of Catholicism (see *L* III, pp. 226–7). His lament about his slow spiritual maturing is similar to Milton's thought in the sonnet, 'How soon hath Time'.

l. 7. See Matt. 13.

l. 14. *learnt how late, the truth*. After he had become a Catholic in 1866 G. M. H. wrote: 'The silent conviction that I was to become a Catholic has been present to me for a year perhaps, as strongly, in spite of my resistance to it . . . as if I had already determined it' (*L* III, p. 27). His note on 12 March 1865, 'A day of the great mercy of God', may relate to his new conviction.

69   *Continuation of R. Garnet's [sic] 'Nix'*. *C* II (printed *J*, p. 64), June–July 1865. Hopkins read Richard Garnett's 'Nix' in *The Children's Garland from the Best Poets*, edited by Coventry Patmore (1862). G. M. H. inaccurately quoted this version of the poem, which differed slightly from the original version, in his Platonic dialogue, 'On the Origin of Beauty' (printed *J*, p. 111). See Appendix *B*, p. 409.

70   *'A noise of falls'*. *C* II (printed *J*, p. 66), August 1865. The poem follows sketches and descriptions of the sky observed while on holiday in Devon and at Hampstead.

'*O what a silence*'. *C* II (printed *J*, p. 66), August 1865, follows 'A noise of falls'. ll. 12–17 occur four entries later in the diary.

l. 3. *skylark*. See 'The Sea and the Skylark', pp. 131 and 350. *traverse*: zigzagging upwards.

71 '*Mothers are doubtless*'. *C* II (printed *J*, p. 67), August 1865.

l. 4. Very faint. 'come to' amended to 'hold + ? (cancellation?)'.

Daphne. *C* II (printed *J*, p. 68), dated 1 September 1865. This poem and the dialogue of the next few pages are all fragments of Hopkins's unfinished play, 'Castara Victrix'.

l. 15 *knopt*. Studded with knobs, here fruit; also used of flower-buds.

72 *Fragments of Castara Victrix*. *C* II, begun August 1865 (see *J*, p. 65). These speeches were written in Sept. 1865 (printed *J*, pp. 68–70). G. M. H. listed the characters: '*Castara Victrix* or *Castara Felix*. Silvan, the king, and his two sons Arcas and Valerian. Carindel. The fool. Carabella. Pirellia, Piers Sweetgale. Daphnis. Daphne.

The melancholy Daphne doats on him.' (*J*, p. 65.)

74 '*My prayers must meet a brazen heaven*'. *C* II (noted *J*, p. 70), 7 September 1865. N. H. MacKenzie suggests Deut. 28: 23 and Lev. 26: 19, the effects of sin will be no rain—heavens hot brass without rain clouds, and the earth parched hard. Cf. Manley Hopkins, '. . . Though the earth be iron, the heaven seem brass' ('Prayer'). G. M. H. used similar terms a number of times; 'sky anointed with warm brassy glow' (*J*, p. 212) and 'The sun just risen/Flares his wet brilliance in the dintless heaven' (*J*, p. 58).

Shakspere. *C* II (noted *J*, p. 70), 13 September 1865. There had been numerous celebrations of the tercentenary of Shakespeare's birth in 1864.

'*Trees by their yield*'. *C* II (noted *J*, p. 71), 28 September 1865. Above the poem Hopkins wrote 'A verse or more has to be prefixed'.

ll. 1–2. See Luke 6: 44.

75 '*Let me be to Thee*'. *C* II (noted *J*, p. 71), 22 October 1865. This again appears to refer to G. M. H.'s conviction of the truth of Catholicism.

76 *The Half-way House*. *C* II (noted *J*, p. 71), October 1865, follows previous poem. The title may come from Newman's statement in the *Apologia* that Anglicanism is the half-way house between atheism and Roman Catholicism (see A. Sulloway, p. 18). It is probably significant that the sentence occurs in Newman's discussion of the doubts about Anglicanism, very similar to those Hopkins felt, that led to his leaving that church and becoming a Roman Catholic (*Apologia* (Oxford, 1967), p. 185). l. 4 'Love,': comma editorial.

l. 7. *Egyptian reed*. See Isaiah 36: 6. The image suggests G. M. H.'s resolve to leave the Anglican Church (the national religion).

l. 8. *vine*. See John 15: 1–10.

ll. 17–18. Hopkins wrote to E. H. Coleridge (1 June 1864), 'The great aid to belief and object of belief is the doctrine of the Real Presence in the Blessed Sacrament of the Altar. Religion without that is sombre, dangerous, illogical, with that it is—not to speak of its grand consistency and certainty—*loveable*. Hold that and you will gain all Catholic truth!' (*L* III, p. 17). 'Catholic' is used broadly here and includes Tractarians. See *L* III, p. 92 (prose, pp. 224–5).

A Complaint. The family's copy is now in *C* (Bodl.). Text from *A*, a revised version written on paper embossed 'Oxford Union Society' and with a watermark of 1865. It is this last copy to which Hopkins referred in a letter to R. B. (*L* I, p. 87, 14 Aug. 1879), 'I had quite forgotten the sonnet you have found, but can now recall almost all of it; not so the other piece, birthday lines to me sister, I fancy.' G. M. H. had forgotten Milicent's birthday (17 October) in 1863 (*L* III, p. 84) and it may have been on this occasion that the poem was written, although, considering the watermark, late 1865 is also a strong possibility.

77 *'Moonless darkness'*. *C* II (noted *J*, p. 71), 25 December 1865.

l. 9. MS 'Xmas'.

'The earth and heaven'. *C* II (noted *J*, p. 71), 5 January 1866.

l. 9. 'favourite . . . gale' cancelled and not replaced.

78 *'As it fell upon a day'*. *C* II (printed *J*, p. 71). Hopkins wrote just above the poem: 'Katie, age 9 (8 January 1866)'. He may have intended the poem for her.

'In the staring darkness'. *C* II (printed *J*, p. 72), follows previous poem. Beneath the verse G. M. H. wrote: 'Grace (8). (same day.)'. Perhaps he wrote the poem for her.

79 *'The stars were packed so close'*. *C* II (printed *J*, p. 72), January 1866. See 'The Elopement', ll. 19–22 (p. 94).

The Nightingale. Autograph in *C*, dated 'Jan. 18, 19, 1866'.

80 *The Habit of Perfection*. Two autographs in *A*: (i) dated 'Jan. 18, 19, 1866', subtitled '(The Novice)'; (ii) text; later, perhaps contemporary with the text of 'Heaven-Haven'. There is also a transcription in CH by Fr. Bacon SJ, entitled 'The Kind Betrothal'. It varies considerably from the autographs.

81 *Nondum*. Autograph in CH, dated Lent 1866. Transcription in *A*. Title means 'not yet'.

ll. 1–6. See Tennyson, 'In Memoriam', LVI.

82 ll. 7–12. Cf. Tennyson, 'In Memoriam', CXXIV.

ll. 19–24. See Genesis 1: 1–13.   l. 27. *deeps calls to deep*. See Psalm 42: 7.

83 *Easter*. MS in CH, undated. 1866? Commas at end of ll. 20 and 21 are from transcript in *A*.

l. 1. *nard*. Spikenard, an expensive oil sold in alabaster boxes. See Mark 14: 3–8 and John 12: 3–8.

l. 21. MS reads 'disshevelled'.

84 *Lines for a Picture of St. Dorothea*. n.d. Text from a manuscript at Stonyhurst which originally belonged to Fr. Bacon, SJ. There is another manuscript at Campion Hall. The poem is a later version of 'For a Picture of Saint Dorothea' and is the earliest example of G. M. H.'s use of sprung rhythm.

Title. Perhaps G. M. H. had in mind Mrs Jameson's remark in *Sacred and Legendary Art* that 'the principal incident of her [St Dorothea's] legend is so picturesque and poetical that one is surprised not to meet with it oftener; in fact, I have never met with it; yet the interview between Dorothea and Theophilus, and afterwards between Theophilus and the angel, are beautiful subjects: the first scene has a tragic interest, and the latter an allegorical significance ... ' (London, 2nd edn., (1850), p. 338). See N. H. M. 'Hopkins and St. Dorothea', *Vital Candle*, ed. John North and Michael Moore (Waterloo, 1984), pp. 21–39.

ll. 7–12. See Philip Massinger, *The Virgin-Martyr*, v. i, to which Mrs Jameson alludes.

ll. 10–11. *set*: develop fruit (OED).

ll. 11–12. *spring*. See *J*, p. 134, 'Hedges springing richly', growing rapidly, as in spring.

l. 20. *sizing*. It appears to increase in size.

l. 21. *mallow-row*. See *J*, p. 66, 'Mallowy red of sunset and sunrise clouds'. G. M. H. puns upon the fact that mallow is a plant as well as a colour.

l. 23. *sphered*. Placed 'above common reach' (OED); but overtones of heaven and astronomy remain.

85 ll. 35–6. Cf. 'The Wreck of the Deutschland', stanza 18 for G. M. H.'s similar reaction to the fate of the tall nun.

l. 39. *wand*. the palm, symbol of martyrdom. Cf. 'The Wreck of the Deutschland', stanza 22 for another statement that the martyr's role is to lead men to God.

Summa. Text from autograph in *A*, 'written when G. M. H. was an undergraduate' (R. B.). Hopkins burnt the original when he became a Jesuit (See *L* I, p. 24).

Summa: 'the summary of what is known of a subject' (OED).

ll. 6–8. See *L* III, pp. 19–20.   ll. 9–16. See *L* I, pp. 27–8.

86 *Jesu Dulcis Memoria*. MS (CH), written in pencil and ink on two sheets, undated, no title, a number of variants. On one page G. M. H. drafted in pencil a translation of the Latin hymn as it appears initially in the Roman breviary. The verses are part of a much longer hymn originally attributed to St Bernard and now thought to have been written by an English Cistercian of the twelfth century (see *The Oxford Book of Medieval Latin Verse*, edited by F. J. E. Raby, pp. 347–53, 493–4). Using other verses from this source there are in the breviary two more hymns, 'Jesu, Rex admirabilis' and 'Jesu, decus angelicum'. G. M. H. has on the verso and other sheet translated five further stanzas, mostly from the third hymn but in no special order. Because

Hopkins's intended sequence is not clear and the later verses cannot be appended to the first hymn, the stanzas are now printed in the order accepted for the original Latin hymn, with the addition of the final verse, which this version omits. G. M. H. has not translated the first line of stanza 8. For the history of the hymn see *Dictionary of Hymnology*, edited by John Julian (London, 1892). Stanzas 1, 2, 3, 5, 10 form 'Jesu dulcis memoria' in the breviary. Stanza 4 is found in 'Jesu, Rex admirabilis'. Among Hopkins's variants are:

ll. 5–7.   Song never was so sweet in ear,
                  Word never was such news to hear,
                  Thought half so sweet there is not one

                  There's no such touching music heard,
                  There's never spoke so glad a word,
                  So sweet a thought there is not one.

ll. 13–16   Jesu, like dainties to the heart
                  Daylight or running brooks Thou art,
                  And matched with Thee there's nothing glad
                  That can be wished or can be had.

87 *Inundatio Oxoniana*. Text from autograph fair copy in CH, undated. On 18 January 1866 *The Times* reported: 'In consequence of the excessive rainfall of the past three months, and of the rapid thaw on Friday last, the waters round Oxford are higher than they have been for some years. Christ Church and Magdalen Meadows are an unbroken sheet of water, and the new pleasure walk made by the University from Marston-lane to the Museum is also partially submerged, though it was thought to have been raised sufficiently high. Port Meadow and all the low lands in the valley of the Thames and Cherwell are flooded' (p. 5). The details of the poem suggest that it was probably written in 1866, rather than in 1865 when the flood was the result of a rapid thaw (see *The Times*, 7 Feb. 1865, p. 7). l. 20, comma editorial (W. H. G.). l. 34, 'limos . . . sequaces', pencil change from singular.

Translation: The spring pastures have for long suffered the glowering clouds with their incessant heavy rain, and the abundance of the water-filled sky pelts down; having destroyed the bounds of the despised river bank and driven by the winds (it) becomes a sea across the broad fields. Here and there, in places amid the glistening waters grass rises more cheerfully, a portion of which, raised by gentle ridges, flourishes; but already empty wetness has hidden the rest. The Isis can scarcely lead its undistinguishable waters in their proper channel; the unshackled winds blow on the watery tracts, and it is delightful to scud (across) new seas with a following South wind; men sail the impassable groves and enter the shady willows; among the tops of shattered woods a drowned poplar gives passage to unaccustomed keels.

But if at last for five days the sun will have ridden in circuit through a clear sky, if so often the fiery aether will have smiled, the receding waters depart. Often then a sea smell is borne across the plain, lightly it will fill the moist breezes; it steals into the middle of the city and encircles the inmost parts of homes, so that you will be quite unable to divert the disagreeable sea-weed. From this cause what a number of illnesses there are, how often men regret having approached; you travel to our homes (that are) exposed to the neighbourhood fever. But it is possible to have moved away and straightway escape the feverish heats; we do not live continually in these shadows. The majority listed as abiding in the area will stay unharmed more securely near (their) forebears' homes, the indigenous people accustomed to the peril: undoubtedly the infections have been driven off from them by a kind power, for although the West wind may renew its warm showers on our fields through the night, it vexes the flattened acres with impunity and a harmless smooth sea is drawn over the fields. No indeed, scarcely could as many lilies, their offspring, grow red in dry meadows as you will see tossing their heads near familiar streams, had the destined waters not played a little in the abundant flood, if the choppy waves had not been able to spread themselves first and leave behind the attendant muds. Thus the tree extends its pleasant twigs more vigorously over the riverside, on this account the willows burgeon; and so the horseman seeks the untrodden plains and rides with greater pleasure into the tall grass.

88 *Ecquis binas*. Autograph in CH, probably contemporary with 'Jesu dulcis memoria'. This medieval Latin hymn was included in *The Rhythm of Bernard de Morlaix*, which Hopkins possessed in a translation by the Revd J. M. Neale. The Latin, as far as G. M. H. translated, runs:

> Ecquis binas columbinas
> Alas dabit animae?
> Ut in almam crucis palmam
> Evolvet citissime,

(*Sacred Latin Poetry*, edited by R. C. Trench (London, 1872))

Above the poem occur the lines:

> How well Thou comfortest!
> And art the soul's sweet guest,
> Guest and refreshment sweet

Elegiacs: 'Tristi tu, memini'. Autograph in CH. Undated drafts with many variants. Written over pencil draft of the first few lines of a letter to R. B. about Dolben's death (30 August 1867, *L* I, p. 16). Shares page with early drafts of 'After *The Convent Threshold*' (Latin).

Translation: I remember, maiden, when your lot was sad, nor at that time had (your) first love stood firm. Now I am going away: again you will be abandoned alone; and so this is a sadder time; yet that (earlier time) was sad enough. My presence is welcomed as the tender of unexpected fire; I am the bright star between empty wintry periods.

A variant runs:

> Tristis eras dum me venturum, Cythna, putares.
>   Et veni et redeo: jam quoque tristis eris.
> Adsum gratus ego necopini apparitor ignis,
>   Inter ego gelidas signa serena nives.

Translation: You were sad, Cythna, as long as you thought I was only likely to come. I have come and I am going back: now too you will be sad. I am welcome here as one who tends an unexpected fire, I am the bright constellations/signs of fair weather between frozen snows.

'Alget honos'. MS in CH, written below 'Tristi tu, memini'.

Translation: The honour of garlands grows cold suspended in trees; her beauty, which was forsaken in a lonely marriage-bed, grows cold!

'Quo rubeant'. MS in CH on same sheet as 'tristi tu, memini'. Question mark editorial.

Translation: Why need the lovely roses or the fruitful summer show their red colour? That is the red you will have in the cheeks of a maiden. An earlier version reads:

> Namque rosisve color qui pomiferove auctumno
> Insit inest teneris virginis ille genis

89 Elegiacs: after *The Convent Threshold*. Autograph in CH, undated, no title. See Appendix *A* for Christina Rossetti's poem, sections 1 and 9 of which this translates. See too 'A Voice from the World'.

ll. 19–20. See Rev. 15: 2.

Translation (W. H. G.): The stream of a brother's blood flows between us, Aulus, and the fresh blood of our slain father is fixed between us. O you who have ever been the best and dearest man in the world to me, that blood now keeps me far sundered from you. There is a road which leads aloft through the stars, and a golden staircase which replaces day by day and night by night: setting foot on this road I shall mount up to the farthest heights of heaven and up to the glassy halls and to the glassy sea. The feet which you say surpass the white lilies in beauty are deeply crimson with a blot of sin. My feet have a crimson stain, and with its sombre gouts I stand for all to see (*exsto*) as a token of my own guilt—the joys that were mine, and the tears after the joys, and the love that fell to the ground and was not raised up again. And yet they (my feet) have not so much blood on them, nor is it so ingrained in them but that doubtless it can be washed away if water be brought: yet—if only I could open the unseen depths of my heart!—in the unseen depths of my heart this guilt lies hidden. But the sea that glows red with mingled flame and glass—that (sea the prophet saw) was molten glass and clear fire—ah, I pray that it may supply healing balms for my maimed feet, antidotes for my marks of shame and for the snare set for me. And since heaven has been shown to us with the pathway built up to it, O set foot with me on the road that leads up to the stars.

. . . .

I will tell you of the dreams I dreamed last night (and) in a twilight when it was doubtful whether it was night or day. At that time my plenteous hair was wet with cold dew: the dew had gathered distilled from the cold ground. You came there and you asked whether I had been touched by a vision of you, whether that sleep of mine, too, was still mindful of you. The heart that once used to leap at that likeness of you lay now a mass of inert dust at your importunate words. Yet I understood your questions, nor was I unable to reply, and I said these few words in a heavy stupor: 'Sad is our bridal bed, its valance is filled with woe; it rests on cold unyielding stones. Seek for yourself a pleasant bed, a new marriage; place your soft limbs, I give you leave, on an attractive couch. You have a second spouse who will cherish you better than I, you have a love that is sweeter than mine.' On hearing this you beat your hands together in wild excitement and your limbs seemed all of a sudden to give way and shake. These were the last things I saw; at the same time I rolled headlong into the inmost parts and empty coffers of the solid earth. But I did not think it was in applause, as at a holiday dance, that you beat your hands together, or that your limbs were trembling from too much wine.

90 *St. Dorothea* (*lines for a picture*). Text from autograph in *A*. R. B. has dated it 1866–8, 'Balliol College / Oxford'. Handwriting suggests late 1867–8. See notes, pp. 316 and 323.

91 l. 47. Second dash editorial (R. B.).

92 '*Not kind! to freeze me*'. Text from MS (CH). Written in pencil on one of seven sheets on which G. M. H. drafted a translation of Horace's ode, 'Odi profanum volgus'. The sheet contains the beginning of a letter to G. M. H.'s aunt, Laura Hodges (with whom he stayed in January 1868). Like 'Odi profanum volgus', the poem was probably written early in 1868 during Hopkins's final term of teaching at the Oratory, Birmingham. It is, as N. H. M. and R. Kilpatrick have shown (*Classical and Modern Literature*, vol. 5, no. 1 (Fall 1984)), a spirited translation of Horace, *Odes*, Book II, xvii, ll. 1–4.

Horace: 'Persicos odi'. Autograph in CH, undated, on four pages of notepaper. One of the sheets contains part of a letter mentioning the visits of Fr. Ignatius Ryder to the Oratory, and concludes 'I do not expect to be long here: if I get a vocation to the priesthood I should go away . . .' (see *J*, p. 534 and *L* III, pp. 52–5). Probably translated at the Oratory between November 1867 and Easter 1868 (H. H.). I am indebted to R. G. M. Nisbet and Margaret Hubbard for the following notes based on their *Commentary on Horace: Odes*, Book I (Oxford, 1970), pp. 421–7. The poem, placed at the end of Book I, is a characteristic request for a simple style of life without luxury and ostentation.

l. 1. *child*: wine-server. There was a tradition in Greek literature of addresses to slaves.

Persian-perfect. Persia was associated with luxury.

l. 2. *bast*. A membrane below the bark of lime trees, used as string to bind together elaborate garlands.

l. 5. *myrtle*. A very common plant. Flowers were often interwoven to decorate garlands of it.

l. 6. Myrtle was said to dispel the fumes of wine. G. M. H. does not bring out Horace's implication that he too, like the wine-server, is still handsome enough to show off a simple garland.

l. 8. *tackled*: fastened to a pergola.

Horace: 'Odi profanum volgus'. Autograph in CH. Pencil drafts on seven sides. Stanza 1 follows 'Not kind! to freeze me'. Probably translated early in 1868. Among the variants are:

l. 2. Grace guard your tongues! ...   l. 4. *make*. Earlier read 'bid'.

ll. 9–10 read originally:
> Say man than man may more enclose
> In rankèd vineyards

93   l. 21 was rewritten 'Sleep that comes light and not afraid', but the 'linkage of this line with stanza 5 seems to us too elliptical for adoption in the text' (W. H. G.).

I am indebted to Gordon Williams for the following notes based on his commentary on *The Third Book of Horace's Odes* (Oxford, 1969).

ll. 1–2. Two commands were given at the beginning of solemn rites: uninitiated listeners were warned to stay away and those remaining were asked to be silent.

ll. 9–16. These lines describe differences between men during life, but l. 16 shows that death takes all.

l. 17. The sword which Dionysius suspended above Damocles' head.

l. 18. Sicilian banquets were proverbial.

ll. 21–32. Unlike the rich man with many crops or ships, the simple countryman can sleep without worry.

l. 24. *Tempe*. A valley renowned for its beautiful, shady trees and singing birds.

l. 28. Stars signalling the beginning and end of stormy winter.

l. 31. The dog-star was associated with heat-waves.

ll. 33–4. Hyperbole suggesting that the many luxurious villas built into the sea encroach upon the fish.

ll. 35–6, 37–40. Anxious, rich men try in vain to divert themselves with luxurious yachts.

l. 41. *Phrygian stone*. A prized marble. l. 42. Expensive clothes.

l. 43. *Falernian-grown*. Wine.   l. 44. *oils of Shushan*. Persian balsam.

94   *The Elopement*. The Revd D. A. Bishoff, SJ writes: 'Early in 1868, two of the fifth form of the Oratory School, Edgbaston, Birmingham, joined with one of the junior masters, J. Scott Stokes, in editing a weekly journal called *The Early Bird* or *The Tuesday Tomtit*. Each issue was limited to three handwritten copies, the first appearing on 18 February 1868. It suffered an early death. One of the issues, however, carried these verses by G. M. H., then a junior master at Dr J. H. Newman's school; they were followed by a parody, "The Robbery", written by R. Bellais and W. Sparrow. The original handwritten copies have disappeared. The only record of these verses is

found in an anonymous essay, "Early Magazines", *The Oratory School Magazine*, No. 13, Nov. 1895, pp. 5–8' (4th edn.).

l. 1. *rud red*. The colour of red ochre. See 'The Woodlark', l. 31 (p. 122).

l. 19. Magazine omits hyphen, 'to-night' (N. H. M.).

l. 20. Magazine reads 'guess', but a similar description in *J* (p. 72) has 'press' (W. H. G.) (this is 'The stars were packed so close', p. 17).

l. 21. *hurdle*: fence. See 'The Starlight Night', l. 13 for a similar idea.

l. 22. *liberties*. The district over which a person's or corporation's privilege extends (now archaic).

l. 30. Magazine places stop at end of line (N. H. M.).

95 *Oratio Patris Condren*. No date. Pencil draft and ink fair copy in *H*. Intermediate fair copy (Latin and English), perhaps intended for display, pasted into *B*. Text from late copy in *H* on verso of a cancelled draft of 'S. Thomae Aquinatis Rhythmus'. The poem is a translation of 'a prayer by Fr. Condren of the French Congregation of the Oratory of St. Philip Neri'. The principal variant in *B* is ll. 5–6:

> In those most perfect ways Thou wendest,
> In the virtues of that life Thou spendest,

l. 4. *fulness* in all MSS.

Ad Matrem Virginem. Autograph, undated (CH). Handwriting suggests that it dates from Christmas 1870.

Translation: 'To the Greater Glory of God (A.M.D.G.). A Communion Hymn to the Virgin Mother at the Feast of the Nativity'
'Mother of my Jesus, Mother of mighty God, teach me about Him, the small sweet God.

How much did you love Him whom you conceived, the inconceivable, the terrible Lord, but as the Word made flesh brought into smaller compass in you? And He too does not despise even my heart: my heart, though so unworthy to receive so great a sign, oh unworthy to carry Him who was with me this morning, he enters, O Mary, in the Eucharist. He himself wishes to enter: I do not wish to withhold myself. Shining example, teach me to love.

Tell me—so that He may be loved the more—what manner of being He seemed when He lay hidden in your womb and had not yet appeared to sight, when your voice made Elizabeth happy, a mother made happy through a mother, a boy cousin made happy through a boy cousin. Teach me to rejoice, O rose, in your spring, O branch, in your flower, O fleece, in your dew, O ark, in your law, O throne, in your king, O army, in your commander, O moon, in your light, O star, in your rays, O mother, in your child. For I am still puffed up and rank with the foul world; I have saddened the Holy Spirit and caused grief to my Guardian when to my God I showed Jesus wounded and battered in my flesh.

Just how did you feel when you saw at long last, in full view, the Lord himself as a little baby on the hay, beheld Him trembling who fixes firmly the universe, and rolled in swaddling clothes Him who, when not yet born of you, unrolled in serenity the everlasting years? What did you say then, and what did you hear? Although He could not speak, He yet spoke. Allow me to embrace Him, grant me a little of your love, and kisses from your mouth. He who wants to give himself for me, to speak to me although He cannot yet use words, to dwell with me, O grant that I may gaze upon Him, O Mother of mighty God, Mother of my Jesus.

Glory to God for ever (L.D.S.).

96 ll. 29–30. See Luke 1: 39–44.

97 *'Haec te jubent'*. Autograph in *H*, undated, with corrections and variants on three scraps of paper, on the back of one of which is a paragraph on 'words and signs'. The poem seems to be a ' "presentation piece" addressed to a new Jesuit Provincial' (W. H. G.). Probably written in May 1871 when the Provincial, Fr. Robert Whitty, visited St Mary's Hall (N. H. M.).

An uncancelled, earlier variant reads:

> Sed candidatus ille quem cernis chorus
> Ipso colore prospera
> Et auguratur et tibi ore optat meo . . .

Translation (W. H. G.): These regions give you greeting, as far as they can, drenched as they are with too much rain, (and) the mower, Father, abandoning the soaked meadows, gives you much greeting. But our band clad in white, which you see, appears, by its very colour, to predict happy things and by my mouth wishes for you both kindnesses and joys. I therefore pray that this your unshorn flock, with its shepherds, may turn out well, and also may the province which is being committed to your right hand as a new spiritual charge.

G. M. H. was responsible for delivering a sermon during the Provincial's visit.

ll. 10–12. Fr. Whitty had been appointed Provincial in 1870.

98 (*May Lines*). Text from undated MS in *H*, headed A.M.D.G. & B.M.V. with L.D.S. written below the poem. The handwriting suggests it dates from 1873. The epigraph, 'Ab initio . . .', is from Ecclesiasticus 24: 14.

Translation (W. H. G.): 'From the beginning and before the world, was I created, and unto the world to come I shall not cease to be' (Douay).

O doubly predestined, you who from all eternity have been the Mother of Christ, (predestined) after the foreseeing of the merits of the Innocent One, a second time after the sins of our race—though the former privilege is the purer crown, yet it is the latter which the more readily brings home to the heart the gifts of God. Assuredly I should marvel at you simply as God's mother, but then I should not take such delight in your birth; I should confess you a virgin made mother, but not as the one who is, among all mankind, for ever unsullied. But, as your two-fold glory, there will always be those things which stand fast and those which have fallen away—both the redeemed sins of mankind and the foreseen merits of the Innocent One.

l. 7. *Iterum*. Originally 'Alterum' (cancelled).

l. 15. MS *partu*, in error for *pastu?*

Ad Mariam. No manuscripts extant. Printed in *The Stonyhurst Magazine*, February 1894. Attributed to G. M. H.; W. H. G. suggests it was written between 1870 and 1873. The printed version has no breaks between stanzas and places a full stop after l. 2 'May' and l. 18 'have seen' (N. H. M.). The metre is that of Swinburne's 'When the hounds of spring' from *Atalanta in Calydon*.

99 l. 23. *Aidenn*: Eden (Hebrew).

100 *O Deus, ego amo te.* Three undated, fair copies in *H*, all slightly different. Two titled 'St. Francis Xavier's Hymn' are headed A.M.D.G. Handwriting of two of the manuscripts suggests that they date from the early 1870s. Text from the third version, written later below the end of 'S. Thomae Aquinatis Rhythmus'. Translation in *A*. See the Welsh version, 'Ochenaid', p. 125.

l. 7. *nails and lance.* MS repeats 'lance' in mistake.

ll. 7–8. variant:
> For me didst bear the nails, the lance,
> And the shaming out of countenance,

Rosa Mystica. Autograph in CH. Probably one of the two or three presentation pieces mentioned in a letter to R. W. Dixon (5 October 1878).

102 l. 40. Question mark editorial.

'Quique haec membra'. Text from an undated fragment in *H* (pencil); handwriting suggests that it was written in the mid 1870s.

l. 11 variant (uncancelled):
> Quamque rogare alium peccat gens credula vitam

Translation: . . . And you who wish these limbs to be a prey to many evils, do not wish them to be too much so, for I do not contend against these woes that cleave to my lot, nor am I ashamed to be mindful of them; only let clemency attend my moderate punishment and let the cross which you bid me carry be yours: because I am to be tortured, I pray that it may be with you, and that I may be tortured according to your skill. But have pity on your so many thousands of Indians—have pity now on your folk; and that a stranger may ask for that welfare wherein they hasten and sin, grant this, God, in the meantime.

On St. Winefred. Text from autograph in *H*, headed A.M.D.G. with L.D.S. written below. Hopkins also wrote the poem in Latin (see below). For his description of St Winefred's Well see *J*, p. 261 (8 October 1874), prose, p. 221.

103 *In S. Winefridam.* MS (Fourier Library, College of Notre Dame of Maryland, Baltimore, Maryland); undated, numerous alterations. Headed, like the English version, A.M.D.G. with L.D.S. below. See 'Quin etiam', below.

Translation: 'To St Winefred who, in addition to the favour of her miracles, bestows care on a bath and a mill'.

Graciously her summer hand tempers her bath for tired limbs, and graciously her winter hand tempers it for stiff ones. Why, having deemed our toils worthy of her right hand (aid), she is even of service to the ever-turning mill, and is not ashamed to be so. Evidently, though she is high in Heaven she does not despise honest dirt, although she enjoins maidenly grace as well.

The original ran:

> Apparat aestiva fessis sua balnea membris,
>   Apparat hiberna balnea rite manu;
> Quin etiam nostri partem dignata laboris
>   Utilis assiduae, nec pudet esse, molaest;
> Ut quae expers maculae sordes non temnat honestas
>   Et quae virgineum suadeat ipsa decus.

Fragments on St. Winefred. From two sides of a sheet in *H* on which G. M. H. has started a letter in Latin to a fellow Jesuit excusing a tardy reply.

(i) 'Iam si rite'. See *J*, p. 261 (8 October 1874), prose, p. 221.

l. 2. *sacro.* originally 'Dei'.

Translation: Now if I duly follow the traces of the ancient deed, these are outstandingly full of holy power

(ii) *'Quin etiam'*. A variant of 'In S. Winefridam', ll. 3–6, above.

Translation: Why, not having disdained our labours, she is even of use to the busy mill, nor does it shame her so to be, as is natural for one who both endures honourable dirt and yet in herself encourages maidenly modesty.

(iii) *'Atque tribus primum'*.

l. 4. Cf. 'The Wreck of the Deutschland', stanza 4, ll. 5–8.

Lines missing after l. 6.

l. 8. Lacunae either side of 'inclusas'.

l. 9. *quae fiunt.* Originally 'nobis data'.

l. 10. *datur.* Originally 'patet' (uncancelled).

Translation: And first (take) the fact that the river issues from three springs: this is the nature, as we believe, of our threefold God; and take the fact that these springs join together and increase with a clear level surface: there you have the singleness of heart which you nourish, gracious faith. What of the fact that the spring, rising from a hidden source, makes its way into the sunlight and into the sight of men?
     The stone edging of the well which you see distinctly shaped in five foils (in order that) with this kerb it may engarland its enclosed water—this is because the outward form (appearance), which announces all things which are created, enters the mind by the five ways by which it is free to enter.

104 *'Miror surgentem'*. Text from an undated fragment in *H* (pencil); New Year's Day (1876? N. H. M.).

Translation: I wonder at Orion rising through the clear night, even though the bright moon is close at hand and presses more heavily on the small stars nor allows them to shine with her. Yet I marvel how this Orion grows up the sky and how it gleams with its own fire, which a force that is not its own makes bright in the heavens, while its soft lustre comes and goes: why, you would think that some winds had the power to whirl its seven star-points round and round. I marvel too that the breezes and the tepid South wind are wafted so pleasantly, and that winter and the first Kalends which the new year keeps are so warm, for from that day which has just set, the fairest in the year, we say the days (of the year) take their start. O heavenly Jesus, you who gather up in your hand us men and these lofty stars, all things come from you: I pray that the year too may take its beginning from you; it will (then) be a good year. All things are in you: may our race (also) live in you because we are your limbs: all the breezes that we enjoy and the sky to which we look up have, I assert, been granted to us (for use), but many, indeed, have no gratitude; yet even if gratitude is lacking, gracious nature, bountiful nature is also at hand to help all men; nature, to whom that provident hand of yours is stretched out everywhere.

S. Thomae Aquinatis Rhythmus. Four drafts with numerous variants in *H*: (i) an early draft of the complete text; (ii) a much revised draft of stanzas 5–7 (cancelled); (iii) a fair copy of the first 3½ stanzas; (iv) text, a late draft with alterations, ending above 'O Deus, ego amo te'. On an attached page G. M. H. noted: 'This is what I sent to Mr. Orby Shipley—

> Godhead, I adore thee fast in hiding; thou
> God in these bare shapes, poor shadows, darkling now: etc.'

This alters the first two lines of (iv), which read:
> Godhead here in hiding, whom I do adore
> Masked by these bare shadows, shape and nothing more,

Title from (iii).

105 l. 21. Uncancelled variant:

Bring the tender tale true of the Pelican;

l. 25. 'shrouded' and 'veilèd' are bracketed as equal alternatives.

106 *Author's Preface.*

this book: MS *B*.

Running Rhythm. G. M. H. used also the term 'Standard Rhythm'.

Stress, Slack. In the following notes the stress = ' and the slack = × or ×× etc.

Logaoedic Rhythm. e.g. in Swinburne's famous *Atalanta* chorus and G. M. H.'s imitation of it in 'Ad Mariam':

× × ╱ × │ ╱   ×   × │ ╱   × │ ╱   ×

'When a sister, born for each strong month-brother,' (l. 1) (W. H. G.).

107 *Counterpoint Rhythm.* First heard in 'God's Grandeur':

╱ × │╱ × │ ×   ╱ │ ×   ╱   ×   ╱

'Generations have trod, have trod, have trod,' (l. 5) (W. H. G.).

Milton is the great master. For G. M. H.'s examples of Counterpoint Rhythm in *Paradise Lost* and *Paradise Regained* see *L* I, p. 38 and *L* II, p. 15 (W. H. G.).

monosyllable . . . First Paeon. E.g. in 'The Wreck of the Deutschland' (only the stressed syllables are marked in MSS):

(*a*) Monosyllabic feet:

       × | ⁄ | ⁄ | ⁄   ‖×   ×   ⁄ | ⁄ | ⁄
    'The sour scythe cringe, and the blear share come,' (l. 88).

(*b*) Paeons:

       ⁄   × ×   × | ⁄   × × × | ⁄ ×   × × |, ×   × | ⁄   ×
    'Startle the poor sheep back! is the shipwrack then a harvest, does tempest
    ⁄ × × | ⁄   × ×
carry the grain for thee?' (l. 248).

In both lines the caesura really breaks the third foot. Note the extended fifth foot in· (*b*) and cf. line 6 of the same stanza. (W. H. G.)

107 *rove over*. See *L* I, p. 86; prose, p. 238.

108 *Echos, second line*. See p. 155.

a principle needless to explain here. See note to 'Hurrahing in Harvest', p. 354.

counterpointed rhythm. See note to 'Harry Ploughman', p. 382, and *L* I, p. 43: 'Please remark the difference between ∞, which means a counterpoint, and ⌢, a circumflex, over words like here, hear, there, bear, to express that they are made to approach two syllables—he-ar etc. No, it should be ∼, not ∪ .' *nursery rhymes*. See *L* II, p. 14 (5 Oct. 1878).

this book: MS *B*.   *Nos 1 and 25*. 'Pied Beauty' and 'Peace'.

110 *The Wreck of the Deutschland*. Transcripts in *A* and *B* and one by Fr. Bacon made before G. M. H. corrected *B* (it has now disappeared (N. H. M.)). Text from corrected transcript in *B*, except, from *A*: title and dedication, initial capitals in 'God' (l. 12) 'We' (l. 198) 'Way' (l. 206) 'English' (l. 276), 'though' instead of 'tho' ' (ll. 38, 55), hyphen in 'all-fire' (l. 184) and 'wild-worst' (l. 192), semicolon (l. 242), 'stanching' (l. 253), 'Double-naturèd' (l. 266); u.c. 'What' (l. 127) editorial, and from Fr. Bacon's transcription, 'flanks' instead of 'planks' (l. 31). (Subtitle is from *B*.) 'Part the first': italics editorial, to match '*Part the second*', stanza numbers regularized.

Metrical marks in *B*: l. 13 hoŭr   l. 112 endūred   l. 272 Ā released.

Hopkins wrote to R. W. Dixon about the poem's history, 'when in the winter of '75 the Deutschland was wrecked in the mouth of the Thames and five Franciscan nuns, exiles from Germany by the Falck Laws, aboard of her were drowned I was affected by the account and happening to say so to my rector he said that he wished someone would write a poem on the subject. On this hint I set to work and, though my hand was out at first, produced one. I had long had haunting my ear the echo of a new rhythm which now I realized on paper. To speak shortly, it consists in scanning by accents or stresses alone, without any account of the number of syllables, so that a foot may be one strong syllable or it may be many light and one strong. I do not say the idea is altogether new; there are hints of it in music, in nursery rhymes and popular jingles, in the poets themselves, and, since then, I have seen it talked about as a thing possible in critics ... But no one has professedly used it and made it the principle throughout, that I know of. Nevertheless to me it appears, I own, to be a better and more natural principle than the ordinary system, much more flexible, and capable of much greater effects. However I had to mark the stresses in blue chalk, and this and my rhymes carried on from one line into another and certain chimes suggested by the Welsh poetry I had been reading (what they call *cynghanedd*) and a great many more oddnesses could not but dismay an editor's eye, so that when I offered it to our magazine the *Month*, though at first they accepted it, after a time they withdrew and dared not print it' (*L* II, pp. 14–15, 5 October 1878). Accounts of the wreck given in *The Times* on 11 and 18 December 1875 can be found in Note F, pp. 439–43 of *L* III.

In *A* the poem is preceded by a note transcribed by R. B.: 'Be pleased, reader, since the rhythm in which the following poem is written is new, strongly to mark the beats of the measure, according to the number belonging to each of the eight lines of the stanza, as the indentation guides the eye, namely two and three and four and three and five and five and four and six;* not disguising the rhythm and rhyme, as some readers do, who treat poetry as if it were prose fantastically written to rule (which they mistakenly think the perfection of reading), but laying on the beats too much stress rather than too little; nor caring whether one, two, three, or more syllables go to a beat, that is to say, whether two or more beats follow running—as there are three running in the third line of the first stanza—or with syllables between, as commonly; nor whether the line begin with a beat or not; but letting the scansion run on from one line into the next, without break to the end of the stanza: since the dividing of the lines is more to fix the places of the necessary rhymes than for any pause in the measure. Only let this be observed in the reading, that, where more than one syllable goes to a beat, then if the beating syllable is of its nature strong, the stress laid on it must be stronger the greater the number of syllables belonging to it, the voice treading and dwelling: but if on the contrary it is by nature light, then the greater the number of syllables belonging to it the less is the stress to be laid on it, the voice passing flyingly over all the syllables of the foot and in some manner distributing among them all the stress of the one beat. Which syllables however are strong and which light is better told by the ear than by any instruction that could be in short space given: but for an example, in the stanza which is fifth from the end of the poem and in the 6th line [stanza 31] the first two beats are very strong and the more the voice dwells on them the more it fetches out the strength of the syllables they rest upon, the next two beats are very light and escaping, and the last, as well as those which follow in the next line, are of a mean strength, such as suits narrative. And so throughout let the stress be made to fetch out both the strength of the syllables and the meaning and feeling of the words.'

* In Part the second the first line of each stanza has three stresses. G. M. H. also noted that 'There are no outriding feet in the *Deutschland*' (*L* I, p. 45).

ll. 1–2. In *A* and *B* these originally read:

> God mastering me;
> Giver of breath and bread;

ll. 5–6. See Job 10: 9–11 and Psalm 138 (Douay; AV 139) (Philip M. Martin, *Mastery and Mercy* (Oxford, 1957), p. 30).

l. 8. See *S*, p. 158 on elevating grace, which Hopkins describes as 'truly God's finger touching the very vein of personality, which nothing else can reach and man can respond to by no play whatever, by bare acknowledgment only, the counter stress which God alone can feel, ... the aspiration in answer to his inspiration'.

ll. 9–16. The incident described is sometimes thought to have occurred during G. M. H.'s first experience of the Ignation *Spiritual Exercises* in the long retreat given shortly after he entered the novitiate (see Introduction, pp. xx–xxi), but referring it to Hopkins's conversion to Catholicism makes ll. 18–19 easier to understand. See P. Milward, *Commentary on G. M. Hopkins' 'The Wreck of the Deutschland'*, p. 23, and R. Boyle (*Immortal Diamond*, ed. Weyand and Schoder, p. 335).

l. 20. *that spell*: during that time.

l. 21. *the heart of the Host*. 'God', but perhaps also 'heart' = 'meaning'. In a letter to E. H. Coleridge (1 June 1864), G. M. H. said 'The great aid to belief and object of belief is the doctrine of the Real Presence in the Blessed Sacrament of the Altar. Religion without that is sombre, dangerous, illogical, with that it is—not to speak of its grand consistency and certainty— *loveable*. Hold that and you will gain all Catholic truth' (*L* III, pp. 17, 92).

ll. 21–3. For the image of the homing pigeon cf. 'The Handsome Heart', ll. 5–6.

l. 24. *tower from the grace to the grace*. The underlying idea would seem to be explained in notes entitled 'On Personality, Grace and Free Will' (*S*, pp. 146–59). Here Hopkins suggests that each man's life may be lived on a series of different degrees of moral goodness, each state called a 'pitch'. God's grace allows man to taste the degree of goodness above the one in which he is already living. In order to rise to the higher pitch man has merely to accept the opportunity. To stay at the new pitch further grace is required.

111   ll. 25–8. These lines with their traditional symbol of mortality, the hour-glass, may suggest man, and more specifically, the poet as he would be without God's grace, gradually disintegrating physically and perhaps morally.

ll. 26–7. *at the wall/Fast*. The sand at the wall of the hour-glass, which at first appears motionless.

ll. 29–32. The second image suggests that man is maintained spiritually by grace as the water in a well is replenished constantly and unnoticeably by the streams that run down the sides of the hill ('voel'). Peter Milward notes (*Commentary*, p. 35) the relevance of *S*, p. 154: 'grace is any action, activity, on God's part by which, in creating or after creating, he carries the creature to or towards the end of its being, which is its selfsacrifice to God and its salvation . . . so far as this action or activity is God's it is divine stress, holy spirit, and, as all is done through Christ, Christ's spirit.' See John 4: 14.

l. 33. *kiss my hand.* salute/greet; an ancient gesture of 'adoration and salutation', although in the Bible associated with pagan worship.

l. 35. See *J*, p. 254, 'As we drove home the stars came out thick: I leant back to look at them and my heart opening more than usual praised our Lord to and in whom all that beauty comes home.'

ll. 38–40. Although the wonder and splendour of nature come from Christ, in order to feel his presence with its significance ('mystery'), a special state of mind is necessary (as in *J*, p. 254). In this receptive state an 'instress', impression, of Christ's presence may be felt (cf. 'Hurrahing in Harvest'). To complete the communication with Christ requires that the instress be consciously accepted ('stressed'); see note to l. 8.

ll. 41–8. Hopkins seems here to be talking about 'actual grace', defined in the *Catholic Encyclopedia* (1909) as 'a supernatural help of God for salutary acts granted in consideration of the merits of Christ'; that is, 'a passing influence of God' helping man towards those actions necessary for his eternal salvation, given to man through and because of Christ (vol. 6, p. 690). See *S*, p. 154, quoted in note to ll. 29–32. But see also *S*, pp. 98–100 on the Holy Ghost.

ll. 49–56 emphasize that it is Christ's Incarnation that made this grace possible. J. E. Keating points out that 'all grace since the fall of man is bestowed through the merits of Christ' (*Wreck*, p. 63).

ll. 56–64. In crises men turn to God.

112 ll. 58–9. 'As to final impenitence, it is absolute; and this is easily understood, for even God cannot pardon where there is no repentance, and the moment of death is the fatal instant after which no mortal sin is remitted' (*Cath. Encyc.* (1910), vol. 7, p. 415). See *S*, p. 247 for Hopkins's statement of this.

ll. 59–62. Man's repentance and acceptance of God can fill him with a feeling of utter revulsion at his own sinfulness or overwhelming gratitude for forgiveness. Presumably both emotions are presented but one dominates (the line may originally have read, as in *A*, 'sour and sweet').

ll. 65–72 continue the description in the opening stanza of God's two very different attributes: his stern mastery and his mercy. Men whose malice is stubborn ('*dogged in den*') require harsh experience to make them acknowledge God and ask for his mercy.

ll. 70–2. Cf. Psalm 17: 10–15 (Douay; AV 18: 9–14).

l. 77. *Paul.* His Jewish name was Saul. He persecuted Christians until God temporarily blinded him and Jesus spoke to him, converting him (see Acts 9: 1–20 and prose p. 286).

l. 78. *Austin*: St Augustine (345–430). Augustine was gradually converted to Christianity by the sermons of Ambrose, Bishop of Milan.

Part the second.

l. 82. *the flange and the rail.* Railway accidents were causing hundreds of deaths in the 1870s (N. H. M., *Reader's Guide*, p. 38). See *S*, p. 247.

ll. 85–6. *Dust.* See Genesis 3: 19 'for dust thou art, and unto dust shalt thou return', and Isaiah 40: 6–8.

l. 88 *sour*: extremely unpleasant.

scythe cringe. Plants cut with a scythe appear to buckle or bow as they fall. Death is called 'the grim reaper'. *cringe*: transitive, 'cause to cringe or buckle'.

blear share: blind and indiscriminate ploughshare.

113 l. 93. *feathers.* See Psalm 91: 4, 'He (God) shall cover thee with his feathers, and under his wings shalt thou trust'.

l. 95. *bay.* Architectural: recess or compartment; see *J*, p. 193, 'opposite bays of the sky'.

l. 96. *vault*: cover.

round. Perhaps in the sense of a loop of rope round a reel.

reeve: fasten.

l. 107. *combs*: crests or ridges of the sandbank.

night. Both the time of day when the *Deutschland* struck the bank, and in part the reason why the ship was on its mistaken course.

l. 108. *dead*: 'precisely' and 'doomed'.   l. 111. *whorl*: propeller.

114 l. 128. *burl*: fullness.

buck. Perhaps the rising of the wave to a crest, also the action of the wave in dislodging anyone or anything not securely fastened to the ship. 'One brave sailor, who was safe in the rigging, went down to try to save a child or woman who was drowning on deck. He was secured by a rope to the rigging, but a wave dashed him against the bulwarks, and when daylight dawned his headless body, detained by the rope, was seen swaying to and fro with the waves' (*The Times*, 11 December 1875; reprinted *L* III, p. 443).

ll. 143–4. An ecstatic joy Hopkins associates with youth.

l. 147. *hawling.* Perhaps from 'hawle'= hail; to throw or send down hard, like hail in a storm; or 'haul', to pull or drag forcibly (Keating, p. 81).

l. 148. *rash smart sloggering*: stinging, hard.

l. 150. *fetch*: a far-reaching effort (OED).

l. 151. *the tall nun.* *The Times* described her as 'the chief sister, a gaunt woman 6ft. high, calling out loudly and often "O Christ, come quickly!" till the end came' (11 December 1875), and 'One, noted for her extreme tallness, is the lady who, at midnight on Monday, by standing on a table in the saloon, was able to thrust her body through the skylight, and kept exclaiming, in a voice heard by those in the rigging above the roar of the storm, "My God, my God, make haste, make haste" ' (13 December 1875).

115 l. 153. There were five nuns.   l. 156. *wide of its good*: evil.

l. 157. *Gertrude*. St Gertrude the Great (1265–1301 or 1302), who died near Eisleben, the German town where Martin Luther was born. Entering the convent at the age of five, she became a pre-eminent example of a simple Benedictine nun. She was a mystic, gifted with frequent visions (*Cath. Encyc.* 1909).

lily, and l. 158 *Christ's lily*. The symbol of purity appropriately used of St Gertrude.

ll. 157–60 state that good and evil are closely entangled on earth. The Ignatian *Spiritual Exercises* warn at a number of points of the watchfulness necessary to distinguish good from evil disguised as good.

l. 158. *beast of the waste wood*. See Psalm 79: 14 (Douay; AV 80: 13). Martin Luther (1483–1546) was outspokenly critical of the ways in which the wealth and power of the Papacy were maintained. He pictured the upper hierarchy of the Catholic Church in Rome as a pack of wolves preying upon the German people and called the lower hierarchy in Rome 'vermin'. The political and religious rebellions which grew out of hand from his questioning of the status quo brought bloodshed and the destruction of much Church property.

l. 160. See Genesis 4.

l. 165. *Orion of light*. See Job 9 and Amos 5. Commentators say that the creation of Orion was considered to be a special example of God's power (see, for example, *The International Critical Commentary* on Amos and Hosea (Edinburgh, 1905, 1960), p. 115, and on Job (1921, 1964), pp. 86–7.

l. 166 *unchancelling*. Nonce-word. Perhaps G. M. H. is suggesting that God is bringing the nuns into public prominence through their deaths. ('Chancels' are door-screens through which cloistered nuns can greet visitors. The word is also used of the part of a church reserved for the clergy and choir.) A large number of people visited the convent at Stratford-le-Bow where the nuns were prepared for burial. The sisters were considered by many Catholics in England to have been martyrs.

l. 169. *Five*. A holy number because of Christ's five wounds in the crucifixion. It became a cipher for Christ.

finding: emblem. Meditating on Christ's sacrifice is an important part of reaching and maintaining Christian belief, of 'coming to or finding' God.

sake. See G. M. H.'s explanation in *L* I, p. 83 (prose, p. 237)

l. 174. *before-time-taken*: predestined. See 1 Peter 1: 18–21 and *S*, pp. 196–7 (prose, p. 288).

prizèd and priced. See Zacharias 11: 13 (Douay).

l. 176. The red rose is a symbol of martyrdom.

l. 177. *Francis*. Francis of Assisi. The nuns were Franciscans.

ll. 178–81. Francis was given to intense meditation. On one such occasion

Christ suddenly appeared to him in a vision of a crucified seraph. Francis then found that he had on his body five marks similar to the wounds Christ had received in his crucifixion: *scape* (= outward sign) of *love* (= Christ) *crucified*, a proof (= *seal*) of the vision. *The Little Flowers of St. Francis* (tr. T. W. Arnold, Chatto and Windus, 1926) adds that Christ told Francis that every year after his death on the anniversary of his death he would go to Purgatory in order to lead from there all Franciscan orders and others devoted to Francis 'to the end that (he) mayest be conformed to (Christ) in death, as . . . in life' (p. 186).

l. 184. See Rev. 1: 14–16.

116 l. 186. *forehead*. St Beuno's is situated on a hillside, surrounded by fields.

l. 192. *The cross to her*: holding the crucifix against herself.

christens her wild-worst Best. Through her faith in Christ the nun makes her ordeal the opportunity for the best experience of her life. G. M. H. leaves ambiguous here the suggestion that the nun identifies herself with Christ on the cross. He rejects it in ll. 209–16.

l. 194. *arch and original Breath*. See OED, *arch-* (prefix) = 'initial', or, less reduplicatively, 'sovereign'; the Spirit of God as in Genesis 1: 2.

l. 196. *body of lovely Death*. Christ crucified.

l. 198. See Matt. 8: 23–7.

ll. 201–4. The drooping, grey clouds of winter are 'peeled' back to reveal the blue and white skies of spring.

l. 208. See 1 Cor. 2: 9.

ll. 209–16 reject first the idea that the nun was asking for ease. That desire is born of dull, repetitive tasks and sorrow long-drawn-out. Nor is it likely that the nun's identification with Christ was particularly close. That requires quiet meditation.

117 ll. 217–24. While the language of this stanza suggests that the nun had a vision of Christ, stanza 29 makes it clear that the nun did not see Christ as a figure but, like G. M. H. in 'Hurrahing in Harvest', realized that Christ is present in nature. She perceived that the storm occurred for a divine purpose.

l. 226. *single eye*. Focused on God to the exclusion of everything else; see Matt. 6: 22.

l. 230. See John 1: 1–3.

l. 231. *Simon Peter*. See Matt. 16: 16–19. Simon Peter recognized Jesus as the Son of God.

l. 232. *Tarpeïan-fast*. Rome's Capitol 'on the Tarpeian rock, her citadel / Impregnable' (*Paradise Regained* iv. 49–50). As J. E. Keating points out, G. M. H. is probably thinking of the Papacy ('Rome') and the nuns' adherence, despite their persecution, to Catholicism ('the true church built on the rock', Matt. 16: 18).

blown beacon of light. See Phil. 2: 15. Perhaps too the image of fires, built on headlands as warnings to ships on stormy nights, that blaze higher the stronger the wind (N. H. M., *Reader's Guide*, p. 52).

ll. 233–40. The *Deutschland* was wrecked on 7 December. 8 December is the Feast of the Immaculate Conception of the Blessed Virgin Mary.

l. 238. *so to conceive thee is done*. We think of Christ as incarnated.

ll. 239–40. By thinking of Christ, recognizing his presence, and calling his name out so that others could hear it, the nun has made Christ's presence felt again in the world, 'brought it to birth' as Mary did the Incarnated Christ. See Lesson 9 on Luke 11: 28 in the Roman Breviary: 'Blessed, too, were all who conceived that same Word spiritually, by the faith that comes from hearing, and who by their good works strove to bring it to birth and, as it were, to nourish it, in their own hearts and in the hearts of their fellow men' (Keating, pp. 98–9).

ll. 241–8. This stanza brings a reconciliation between the picture of God as sternly masterful, even destructive, and God the merciful.

ll. 241–2. Cf. John 16: 21.

l. 244. See G. M. H.'s notes for a sermon on death (*S*, pp. 247–52); prose, pp. 295–301.

ll. 246–7. Cardinal Manning's funeral sermon for the nuns was reported to include the statement that 'there was reason to believe that the sight of the calm resignation of these holy Sisters proved a useful example to some who shared their fate. Who could tell how many acts of contrition, of faith, and of submission there were during those hours of agony? The example of those sisters was like an articulate voice preaching to others' (Keating, p. 100). G. M. H. accords this power to the cry of the tall nun rather than to the behaviour of all the sisters.

118 l. 250. *Yore-flood*. Perhaps an allusion to Genesis 1: 2 or to Noah's flood (Genesis 6–8) or simply to the ocean.

year's fall. Perhaps the annual rainfall.   ll. 251–2. Cf. Job 38: 8–11.

ll. 253–4. God restrains the restlessness of men's minds and gives their lives a firm foundation. Cf. ll. 29–32.

l. 256. *bodes but abides*. foresees but waits, leaving man with free will.

ll. 257–64. See *S*, p. 252; prose, p. 300.

ll. 259–60. See *S*, p. 190, where, transcribing passages from the *Spiritual Exercises*, G. M. H. writes 'after Christ had expired on the Cross . . . his blessed soul . . . united to the Divinity descended into hell, whence releasing the souls of the just.' The Creed also mentions the descent into hell.

l. 261. *pent in prison*. See 1 Peter 3: 18–19.

ll. 262–4. These people, who had almost lost any chance of salvation, were the furthermost from salvation that Christ, sent by his merciful Father, reached and gathered in his Passion and Resurrection (see *S*, pp. 140–1).

For a summary of various interpretations of these lines see Keating, pp. 104–6.

l. 268. *Mary-of-flame*. See the Roman Breviary, feasts of the Circumcision and Purification, 'Rubum quem viderat', 'The bush which Moses had beheld unburnt we have recognized as thy praiseworthy unstained virginity; Mother of God intercede for us' (Keating, p. 107). Cf. Rev. 12: 1 and G. M. H.'s commentary on it in *S*, pp. 170–1.

l. 272. *lightning of fire hard-hurled*. See Matt. 24: 27 and Luke 9: 54–6.

l. 277. *easter in us*: come to spiritual rebirth in us. See in the Roman Breviary the pre-Christmas antiphon 'O Oriens', 'O dayspring, splendour of eternal light and sun of justice, come and illumine those who sit in darkness and in the shadow of death' (Keating, p. 109).

crimson-cresseted east. The rising sun is a symbol of a new and holier day. See 'God's Grandeur', ll. 12–14. A 'cresset' was a metal or stone vessel for holding oil, coal, etc., for light. It could be used as a beacon, or placed in a church beside the cross.

119   l. 278. *rare*: special, 'of uncommon excellence' (OED).

The Silver Jubilee. Autographs in *C, A, B*; transcription in the Silver Jubilee Album compiled for the occasion and housed at St Beuno's. The poem was also printed in 1876 with a sermon by Fr. John Morris, SJ. MS *A* was written from memory and corrected in a letter of 15 February 1879 (*L* I, p. 65). Text is from *B*, except layout from *A*. G. M. H. also contributed poems in Latin and Welsh, see pp. 120–1 (mentioned in *L* III, pp. 139–40).

l. 5. G. M. H. notes in a letter (*L* I, p. 65, 29 January 1879) that it was in fact twenty-six years, not twenty-five since the restoration of the Catholic hierarchy in Britain.

120   *Ad Episcopum Salopiensem*. Fair copy in *H*, dated April 1876; G. M. H. noted below it: 'They said the beginning was unintelligible and struck out the first nine couplets, so that I had to make the address begin—

> Quod festas luces juvat instaurare Beatis
>   Natalesque suis mos cumulare rosis,
>   His, pater, indiciis etc—

with some other slight changes.' Text is from the shorter version, also in *H*, headed A.M.D.G. with L.D.S. written below. G. M. H. abbreviated the original title by cancelling 'annum agentem et sui praesulatus et restituti apud Anglos episcoporum ordinis vicesimum quintum, qui jubilaeus dicitur' ('in the twenty-fifth year both of his prelacy and of the restoration of the episcopal hierarchy in England, which year is called the Jubilee').

Discarded opening:

> Vertitur in gyrum toto pulcherrima gyro,
>   Attamen est quo sit pulchrior urna loco.

Scilicet hic { hominis vultus habet: ecce recurrunt:
              { faciem spectas: modo verte, recurret;

Non { alium spectas qui venit } ore novo.
     { aliam cernis quae placet }

Miramur rediisse quod ipse redire coegit
  Orbis et in modulum testa rotata suum.
Sic iterat caelum spatiis sua tempora certis
  Quaeque nitere vides astra videbis ait;
Quod si Cassiope magis hac tibi parte venustast
  Hac te Cassiope parte venusta manet.
Indidit hoc nobis varium qui temperat annum,
  Sol ubi prae cunctis igneus unus inest;
Et per versa vices series succedat ut arvis
  Et media his aestas ut sit aprica magis.
At si quid rerum minus ipse notaverat ordo
  Addita non illud signa latere sinunt.
Obscuras olim tulit ambitus ille calendas
  Nostra sed insignes esse rubrica facit.

Translation: (discarded opening) An urn which turns round is most beautiful in its complete revolution, even though it is more beautiful in one particular position (than the others). For here it shows the features of a man: see, they reappear: you do not see another man who comes with a new face. (Original version: For here you see a face; just turn the urn it will come back again: you do not see another face which pleases with new features.) We are surprised that that has returned which the circular movement itself, and the urn revolving according to its own symmetry, has compelled to return. Just so the sky repeats its rhythms in fixed regions, and 'the stars you see shining there you will see again', it says. 'But if Cassiope (a constellation) is more lovely to you in a certain quarter, it is in this quarter that the lovely Cassiope awaits you.' He who rules the changing year, in which there is one sun which burns more fiercely than all others, has caused this to be so for us, and (has ordained) that in turns a changed lot should come upon the fields, and that midsummer should bring them more sun. But in case the very order of the universe had failed to indicate clearly something within itself, additional signs do not allow that thing to escape our notice. That old circling year of ours brought us a first day that was undistinguished, but the title of my poem has turned it into a red-letter day.

(*The approved text*) Because it pleases us to celebrate anew, in their honour, the festivals of the Saints, and it is our wont to heap up their birthdays with the roses due to them, we, Father, your dutiful flock, rejoicing in these tokens and in this custom, honour you on this memorable day; and the paying of those respects which would have been fitting at any other time of your life has now, on this late occasion, come round to us. For, still in your prime, you have reached the twenty-fifth year since the time when the sacred mitre crowned your head. As I believe, that number adds a quarter to the mortal centuries: you add a quarter to the centuries; may you in the same manner be able to halve them. If the famous Pius [Pope Pius IX, 1846–78] attains to the years of Peter, and more, then surely there is one [i.e. you] whose years will be those of the long-lived John. But I am not a soothsayer turning my mind to that time: I only conjecture that the day which is here now is auspicious; may it be—but assuredly it *is* a happy one for your country and for you: that you should be what you are, she, your country, regards as her concern too. With you as a shepherd, as seemed good to god, we English began, as an integrated flock, to form part of the sacred flock (of our Church). Indeed, gracious England even communes thus with herself at your jubilee: 'All may see that it is from that time (1850?) that I have been accounted holy. With these men to aid, after such great disasters, after generations so inhuman,—with their aid, you, O ancient Faith, have been revived for me. Therefore, Heaven keep you, forerunners of events so desired, you the vanguard of my good fortunes. From this time I now number my days; I am now as one clad in white in your calendar, I who through you have been able to please God as His bride.'

See also 'The Silver Jubilee' and 'Cywydd'.

Cywydd. Text from corrected autograph in *H*, dated [April 24] 1876. For G. M. H.'s interest in Welsh see Introduction, p. xxiv. The poem is one of three that Hopkins wróte for an album presented to James Brown, Bishop of Shrewsbury in celebration of the twenty-sixth year of his episcopate. G. M. H. wrote to his father, 'we presented him with an album containing . . . compositions, chiefly verse, in many languages . . . For the Welsh they had to come to me, for, sad to say, no one else in the house knows anything about it; I also wrote in Latin and English' (*L* III, p. 140, 6 August 1876). See 'The Silver Jubilee' and 'Ad Episcopum Salopiensem'.

Cywydd: strictly cywydd deuair hirion: a Welsh poetic form consisting of rhyming couplets of 7-syllable lines following the rules of stress and rhyme as stipulated by *cynghanedd*. See Gardner, *Study* ii, pp. 145–54.

Dedication. Thomas is an error for James Brown (Thomas Brown was Bishop of Newport and Menevia).

Translation (by W. H. G. with the aid of Sir Idris Bell, Dr T. Parry, and Mr M. Harris): Address to the Very Rev. Dr. Th. Brown, Bishop of Shrewsbury, on his reaching his five and twentieth year, which is known as the Jubilee; and the poet complains that earth and sea give greater testimony to the old religion of North-West Wales than man; and he says also that he hopes that this will be changed through the work of the bishop.

Our focal point here is bright and glad with the streamlet of many a fountain, a holy remnant kept for us by Beuno and Winefred. Under rain or dew, you will hardly find a country beneath heaven which is so luxuriant. Weak water brings faithful testimony to our vale, but man bears no such witness. The old earth, in its appearance, shows an eternal share of virtue; it is only the human element that is faulty; it is man alone that is backward. Father, from thy hand will issue a spring from which will flow the beautiful prime good. Thou bringest by faith a sweet healing, the nourishment of religion; and Wales even now will see true saints—pure, holy, virgin.

Brân Maenefa sang this
April the twenty-fourth 1876.

Mr Richard Jeffery suggests to me that Hopkins was unaware of the rules of *cynghanedd* and simply tried to imitate the poets he read. He keeps perfectly to the seven-syllable line and the special uneven type of end-rhyme, and every line has either decorative recurrence of consonants or, occasionally, internal rhyme and alliteration, but without the elaborate rules which standardize the positions of these in the line. Real *cynghanedd* in this metre always links the rest of the line to that part of the line-end not involved in the end-rhyme, by internal rhyme, by internal rhyme coupled to alliteration, or, most typically, by repetition of a whole series of consonants in exact sequence at the beginning and end of the line. If Hopkins read some famous poems by Dafydd ap Gwilym, the earliest of the *cywydd* poets, which add a great deal of extra consonant-decoration to these patterns, it would make it still more difficult for him to recognize the underlying norms.

121 *Moonrise.* Text from draft in *H* dated 'June 19, 1876'. Late transcription by R. B. in *A*. Below the poem G. M. H. experimented with adding another stress to the second half of the first line.

> in the white of the dusk, in the walk of the morning
> in the wake of the yesterday, walk of the morning
> in the yesterday light, in etc.

He added 'And so alter throughout' but no further drafts are extant.

l. 3. *paring of paradisaïcal fruit.* The dull, golden skin; see *J*, p. 209.

l. 4. *Maenefa.* The mountain behind St Beuno's (N. H. M.).

l. 5. *fluke*: the triangular-shaped piece of iron near the tip of the arms of an anchor.

fanged. Originally 'fanged in him': 'of an anchor: to "bite" with its fluke' (OED), to hold.

not quit utterly. Originally 'not free utterly'.

l. 6. caesural mark omitted in MS.

122 *The Woodlark.* Text from draft in *H* dated 'July 5' [1876]. The poem is in the very early stages of composition, with a great many lines missing. The version printed here follows the manuscript order and shows Hopkins in the intermediate stage between the fragmentary lines of description, like those jotted into his diaries and notebooks, and the smooth coherence of his finished poems.

G. M. H. has put dots below words to mark the stresses in ll. 27, 29–34.

ll. 1–11. See *J*, p. 138 (3 June 1866), 'The cuckoo singing one side, on the other from the ground and unseen the wood-lark, as I suppose, most sweetly with a song of which the structure is more definite than the skylark's and gives the link with that of the rest of birds'.

l. 26 cancelled and not replaced.

ll. 27–34 revised several times.

123 l. 36. *oxeye*. See *J*, p. 144, 'Those ox-eye-like flowers in grain fields smell deliciously', and p. 138, 'The meadows . . . containing white of oxeyes'.

l. 38. *fumitory*. See *J*, p. 135, 'Fumitory graceful plant'.

In Theclam Virginem. See note to 'St. Thecla', p. 318.

Penmaen Pool. There are several transcriptions and two autographs. Text is from autograph in *B*, except: l. 1 'leisure' amended to 'pleasure' as in the other manuscripts, and l. 8 'skulls' changed to 'sculls'. *B* is dated 'Barmouth, Merionethshire. Aug. 1876'.

124 l. 10. G. M. H. notes that the 'Giant's Stool' is the mountain, Cadair Idris.

l. 17. *Charles's Wain*: the Plough.

l. 32. Two transcriptions read 'darksome danksome'.

ll. 34–5. Variant: 'Who'll / But praise it?'

125 *Ochenaid Sant Francis Xavier*. Text from undated fair copy in *C* in an unknown hand. It has been ascribed to G. M. H. because of its style (see *Poems*, 4th edn., pp. 324–5). It is a Welsh translation of the same Latin hymn as 'O Deus, ego amo te', p. 61.

Translation (by W. H. G., Sir Idris Bell, Dr T. Parry, and Mr M. Harris): The Sigh of St Francis Xavier, Apostle of the Indians.

Not because Thou hast redeemed me do I love Thee, Lord, in truth, nor because of those who do not love Thee and are condemned to eternal fire.

Thou, thou, who didst embrace me (all of me), my Jesus, on the Cross; from lance, nails, and slanderous tongues didst suffer great agony.

Infinite grief hast Thou endured, and pain and sweat on my behalf, sinner that I am—even unto death for my sake.

Therefore, most loving Jesus, why should I not love Thee steadfastly? Not so as to receive heaven at Thy hands, or any reward; nor lest I receive torment for ever;

But just as Thou didst love me, so shall I love Thee and *do* love Thee, only because Thou art God and art to me a Ruler.

(Margaret Clitheroe). Text from draft in *H* without date or title. Stanzas 6 and 7 lack their final four lines and stanza 8 is missing line 3. The sequence is uncertain apart from the first three stanzas, which G. M. H. numbered. (1876–7?)

Margaret Clitheroe (*c*. 1556–1586). Married to a wealthy butcher in York, she became a Catholic convert, raised her children as Catholics and harboured priests. She became known for her outspoken faith and was imprisoned for it several times. In 1586 the government became more determined to stamp out Catholicism in York and Margaret was arrested and questioned again. She refused to plead in order to prevent her children being forced to give evidence against her. The penalty for silence was to be 'pressed to death'. (*Cath. Encyc.* 1967.) She was declared a saint in 1970. See *L* I, p. 92, and *S*, p. 48.

ll. 1–4. Two ideas seem to be clear here: that Margaret was predestined to martyrdom (see 'The Wreck of the Deutschland', ll. 173–4), and that, although she suffered, her reward was the everlasting bliss of the martyrs.

l. 2. *the chief of bliss.* Earlier read: 'out-of-sight with bliss'.

126 l. 10. *crisp:* see *J*, p. 144, 'Strange pretty scatter-droop of barley ears, their beards part outside like the fine crispings of smooth running water on piers etc.'

l. 18. *clinching-blind.* She was sentenced by Judge Clinch.

l. 27. Comma editorial.   l. 31. The Holy Trinity.

127 l. 42. *Thecla.* See note to the poem, 'St. Thecla'.

ll. 51–61. The punishment was intended to last three days but Margaret died within a quarter of an hour.

'Hope holds to Christ'. Text from a torn scrap of paper in *H*.

l. 1 earlier read: 'Hope holds towards Christ her home-made mirror out'; 'her home-made' was then altered to 'a living'.

128 l. 7. Earlier: 'Her glass can see'.   Gap between ll. 10 and 11.

God's Grandeur. Draft in *C*; two autographs in *A* ('Feb. 23, 1877' and 'March 1877'). Autograph in *F* (see *L* III, pp. 144–5). Text from corrected transcription in *B*, except: layout, u.c. 'Why' (l. 4), comma after 'And' (l. 9), 'eastwards' (l. 12) from autographs. *A*² is marked, 'Standard rhythm, counterpointed'.

Metrical marks in *B*: l. 1 with the grandeur   l. 3 gathers   l. 5 Generations. Hopkins noted that this poem and 'Starlight Night' were 'to be read, both of them, slowly, strongly marking the rhythms and fetching out the syllables' (MS *A*).

ll. 1–3. 'All things therefore are charged with love, are charged with God and if we know how to touch them give off sparks and take fire, yield drops and flow, ring and tell of him' (*S*, p. 195).

l. 2. *shining from shook foil.* 'I mean foil in its sense of leaf or tinsel . . . Shaken goldfoil gives off broad glares like sheet lightning and also, and this is true of nothing else, owing to its zigzag dints and creasings and network of small many cornered facets, a sort of fork lightning too' (*L* I, p. 169). *A*² read 'lightning' instead of 'shining' but G. M. H. rejected this in *B*.

l. 3. *oil.* Crushed from olives. An early version read: 'like an oozing oil / Pressed'.

ll. 13–14. See Genesis 1: 2 and Milton, *Paradise Lost*, i. 19–22. G. M. H. also observed the way in which the sea 'warped to the round of the world' (*J*, pp. 222, 251).

The Starlight Night. Autograph in *F* (see *L* III, pp. 144–5), two autographs each in *C* and *A*, two transcriptions, in *A* and *B* (latter cancelled G. M. H.).

Text from autograph in *B*, layout from earlier manuscripts. *A*¹ is dated 'Feb. 24 1877'. *A*² is marked 'Standard rhythm opened and counterpointed'.

Metrical marks in *B*²: l. 2 sitting in the air l. 3 boroughs l. 6 airy abeles l. 9 patience.

l. 2. See *J*, p. 46, 'Sky peak'd with tiny flames'.

l. 3. Areas of the sky thickly strewn with stars.

l. 4. *dim woods*. Areas of the sky where fewer stars can be seen.

delves: plural of 'delf' (obs.) a mine. *A*¹ and *A*² read 'diamond wells'.

l. 5. *quickgold*. Analogous to 'quicksilver'. *A*¹ and *A*² read 'gold-dew'. See *J*, p. 150, 'the odd white-gold look of short grass in tufts'.

l. 6. *whitebeam, abeles*. Trees whose leaves have silvery undersides.

l. 7. *Flake-doves*. See *J*, p. 261 where starlings are described as 'black flakes hurling round'.

l. 8. R. B. suggested comparing George Herbert's 'Church Porch', stanza 29:

> What skills it, if a bag of stones or gold
> About thy neck do drown thee? raise thy head;
> Take stars for money; stars not to be told
> By any art, yet to be purchased.

129 l. 10. See *J*, p. 249, lime-trees 'starrily tasselled with blossom'.

l. 12. *barn*. See Matt. 13: 30.   l. 13. *shocks*: sheaves.

piece-bright paling. The image is of wooden walls with knot-holes through which star-like points of light can be seen. Cf. 'The stars were packed so close' (p. 79). Cf. too 'He hath abolished the old drouth', l. 14.

'The dark-out Lucifer'. Text from *H*, written above the draft of 'As kingfishers catch fire'.

l. 1. Alternative: 'Dark-out-for-ever Lucifer hates this'.

l. 2. Alternative: 'Entrellises the touch-tree in live green twines'.

touch-tree. See Genesis 3: 3.

ll. 2–3. See Milton, *Paradise Lost*, ix. 494–505 (beauty-bines) and 575–97.

'As kingfishers catch fire'. Transcription in *A*. Text from draft in *H*.

Metrical marks in *H*: l. 1 dragonflies; l. 8 Crying; ll. 9, 11, 12 begin with great colons; l. 10 has a great colon before 'that'; l. 11 eye he.

See *S*, pp. 238–9 (prose, pp. 290–1).

ll. 1–4. See *S*, p. 195: 'All things . . . are charged with God and . . . give off sparks and take fire, yield drops and flow, ring and tell of him'.

l. 3. *tucked string*. Dial. for 'plucked'; originally, 'every string taxed'.

ll. 3–4. Earlier, 'as every sweet string tells, each bell's / Bow answers being asked and calls its name'.

l. 9. Variant in regular rhythm: 'Then I say more: the just man justices'.

justices. Acts in a godly manner aided by God's grace; see note to l. 10.

l. 10. Alternative (regular rhythm): 'Keeps grace and that keeps all his goings graces'. See *S*, p. 154, 'grace is any action, activity, on God's part by which, in creating or after creating, he carries the creature to or towards the end of its being, which is its selfsacrifice to God and it's salvation'. To maintain such behaviour man needs further grace (*S*, pp. 155, 240; prose, pp. 284–5, 292).

l. 11. Variant in regular rhythm: 'In God's eye acts what &c.'

ll. 11–14. Both sorts of grace come to man through Christ. (Cf. *S*, p. 154; prose, p. 285). It is the Christ-like part of man that experiences the initial belief, and it is through emulating Christ's behaviour that man can achieve the second type of grace.

Ad Reverendum Patrem Fratrem Thomam Burke O.P., 'Apr. 23 1877.' Autograph in H; four corrections. Text has the heading 'A.M.D.G.' and the subscription 'L.D.S.'. This 'presentation piece' may have been called for by G. M. H.'s Rector. Thomas Nicholas Burke (b. 1830), Order of Preachers, was the celebrated Dominican orator who had lectured and preached with great success in America from 1870 to 1873. He returned broken in health, but continued his mission throughout Great Britain until his death at Tallaght, Ireland, in 1883. [I am grateful to Dr J. Kuhn for correcting this date.] He visited St Beuno's in G. M. H.'s last year at the college. (W. H. G.)

130 l. 16. *pater . . . stelliger*, i.e. St Dominic (1170–1221): it is said that at the baptismal font a bright star illuminated the infant Dominic's forehead, symbolizing the future greatness of his intellect.

ll. 19–20. The Latinized names are of eminent Dominican theologians, all commentators on Aquinas: *Gudinus* (Antoine Goudin), 1639–95; *Godatus*, MS 'Gobatus'; the poet must have meant Godatus (Pedro de Godoy), the Spanish bishop, died 1677, who was often linked in writing as in life with the French *Gonetus* (Jean Baptiste Gonet), *c.*1616–81; *Cajetanus* (Tommaso di Vio), 1468–1534, cardinal, professor, voluminous thomist exegete, and defender of the *Summa Theologica* against the attacks of Scotus.

ll. 23–4. Burke was at one time novice-master at Woodchester.

ll. 33–4. In 1872 Burke had published four lectures, 'The Case of Ireland Stated', in refutation of the English historian, J. A. Froude. It is noteworthy that the poem is dated April 23—St George's Day.

l. 35. *Guenefrida*, for St Winefred and the healing properties of her well see pp. 102, 103, 120, 161, and 221. (W. H. G.)

Translation:

> To the Reverend Father, Brother Thomas Burke, OP
> on his visit to St Beuno's College.

Seeing an unknown man walking about in our garden, relaxing at our tables, and performing unfamiliar religious offices, I wondered who he could be. In the whiteness of his clothing he was nearer than we are to the guileless doves, and by his dress was such as might call to mind a sheep. Later when I enquired: 'What is the name and order of this monk who is thus disposed to be singular in the midst of our community, whose tonsured head is covered by a cowl of pure wool and whose plain white gown descends to the middle of his foot, a strange double robe falling from back and breast, while a large rosary rubs his left side?'—they replied: 'This is the voice of one crying throughout the world and preparing a way for the Lord through the hearts of men; to this man the Ocean (Atlantic) did service with subdued waves; it has heard him on its western seaboard and on its eastern. But "monk" is not correct: he is one of a Brotherhood, one whom the famous Father who wrote the star acknowledges as his; one skilled at interpreting the oracular words of Thomas Aquinas, if indeed there is still anything obscure in the utterance of him whom Gudinus, Godatus, Gonetus, and Cajetanus strive to make so clear, who has already long endured countless interpreters, and whom each man twists, without hesitation, to suit his own conceptions. Moreover this man (Fr. Burke) was once the guide of tender novices and their shining example on their unaccustomed path. But the whole man is not engrossed in these matters, or, if you like, he is completely engrossed in them, but in such a way that he can be light-hearted amid serious affairs, for he mingles jests with his sacred duties, so that neither his voice nor his facial expression remains always the same.'— These and other things they told me, and added his name, but this latter my Muse (metre?) did not allow me to use, or would have allowed it only with difficulty. Such a man I should now warmly greet; but one doubt keeps me hesitant, namely, whether he would wish to be praised by me, an Englishman—he who (in controversy) lays my countrymen low throughout the world. But whatever happens, my St Winefred gives me healing from all ills: may she give it to him, and soften his hostile breast with love. And whatever else may be of benefit to a faithful people (the Irish), may he owe that also, by way of gift, to our soil.   (W. H. G.)

Spring. Autograph in *A* and one formerly owned by Lady Pooley. Text is from corrected transcription in *B*, except layout, l. 1 initial capital 'Spring', l. 6 'peartree' (one word), l. 13 initial capital 'Mayday' (one word) from *A*. *A* is dated May 1877 and marked 'Standard rhythm, opening with sprung leadings'. In *A* ll. 1, 5, 9, and 13 begin with great colons (see p. 307). In *B* Hopkins marked: l. 5. to hear   l. 8. the racing. In Lady Pooley's MS ll. 1–9 are linked together by a curved line and marked 'staccato' while ll. 10 and 14 are marked 'Rall.' (rallentando).

l. 3. Thrush's eggs are blue.

131   l. 6. *glassy*. 'A notable glare the eye may abstract and sever from the . . . colour of light reflected' (*J*, p. 231, 11 May; prose, p. 215); 'the flesh being . . . sometimes glassy with reflected light' (*J*, p. 154; prose, p. 192).

The Sea and the Skylark. Autograph in *A*. Cancelled transcription in *B*. Text from autograph in *B*, except layout from *A*. The poem was originally titled 'Walking by the Sea'. *A* is dated 'Rhyl May 1877' and marked 'standard rhythm, in parts sprung and in others counterpointed'.

In *A* there are numerous musical signs including staccatos over l. 4 Frequenting there, l. 5. Left hand, off land, and all of l. 7; rallentando above the beginning of l. 14; accents (>) above l. 2 right, ramps, l. 9 frail; a pause (⌒) over l. 6 more, and several different metrical symbols. In *B* Hopkins confined his metrical marks to: l. 1 to end   l. 8 music.

G. M. H. was not pleased with the earlier version (*A*, *B¹*). Parts of the explanation which he sent to R. B. in 1882 apply to the final, revised poem:

ll. 6–7 refer to 'a headlong and exciting (*rash-fresh*) new snatch of singing, resumption by the lark of his song, which by turns he gives over and takes up again (*re-winded*) all day long, and this goes on, the sonnet says (l. 1), through all time, without ever losing its first freshness, being a thing both new and old'.

l. 6. *new-skeinèd score*. 'the lark's song, which from his height gives the impression . . . of something falling to the earth and not vertically quite but tricklingly or wavingly, something as a skein of silk ribbed by having been tightly wound on a narrow card or a notched holder . . . the laps or folds are the notes or short measures and bars of them. The same is called a score in the musical sense of score and this score is "writ upon a liquid sky trembling to welcome it", only not horizontally. The lark in wild glee races the reel round, paying or dealing out and down the turns of the skein . . . right to the . . . ground, where it lies in a heap, . . . or rather is all wound off on to another winch . . . in Fancy's eye by the moment the bird touches earth and so is ready for a fresh unwinding at the next flight' (*L* I, p. 164).

Hopkins noted that when he had written the poem he had been fascinated with cynghanedd, or consonant-chime (*L* I, p. 163, see Introduction, p. xxvii).

l. 2. *trench*: make a deep impression.

l. 3. *flood*: high tide.   *fall*: low tide.

l. 9. The seaside resort of Rhyl in Wales.

l. 11. *cared-for crown*. See Genesis 1: 26–9 and Matt. 6: 29–33. Cf. 'God heeds all things and cares and provides for all things but for us men he cares most and provides best' (*S*, p. 90).

l. 13. *make*: species.

making. Things made, such as the 'frail town'.

l. 14. *dust*. Cf. the Anglican burial service: 'Forasmuch as it hath pleased Almighty God of his great mercy to take unto himself the soul of our dear brother here departed, we therefore commit his body to the ground; earth to earth, ashes to ashes, dust to dust . . .'

first slime. See Genesis 2: 7.

In the Valley of the Elwy. Autograph in *A* and one formerly owned by Lady Pooley. Text is from corrected transcription in *B*, except layout, initial capital 'Valley' in title, and 'Spring' (l. 7) from *A*. Lady Pooley's MS is dated 'May 23 1877'. *A* is marked 'standard rhythm, sprung and counterpointed'.

Hopkins's marks in B are: l. 1 I remember l. 3 very entering l. 7 morsels l. 9 waters, meadows l. 10 the air l. 12 swaying. Lady Pooley's MS has 'Rall.' marked before ll. 9 and 14, and Sf (sforzando) before l. 10. In A l. 8 'seemed' and l. 11 'Only' are preceded by great colons.

G. M. H. wrote of the poem: 'The kind people of the sonnet were the Watsons of Shooter's Hill, nothing to do with the Elwy. The facts were as stated . . . The frame of the sonnet is a rule of three sum *wrong*, thus: As the sweet smell to those kind people so the Welsh landscape is NOT to the Welsh; and then the author and principle of all four terms is asked to bring the sum right' (*L* I, pp. 76–7). He noted that the companion poem to the above was 'Ribblesdale'.

132 ll. 5–6. See *J*, p. 233 (22 July 1873); prose, p. 216 (N. H. M.).

The Windhover. Two autographs in *A*. Text from corrected transcription in *B*, dated 'St Beuno's. May 30 1877', except layout, commas after 'swing' (l. 5) and 'plume' (l. 9), capital 'O' (l. 11) from *A*. *A*[1] is marked, 'Falling paeonic rhym (*sic*), sprung and outriding'.

Hopkins's metrical marks in *B*: l. 2 dauphin l. 3 rolling . . . him l. 4 there l. 6 heel . . . the hurl l. 8 achieve of l. 9 oh, air l. 11 lovelier . . . dangerous l. 12 of it.

Windhover. A kestrel or falcon, which hunts by hovering on the wind with quivering wings and occasional rapid bursts of larger wing-movement. It can glide sideways from one level to another and dives to snatch its prey.

ll. 4–5. *A*[1] reads 'Hung so and rung the rein of a wimpled wing / In an ecstacy'. Wings quivering in a hover.

l. 5. All MSS read 'ecstacy'.

ll. 5–6. One of the kestrel's most beautiful movements is a rapid side-slip, gliding in a curve from one level to another.

ll. 7–8. *My heart in hiding.* Discretion would be necessary to get close to the bird, but the phrase also suggests a stark contrast between the bird with its activity and freedom and proud courage and the still, earthbound observer, his movements arrested by his admiration for the bird's skill.

l. 8. *achieve*: achievement. *A*[1] read: 'Stirred for a bird,—for the mastery of the thing!'

ll. 9–11. This is ambiguous, but many critics suggest that the poet has a sudden vision combining the beauty, courage, and skill of the bird into a perception of its essential nature (inscape) and, beyond that, envisages Christ, who is a billion times lovelier and mightier.

l. 10. *buckle*: come together. See N. H. M., *Reader's Guide*, pp. 76–84.

l. 11. *O my chevalier.* Perhaps addressed to Christ, to whom G. M. H. dedicated the poem when correcting *B*. However, the dedication may be explained by the fact that Hopkins considered this poem the best he had written.

l. 12. *sillion*: a strip of arable land usually worked by a tenant farmer. A plough, rusty at the end of winter, soon becomes shiny when used.

l. 14. Apparently cold, black embers often have a hot, glowing centre.

Figuratively in the sestet light is suggestive of Christ, whose presence can be perceived even in dull daily toil (represented by ploughing) and whose sacrifice is recalled by the self-destruction of the embers. (The observation of Christ's presence in the world is frequently a subject in the poems Hopkins wrote during 1876–8.) The final six lines stress the relationship between God and the world (see also 'As kingfishers catch fire'). Man can contribute to this relationship by keeping in mind Christ's example of obscure toil and self-sacrifice and by dedicating his own physical work (also ll. 12–13) and inner spiritual life (ll. 13–14) to the glory of God (see *S*, 238–41, prose pp. 290–2).

Pied Beauty. MS *A* dated 'St Beuno's, Tremeirchion. summer 1877' and marked 'Curtal-sonnet: sprung paeonic rhythm'. Text from corrected transcription in *B*, except layout from *A*, possessive 'finches'' (l. 4), omission of comma after 'freckled' (l. 8) and colon in place of semicolon after 'change' (l. 10).

Metrical marks in *B*: l. 9 sōur. *Curtal*: see Author's Preface, p. 109. *A* has great colons before the opening of ll. 1, 5, 7, and 'finches' (l. 4), 'trades' (l. 6), 'strange' (l. 7), 'slow' (l. 9), 'change' (l. 10), 'him' (l. 11).

l. 2. *brinded*: brindled, streaked.

l. 3. Trout lose their rose-coloured marks when they die (N. H. M.).

133 l. 4. *Fresh-firecoal*. 'Chestnuts as bright as coals or spots of vermilion' (*J*, p. 189).

finches' wings have conspicuous light-coloured bands.

l. 7. *counter*: unusual.   *spare*: undecorated.

l. 8. *fickle*: changeable.

frecklèd: variegated, here slightly eccentric?

l. 10. *fathers-forth*: creates.   *past change*: eternal.

The Caged Skylark. Autograph in *A*, corrected *L* I, p. 42 (8 August 1877). Text from corrected transcription in *B*, except layout, initial capitals 'Caged Skylark' in title, from *A* (G. M. H. checked the punctuation in *B* carefully). He dated *B* 'St. Beuno's. 1877'. *A* is marked 'falling paeonic rhythm, sprung and outriding'.

Metrical marks in *B*: l. 4 drudgery   l. 8 barriers   l. 10 babble and   l. 13 uncumberèd   l. 14 footing it.

l. 2. *bone-house*. Perhaps from 'bānhūs', Old English for 'body', but the rib-cage provides the closest visual analogy.

l. 3. *beyond the remembering his free fells*. *A* reads 'beyond recollection of free fells'.

fells: moors, hilltops.

l. 5. *turf.* Piece of clover frequently placed in a skylark's cage.

ll. 9–11. The free bird.

l. 10. 'The lark descends, still singing. When yet at a height, the song ceases and the bird drops abruptly, recovering itself a foot or so above the grass and skimming forward before alighting' (T. A. Coward, *The Birds of the British Isles and Their Eggs*, London (1939), p. 93).

ll. 12–14. The belief that after the Day of Judgement, when those who believe in God have repented of their sins and are granted eternal life, they will have perfect bodies that will hinder their spirits no more than a rainbow damages the feathery seeds of thistles or dandelions (*meadow-down*).

'To him who ever thought'. Text from sheet in *H* on which there are two drafts and a fair copy, each line of which is underlined. Probably 1877. Transcription in *A* by R. B.

Sister Mary Jeremy noted (*TLS*, 14 November 1952) that Hopkins's poem is based on a passage in the *Revelations of St. Gertrude* (1865):

'Having heard a preacher declare that no person could be saved without the love of God, and that all must at least have so much of it as would lead them to repent and to abstain from sin, the Saint began to think that many, when dying, seemed to repent more from the fear of hell than from the love of God. Our Lord replied: "When I behold anyone in his agony who has thought of Me with pleasure, or who has performed any works deserving of reward, I appear to him at the moment of death with a countenance so full of love and mercy, that he repents from his inmost heart for having ever offended Me, and he is saved by this repentance." '

(*The Life and Revelations of St. Gertrude*, 1949 reprint, Newman's Press, Maryland, p. 201.) For St Gertrude see note to 'The Wreck of the Deutschland', l. 157.

l. 6 originally read: 'Will grieve his ever sinning and be freed'; revised to: 'Repent he sinned and so his sins be freed'.

freed. Perhaps 'to clear from blame or stain; to show or declare to be guiltless; to absolve, acquit' (obsolete), OED, as in Romans 6: 7, 'For he that is dead is freed from sin'.

Above the poem is a single line related in thought:

Matchless mercy in disasterous, a disastrous time

134 *Hurrahing in Harvest.* Draft in *C*, autograph in *A* dated 'Vale of Clwyd Sept. 1 1877' and marked 'sprung and outriding rhythm; no counterpoint. Take notice that the outriding feet are not to be confused with dactyls or paeons, though sometimes the line might be scanned either way. The strong syllable in an outriding foot has always a great stress and after the outrider follows a short pause. The paeon is easier and more flowing' (see *L* I, p. 45, and *L* II, pp. 85–7; prose, pp. 228 and 246–8). Text from corrected transcription in *B*, except layout, most of the punctuation from *A*.

Hopkins's metrical marks in *B* are: l. 1 now; . . . barbarous l. 8 a͞/Rapturous . . . greeting l. 9 azurous l. 10 Majestic . . . stalwart l. 14 for him . . . for him.

G. M. H. said 'the Hurrahing Sonnet was the outcome of half an hour of extreme enthusiasm as I walked home alone one day from fishing in the Elwy' (*L* I, p. 56 [16 July 1878]).

ll. 2–3. *wind-walks . . . / . . . silk-sack clouds.* Fluffy clouds blown across the sky as if making their way along a path. Cf. 'That Nature is a Heraclitean Fire', ll. 1–2. See *J*, p. 204 (prose, p. 204), 'bright woolpacks . . .'.

ll. 3–4. *wilful-wavier / Meal-drift.* High, wispy clouds.

ll. 7–8. *a / Rapturous.* Cf. the use of run-on rhyme in 'The Loss of the Eurydice'.

ll. 11–14. Cf. 'The Wreck of the Deutschland', ll. 38–40.

The Lantern out of Doors. Autograph in *A*, marked 'Standard rhythm, with one sprung leading' (l. 9) 'and one line counterpointed' (l. 14). Hopkins subsequently altered l. 5 so that it now opens with a spondee. Text from corrected transcription in *B*, except layout from *A*. G. M. H. dated *B* 'St. Beuno's. 1877' and noted that the poem's companion is 'The Candle Indoors'.

Metrical marks in *B*: l. 13 them, heart l. 14 rescue, and.

l. 8. *distance buys them quite.* Jesuits were regularly moved from one community to another and rarely had the same daily companions for long (see Introduction, p. xxvii).

ll. 9–10. See *L* I, pp. 66–7; prose, p. 235.

ll. 12–14. Cf. *S*, p. 89, 'God knows infinite things . . . ourselves', prose, pp. 278–9.

l. 14. *ransom.* Christ's sacrifice for man's salvation.

135 *The Loss of the Eurydice.* Transcription in *A* marked: 'written in sprung rhythm. The 3rd line has three beats, the rest 4. The scanning runs on without break to the end of the stanza, so that each stanza is rather one long line rhymed in passage than 4 lines with rhymes at the ends' (see *L* I, p. 86). Text from autograph in *B* dated 'Mount St. Mary's, Derbyshire. April 1878'. Slurs linking the end of third line to the beginning of the fourth to complete the rhyme occur in stanzas 6, 17, 23. See prose, pp. 107–8.

l. 6. *furled them, the hearts of oak.* The image is of 'a stroke or blast in a forest of "hearts of oak" (. . . sound oak-timber) which at one blow both lays them low and buries them in broken earth' (*L* I, p. 52). *hearts of oak*: also brave, doughty, see OED, *heart* 19b.

l. 8. *forefalls.* The closest, sea-facing slopes.

ll. 9–20. The *Eurydice* was a naval training vessel returning from exercises in the West Indies. Contemporary descriptions can be found in *Immortal Diamond*, ed. Weyand, pp. 375–92.

l. 22. *bay*. 'Of heaven', see *L* II, p. 33.

l. 23. *Boreas*. The Greek god, personifying the North Wind, who destroyed the Persian fleet (N. H. M.).

136 ll. 29–32. Carisbrook, Appledurcombe, Ventnor, and Boniface Down are all places on the Isle of Wight.

l. 33. *press*: 'as much sail as wind etc. will allow' (OED).

l. 34. *royals*. Above topgallant sails (OED), used only in fine weather (N. H. M.).

l. 47. *Cheer's death*: despair.

ll. 53–6. ' "Even" those who seem unconscientious will act the right part at a great push' (*L* I, p. 53).

137 ll. 89–92. Many Catholic shrines in England had been destroyed by Henry VIII and their contents taken.

138 l. 94. *wildworth*. Wild flowers: a reference to the men, who are hardy, healthy, and mature (*blown*) but are lost because they are outside the protection of the Catholic Church, see pp. 258, 297.

l. 98. *my master*: Christ or God.

ll. 99–100. Henry VIII had severed the link between the national church of England and Roman Catholicism.

l. 102. The Milky Way (*marvellous Milk*) was called the 'Walsingham Way' because it guided pilgrims travelling at night towards the popular Catholic shrine to the Virgin Mary at Walsingham in Norfolk.

l. 103. *And one*. Duns Scotus, 'champion of her (Mary's) Immaculate Conception' (*L* I, p. 77).

l. 112. *O Hero* (that) *savest*: Christ. *Hero*. See *S*, p. 34; prose, p. 276.

ll. 113–20 refer to the belief that those who have not been damned to hell but only appear to be doomed can be given God's mercy till the Day of Judgement through the prayers of those still living.

139 *The May Magnificat*. Autographs pasted into *A* and *B*. Text from later autograph in *B* dated 'Stonyhurst. May 1878'. *A* is marked 'sprung rhythm: four stresses in each line of the first couplet, three in each of the second'. The poem was requested for hanging in front of a statue but was rejected by G. M. H.'s superiors, perhaps because of its sprung rhythm.

Magnificat. Mary's hymn praising God. See Luke 1: 46–55.

l. 5. *Candlemas*. (2 Feb.) the feast celebrating Mary's purification after the birth of Christ and her presentation of him in the temple.

Lady Day. (25 March) the Feast of the Annunciation.

l. 21. *bugle*: a plant with blue flowers (*Ajuga reptans*).

l. 25. *sizing*: growing in size.

140 ll. 37–8. Apple blossom is white with a touch of deep pink.

l. 41. Bluebells are grey when they first flower. Their colour then gradually deepens to blue (see *J*, pp. 208–9, 231).

ll. 43–4. See *J*, p. 232, 'Sometimes I hear the cuckoo with wonderful clear and plump and fluty notes: it is when the hollow of a rising ground conceives them and palms them up and throws them out, like blowing into a big humming ewer—for instance under Saddle Hill one beautiful day and another time from Hodder wood when we walked on the other side of the river.'

'*O where is it, the wilderness*'. A fragment of a proposed poem mentioned in a letter to R. B., printed *L* I, pp. 73–4 (26 February 1879). See 'Inversnaid', p. 153.

'*Denis*'. Text from a draft in *H* written before 'The furl of fresh-leaved dogrose', fragments of which surround it. On the verso is a draft of 'Binsey Poplars' dated 'March 13 1879'. Early March (?) 1879.

Metrical marks: l. 3 occasion.   Great colons before 'blinking' and 'done' (l. 5).

l. 4. *three-heeled timber*. Traditional, wooden archer's arrow with three stabilizing vanes.

ll. 5–6. *gold*. The centre of the target (*butt*, used loosely).

141 '*The furl of fresh-leaved dogrose*'. Text from draft in *H* (see note to 'Denis'). Early March(?) 1879.

Metrical marks: l. 17 sidled . . . dewdrops . . . diamonds.

ll. 1–4. originally:

> Soft childhood's carmine dew drift down
>   His cheeks the forward sun
> Has swarthed about with a lion-brown
>   Before Spring season is done.

l. 1. *furl*. Noun perhaps drawn from the verb in its sense of 'covering or wrapping round'.

l. 3. *swarthed*: made swarthy, darkened the complexion.

l. 6. MS not indented, in error.

l. 9. See *J*, p. 209, 'The bluebells in your hand baffle you . . . : if you draw your fingers through them they are lodged and struggle/ with a shock of wet heads; the long stalks rub and click and flatten to a fan on one another . . . making a brittle rub and jostle like the noise of a hurdle strained by leaning against' (prose, p. 208).

l. 11 earlier read: 'Or like a hurdleless fleecy flock'.

'*He mightbe slow*'. Text from draft in *H* written above a draft of 'Binsey

Poplars' (see note to 'Denis'). Spring 1879, before 13 March. It may have been written about the Arthur who is mentioned in 'Denis'.

l. 1. *slow*. Originally 'dull'.

l. 2. *feck at first*. Cancelled and not replaced.

'What being in rank-old nature'. Three drafts in *H*, transcription by R. B. in *A*. Text is from the third draft in *H*, written on the reverse of an early draft of 'The Handsome Heart'. The early versions were written between summer 1878 (see N. White in *Review of English Studies*, vol. 20, no. 79 (Aug. 1969) pp. 319-20) and June 1879 (*N. H. M.*).

Metrical marks: l. 1 nature   l. 2 personal   l. 5 crumbling ... thundering.

There are great colons before 'personal' (l. 2), 'billow' (l. 3), 'westerly' and 'blustering' (l. 4), 'Underneath' (l. 6).

ll. 1-2 earlier read:

> What things in nature should have, earlier, that breath been
> Which, personal, tells off these heart's-song powerful peals?—

l. 3 earlier read:

> Some billow, a casquèd billow:

N. H. M. has suggested that the poem was 'inspired by a piece of music, perhaps Purcell or Handel played on some great organ' (*Reader's Guide*, p. 222).

142 *Duns Scotus's Oxford*. Autograph in *A* dated 'Oxford, March 1879'. Text from corrected transcription in *B*, except layout, omission of commas after 'city' (l. 1) and 'Rural' (l. 8), hyphens in 'lark-charmèd', 'rook-racked' (l. 2), grave in 'poisèd' (l. 4), semicolon in place of colon (l. 10), from *A*.

Metrical marks in *B*: l. 2 cuckoo-echoing, bell-swarmèd, lark-charmèd   l. 3 thee   l. 4 encounter in   l. 8 keeping   l. 10 on ... waters   l. 11 haunted ... men   l. 12 rarest-veinèd   l. 13 insight.

l. 1. From almost any direction the college towers and trees of Oxford are conspicuous.

ll. 3-4. Country and town were once equally matched and balanced powers in Oxford.

ll. 5-6 contrast the ugly new, brick-built suburbs with the grey stone of the older, college buildings. Both Ruskin and William Morris were to complain of the new ugliness.

l. 8. *keeping*: harmony.

l. 10. *he*. Duns Scotus, who is thought to have taught at Oxford around the year 1300.

l. 11. G. M. H. wrote, 'At this time I had first begun to get hold of the copy of Scotus on the Sentences ... and was flush with a new stroke of enthusiasm. It may come to nothing or it may be a mercy from God. But just then when I took in any inscape of the sky or sea I thought of Scotus' (*J*, p. 221, 3 Aug. 1872).

l. 12. *realty*: reality.

ll. 12–3. Hopkins remarked to Patmore that Duns Scotus 'saw too far, he knew too much; his subtlety overshot his interests; . . . and the ruck of talent in the Schools finding itself, as his age passed by, less and less able to understand him, voted that there was nothing important to understand and so first misquoted and then refuted him' (*L* III, p. 349). See Introduction, p. xxiii.

l. 14. See *S*, p. 45, 'It is a comfort to think that the greatest of the divines and doctors of the Church who have spoken and written in favour of this truth [the Immaculate Conception] came from England: between 500 and 600 years ago he was sent for to go to Paris to dispute in its favour. The disputation or debate was held in public and someone who was there says that this wise and happy man by his answers broke the objections brought against him as Samson broke the thongs.and withies with which his enemies tried to bind him.' Scotus was born in Scotland, but at the time Hopkins was writing it was thought that he might have been born in Northumbria.

Binsey Poplars. Drafts in *H* and *C*. Autograph in *A*. Text from corrected transcription in *B*, except subtitle, layout from *A*. *H* is dated 'March 13 1879'.

Metrical marks in *B*: l. 8 river and wind-wandering.

Hopkins wrote to Canon Dixon, 'I have been up to Godstow this afternoon. I am sorry to say that the aspens that lined the river are everyone felled' (*L* II, p. 26).

l. 1. *aspen*. An unusually broad type of poplar with fluttering leaves.

l. 6. *dandled*. Used of bouncing a child up and down. Cf. Milton, *Paradise Lost*, iv. 343.

In April 1873 G. M. H. wrote, 'The ashtree growing in the corner of the garden was felled. It was lopped first: I heard the sound and looking out and seeing it maimed there came at that moment a great pang and I wished to die and not to see the inscapes of the world destroyed any more' (*J*, p. 230).

143 *Henry Purcell*. Autograph in *A* dated 'Oxford, April 1879' and marked 'Alexandrine: six stresses to the line'. Text from corrected transcription in *B* (additional punctuation was added by G. M. H.), except layout, dedication, hyphens 'purple-of-thunder' (l. 12) from *A*.

Metrical marks in *B*: l. 1 fair fallen . . . fallen l. 2 To me l. 4 sentence . . . listed l. 5 meaning l. 6 that l. 8 there . . . on . . . the ear l. 9 angels . . . lay me! only I'll l. 10 of him . . . moonmarks l. 11 stormfowl l. 13 palmy l. 14 him . . . motion.

l. 6. *nursle*: to nurse, foster, cherish.

l. 7. *forgèd feature*. Inescapable impress of personality.

l. 8. *abrupt*: frank, unselfconscious.

For Hopkins's detailed explanation of the poem's meaning, see *L* I, pp. 170–1 (4 Jan. 1883) and p. 83 (26 May 1879), quoted above, pp. 258–9, 236–7.

144 *'Repeat that, repeat'*. Text from an undated fragment in *H*.

Metrical marks: great colon before 'landscape' (l. 5).

See *J*, p. 232; notes to 'The May Magnificat', ll. 43–4.

The Candle Indoors. Draft in *C*. Autograph in *A* dated 'Oxford 1879' and marked 'common rhythm, counterpointed'. Text from corrected transcript in *B*, except layout, initial capital 'Indoors' (title), dash (l. 8), hyphen (l. 13) from *A*. The semicolon in *B* (l. 10) and comma (l. 14) were added by G. M. H.

Metrical marks in *B*: l. 5 window l. 6 wondering.

Hopkins noted that the poem's companion is 'The Lantern out of Doors'.

l. 3. *blear-all black*. Darkness that blurs the outlines of all things.

l. 4. Perhaps the rays of light that appear to radiate out from the candle seem to revolve, an optical illusion caused by the movement of the eyelashes. *C* reads 'Or truckling to-fro trambeams ~~finger~~ dally at the eye'.

l. 8 The oblique slash indicates a pause in reading aloud but not the break in meaning that a comma would imply.

In ll. 9–14 the poet addresses himself.

l. 12. *beam-blind*. See Matt. 7: 3–5. The poet reminds himself to be more attentive to his own faults and less critical of others.

l. 13. *liar*: false thing, i.e. salt that is incapable of 'salting'.

l. 14. *spendsavour salt*. See Matt. 5: 13–16, in which Christ warns his disciples not to lose their power of preserving mankind but to spread Christianity through their example and teaching.

conscience: men of good conscience.

The Handsome Heart. Draft in *H*. Two autographs in *A*, dated 'Oxford 1879'. A transcription and an autograph in *B* were cancelled by G. M. H. Text from final autograph in *B*, layout from *A*². All versions except the final one, which uses a 6-stress line, are in pentameters. (See prose, p. ixl.)

Metrical marks in text: l. 5 carriers l. 11 bathed.

For the incident that prompted G. M. H. to write the sonnet see *L* I, p. 86 (14 August 1879), above, p. 238.

R. B. followed *A*¹ except for 'heart' (l. 5) from *A*², and l. 8 from the draft. This more lively version reads:

> 'But tell me, child, your choice; what shall I buy
> You?'—'Father, what you buy me I like best.'
> With the sweetest air that said, still plied and pressed,
> He swung to his first poised purport of reply.

What the heart is! which, like carriers let fly—
Doff darkness, homing nature knows the rest—
To its own fine function, wild and self-instressed,
Falls light as ten years long taught how to and why.

Mannerly-hearted! more than handsome face—
Beauty's bearing or muse of mounting vein,
All, in this case, bathed in high hallowing grace . . .

Of heaven what boon to buy you, boy, or gain
Not granted?—Only . . . O on that path you pace
Run all your race, O brace sterner that strain!

See also Note on the Text and Appendix *C*, pp. ixl and 413.

l. 4. *swung to*. Like the needle in a compass (see prose, p. 286).

145   l. 5. *carriers*: pigeons capable of flying home instinctively.

heart (*A*¹) amended to 'soul', as in draft.

l. 6. *Doff darkness*: open the pigeons' travelling-basket.

ll. 7–8. *Heart to its . . . / Falls*. Following its own fine nature, which it reveals through its actions. Cf. *J*, p. 261, 'naturally and gracefully uttering the spiritual reason of its being'.

ll. 9–10. See *L* I, p. 95 (22 October 1879), above, p. 240.

l. 12. *buy*. By prayers.

ll. 13–11. To fulfil the promise he has shown of living a life of which God would approve.

In his letter of 8–16 October 1879, G. M. H. told R. B. that 'the little hero of the Handsome Heart has gone to school at Boulogne to be bred for a priest and he is bent on being a Jesuit'.

'How all is one way wrought'. Text from undated draft in *H* with many revisions. Several verses are written on a note from W. H. Pater dated 'May 20 [1879]'. It may be the poem G. M. H. said in a letter of 22 June 1879 that he was writing (*L* I, p. 84). The sequence of stanzas printed here follows the manuscript and may well not be the arrangement Hopkins would have chosen had he finished the poem.

Earlier alternatives include:

ll. 1–4.   How all's to one thing wrought!
                  The members how they sit!
                  O what a tune the thought
                  Must be that fancied it.

l. 7.   Since all that makes the man(cancelled)

ll. 9–12.   Who shaped these walls has shewn
                  The music of his mind,
                  Made known in earth and stone,
                  What beauty beat behind.

ll. 14–16.	His hand seemed free to play	
	He did but draw but what he was	(cancelled)
	To draw and must obey.	
l. 20.	That vaulted round his voice.	
	(full stop from earlier version)	
146 ll. 22–3.	This sweetness, all this song,	
	This piece of perfect good,	
ll. 23–4.	Punctuation from earlier versions. See S, pp. 295–6.	
l. 28.	That's cloistered by the bee.	
l. 30.	With that the man shall make:	
ll. 35–6.	But right must choose its side	
	To champion, and have done.	

The opening stanza suggests that a piece of art, here a piece of architecture, must be revelatory of its creator. In the second and third stanzas the idea is examined, with the conclusion that the heart cannot be revealed by a building, which is also too clumsy ('rough hew and rugged rind' in one of the alternatives) to express the subtlety of its creator's mind. Stanzas four and five suggest that artistic individuality arises from the artist's intrinsic, created nature or from inspiration that (as in Plato's *Apology*) utilizes his talent. His product, therefore, (stanzas six and seven) may be artistically perfect but is morally neutral. The final two stanzas concentrate on the artist, stressing that he has moral obligations and that it is his deliberate moral attitude that governs what he makes of that more important product—himself.

Cheery Beggar. Text from draft in *H*. An early draft of 'The Bugler's First Communion' is on the verso. Written at Oxford during summer 1879. Transcription in *A* by Mrs Bridges.

Metrical marks: great colons before 'Magdalen' (l. 1) and the second 'pine' (l. 7); l. 1 Magdalen ... thēre   l. 4 sweet-and-sōur of ... finefloūr of   l. 5 links.

l. 3. In February G. M. H. wrote to his mother, 'the long frost, severer, it is said, at Oxford than elsewhere, has given place to great rains and those to fine weather' (*L* III, p. 151).

ll. 4–5. *fineflour*: pollen.   *goldnails*: stamens.

l. 5. *gaylinks*. Earlier read 'gaylatchets'.

The Bugler's First Communion. Partial draft in *H*, draft in *C*, autograph in *A* dated 'Oxford ~~Aug. 8~~ July 27(?) 1879' and marked 'sprung rhythm, overrove; an outride between the 3rd and 4th foot of the 4th line in each stanza'. Text from corrected transcription in *B*, except l. 3 initial capital 'English' from *A*, l. 34 'for now' (R. B. reversed the words in error), l. 44 capital 'Lord', l. 48 'Forward-like' for R. B.'s error, 'Froward-like'.

Metrical marks in *B*: l. 3 hē   l. 4 surely   l. 6 mē, overflowing   l. 8 to it   l. 12 housel his   l. 16 chastity   l. 20 dexterous   l. 24 wellbeing of a

l. 25 though I͡   l. 28 soldi͡ery   l. 32 hĕir to ... thĕre   l. 36 of him͜
l. 40 Galaha͜d   l. 44 Euchari͜st   l. 48 howeve͜r, and .

Hopkins's parish duties included spending one day a week at the Cowley barracks in Oxford.

147 ll. 5–6. G. M. H. explained that the soldier 'came into Oxford to our Church in quest of (or to get) a blessing which, on a late occasion of my being up at Cowley Barracks, he had requested of me' (*L* I, p. 97, 22 October 1879).

l. 12. *housel*: the communion wafer that, in Catholic belief, becomes the body of Christ.

ll. 17–20. *angel-warder*. See *S*, pp. 91–3 on the idea that each person has a guardian angel (prose, pp. 280–1).

l. 18. *squander*: scatter.   *hell-rook ranks* (that) *sally*.

l. 24. *of a self-wise self-will*. In a forthright manner ('headstrong') does what is most true to man's nature as God intended it. Cf. 'The Handsome Heart' l. 7, 'its own fine function, wild and self-instressed'.

l. 25. *tufts of consolation*: intermittent comforting thoughts.

ll. 29–30. *strains / Us*: inspires us to effort.

l. 30. *freshyouth. A* read 'boyhood'. *C* read: boyboughs fretted in a flowerfall all portending / Fruit of sweet's sweeter ending

The image of a tree was one that G. M. H. frequently used for a youth (see 'The Loss of the Eurydice, l. 6, and 'On the Portrait of Two Beautiful Young People', l. 32, for example). Here Hopkins describes a fruit-tree, suddenly altered/ruffled (*fretted?*) by the loss of its blossom, whose fall indicates that fruit is developing, a metaphor for the growing dedication to Christ within the youths.

148 l. 37. *least me quickenings lift*=least quickenings lift me.

l. 46. *brandle*: shake.

l. 48. *forward-like*: presumptuous.   *like*: belike, probably.

Below *A* Hopkins noted that the soldier had been 'ordered to Mooltan in the Punjaub; was to sail Sept. 30'. On 8 October 1879 G. M. H. wrote to R. B., 'I enclose a poem, the Bugler. I am half inclined to hope the Hero of it may be killed in Afghanistan' (*L* I, p. 92).

Andromeda. Draft in *C*, autograph in *A* dated 'Oxford Aug.12 1879'. Text from corrected transcription in *B*, except layout, comma at the end of l. 12, 'barebill' (one word) and virgule (l. 14), from *A*.

In the Greek myth the princess, Andromeda, was chained to a rock as a sacrifice to a sea monster in exchange for its promise to stop ravaging her father's kingdom. Perseus, a hero with winged shoes that enabled him to fly and who was armed with a sword and the head of Medusa that turned to stone any who looked at it, saw Andromeda and rescued her.

It has been suggested that the poem is an allegory in which Andromeda represents the Catholic Church, which has been attacked through the centuries and seems doomed to be conquered by evil ('dragon food'). The 'wilder beast from West' may be the rise of industrialization, evolutionary theory, or Liberalism, which Newman condemned as the 'subjecting to human judgment (of) those revealed doctrines which are in their nature beyond and independent of it' (*Apologia*, Note A, see M. Moore, *Hopkins Quarterly*, vi, no. 3 (1979)). Perseus represents Christ, who will in time rescue the faithful. Hopkins may have had in mind the Day of Judgement or, more probably, the reconversion of England to Catholicism that he hoped for and frequently mentioned in his poems (see, for example, 'The Wreck of the Deutschland', stanza 34).

See *L* I, p. 87 (above, p. 238), 'Lastly I enclose . . .'.

Morning, Midday, and Evening Sacrifice. Draft in *C*, autograph in *A* dated 'Oxford Aug. 1879'. Text from corrected transcription in *B*, except title, l. 2 where G. M. H. added 'the', ll. 5–6 where he removed commas from before 'beauty' and 'freshness' (*L* II, p. 132), and layout, initial capital 'Nature', colon (l. 9), from *A*.

l. 1. *die-away.* Suggests subtle colouring and smooth shape.

l. 2. *wimpled.* Curved, Cupid's bow of the upper lip.

149 l. 6. *fuming.* Suggestive of the fleeting, almost transparent quality of childhood beauty. See *J*, p. 220, 'fuming of the atmosphere marked like the shadow of smoke'.

ll. 13–14. See *S*, pp. 240–1; prose, pp. 291–2.

l. 17. G. M. H. explained, ' "*In silk-ash kept from cooling*". I meant to compare grey hairs to the flakes of silky ash which may be seen round wood embers burnt in a clear fire and covering a "core of heat", as Tennyson calls it' ('In Memoriam', cvii), *L* I, pp. 97–8.

l. 18. *rind.* The outer covering of the ember.

l. 21. See *L* I, p. 98, 'Come, your offer of all this (the matured mind), and without delay either!'

Peace. Curtal sonnet in alexandrines. Draft in *C* dated 'Oct. 2 1879'. Transcription in *A*. Text from autograph in *B*. Layout from *C*. G. M. H.'s concessionary revision to l. 4 made to R. B. on 24 Aug. 1884 (*L* I. p. 196) has not been incorporated. Hopkins wrote the poem out of the unsettling knowledge that the following day he was to move yet again, from Oxford to a new post as preacher in the industrial, northern town of Bedford Leigh. *Curtal*: see Author's Preface, pp. 109 and 246–8.

l. 2. G. M. H. remarked that in 'under be' he had reversed the order of the words for the sake of the rhythm (*L* I, p. 196).

l. 7. *reave:* rob (*L* I, p. 196).

l. 9. *plumes to.* Matures into, as a fledgeling develops adult plumage.

150 *At the Wedding March.* Two drafts in *C*, autograph in *A* marked 'sprung rhythm' with numerous great colons. Text from corrected transcription in *B*

(dated 'Bedford, Lancashire. Oct. 21 1879'), except initial capitals in title 'Wedding' (from *A*), 'March' (editorial), l. 9 'March' (from *A*).

l. 3. *scions*. Shoots of plants, especially for grafting. See the Catholic mass for a bridegroom and bride: 'Thy wife shall be fruitful as the vine that grows on the walls of thy house. The children round thy table sturdy as olive branches.'

ll. 10–12. See *S*, p. 35, Christ is 'the true love and the bridegroom of men's souls' whose sacrifice offers man eternal life (N. H. M.).

Felix Randal. Autograph in *A* dated 'Liverpool. April 28 1880', and marked 'sprung and outriding rhythm; six foot lines'. Autograph pasted into *B*. Text from corrected transcription, *B²*, except layout, punctuation (ll. 5, 13), hyphen 'big-boned' (l. 2), 'though' (l. 6), from *A*.

    Metrical marks in *B²*: l. 1 Randal . . . farrier . . . then l. 3 pining . . . in it l. 5 him . . . Impatient l. 7 earlier l. 8 to him. Ah well l. 11 child, Felix, l. 12 of l. 13 random l. 14 drayhorse .

l. 6. See *S*, pp. 248–9; prose, pp. 298–9.

l. 7. *sweet reprieve and ransom*. Holy communion recalling Christ's sacrifice.

l. 8. *God rest him*. God forgive him.

all road ever . . . Lancashire dialect, 'for whatever (sins) he committed'.

l. 14. *sandal*. The technical name for a particular type of horseshoe (N. H. M.).

151 *Brothers*. Draft in *C*. Two autographs in *A* dated 'Hampstead. Aug. 1880'. Text from corrected transcription in *B*, except commas (l. 12) from *A²* and 'through' (l. 21), comma (l. 28) from autographs. *A²* is marked 'sprung rhythm; three feet to the line; lines free-ended and not overrove; and reversed or counterpointed rhythm allowed in the first foot'.

See *L* I, p. 86 (14 Aug. 1879); prose, p. 238: 'I hope to enclose a little scene . . .' Jack may have played the part of a herald in a single-act farce called 'A Model Kingdom', which was performed after a two-act play (l. 29). The eleven-year-old might have had to play a few notes on a trumpet (ll. 33–4), which further explains Harry's apprehension, and makes the description of Jack as 'brass-bold' (l. 25) still more appropriate (Fr. Francis Keegan, SJ, *Hopkins Quarterly*, spring 1979, p. 26). See Introduction, p. xxviii.

152 *Spring and Fall*. Autograph in *A* dated 'Lydiate, Lancashire. Sept. 7 1880' and marked 'sprung rhythm'. Autograph in *D*, sent to Dixon in a letter of 22 Dec.–16 Jan. 1881 (*L* II, p. 42), see prose, p. 244. This version is quoted accurately in *L* II, p. 174. Text from corrected transcription in *B*, except title, italics (l. 9), from *A*. Hopkins wrote to R. B. that the poem was 'not founded on any real incident' (*L* I, p. 109). The poet suggests that children intuitively feel the sorrow of death. As adults, he says, they will understand that the source of this sorrow is their own mortality, which since the Fall all men must experience (Genesis 3: 19).

l. 2. *unleaving*: losing its leaves.

l. 8. *A* read 'Though forests low and leafmeal lie'. See *J*, p. 239, 'Wonderful downpour of leaf: when the morning sun began to melt the frost they fell at one touch and in a few minutes a whole tree was flung of them; they lay masking and papering the ground at the foot. Then the tree seems to be looking down on its cast self as blue sky on snow after a long fall, its losing, its doing'.

wanwood. Perhaps from 'wann', Old English for 'dark' but I prefer the associations suggested by 'wan', pale from fatigue or sorrow.

leafmeal. Perhaps a more specific version of 'piecemeal'; the tree's 'cast self' disintegrating as fallen leaves. See the final note to 'Binsey Poplars'.

l. 13. *ghost*: spirit.

Latin Version of Dryden's Epigram on Milton. From the *Stonyhurst Magazine*, vol. i, no. 2, July 1881. The original was first printed anonymously below a portrait of Milton on the frontispiece to the fourth edition of *Paradise Lost*. It reads:

> Three *Poets*, in three distant *Ages* born,
> *Greece*, *Italy* and *England* did adorn.
> The *First* in loftiness of thought Surpass'd;
> The *Next* in Majesty; in both the *Last*.
> The force of *Nature* cou'd no farther goe:
> To make a *Third* she joynd the former two.

153 *Inversnaid*. Transcription in *A*. Text from autograph in *H* dated 'Inversnaid Sept. 28 1881'. Inversnaid is a small settlement in the Scottish Highlands. G. M. H. described his visit to W. M. Baillie (7 September 1887): 'I hurried from Glasgow one day to Loch Lomond. The day was dark and partly hid the lake, yet it did not altogether disfigure it but gave a pensive or solemn beauty which left a deep impression on me. I landed at Inversnaid (cf. Wordsworth and Matthew Arnold) for a few hours . . .' (*L* III, p. 288).

l. 3. *coop*. Perhaps a hollow; 'Rushing streams may be described as inscaped ordinarily in pillows—and upturned troughs' (*J*, p. 176).

comb. Water pouring over a rock so that it forms ridges. See *J*, plate 28, 'At the baths of Rosenlaui'. The picture elucidates both 'comb' and 'flutes'.

l. 4. *flutes*. The architectural meaning applied to ridges of water, used here to suggest the appearance of grooves in the falling plane of water.

l. 6. *twindles*. Perhaps Lancashire dialect 'twins', the foam divides into two or doubles. Alternatively the verb may be a combination of 'twine' (to coil) and 'spindle' (to grow into a long, slender form).

l. 9. *degged*. Lancashire dialect for 'sprinkled', 'bedewed'.

l. 11. *heathpacks*: heather.

flitches. Stiff, browned fronds like thin strips of tree trunk (N. H. M.).

l. 12. *beadbonny ash*. Mountain ash, which has lots of red berries in the autumn.

ll. 13–16. See *L* I, pp. 73–4: 'I have . . . something, if I cd. only seize it, on the decline of wild nature, beginning somehow like this—O where is it, the wilderness . . .' (printed above, p. 140).

Angelus ad virginem. Text from the *Month*, January 1882, pp. 100–11. G. M. H. wrote to his mother (24 December 1881): 'Father Richard Clarke, late Fellow of St. John's, has now become editor of the *Month* our magazine. A learned friend of his, while routing in the British Museum, came, in a MS of the 13th century, on a hymn to the Blessed Virgin in Latin and English, with the music, which is quite easy to read to anyone that knows a little about plainsong. He shewed it to Mr. Furnival and also to Fr. Clarke, both of whom want to publish it, but Clarke will be first in the field. It is appearing in the January *Month*, which I believe is already out. He had written an article on it; the music is harmonised by the organist at Farm Street; the Latin and English text are carefully reproduced; and a modernisation of the latter follows, made originally by me but altered since, perhaps not altogether for the better. The footnotes on the old english are mostly by me. The music is interesting from its great age besides being striking and pretty' (*L* III, p. 161). 'This hymn (MS Arundel 248, leaf 154) was written about 1260 and was the Angelus ad virginem sung by the clerk Nicholas in Chaucer's "The Miller's Tale" . . . The Middle English text appears as "Gabriel's Greeting to Our Lady" in *English Lyrics: 13th Century* (ed. Carleton Brown, 1932), no. 44' (4th edn.).

The notes on the Middle English text are:

'The English metre reproduces the Latin exactly, as will be found when allowance has been made for the *e* of inflection, which is sounded or dropped freely. In the English, however, there is this curious addition. The 8th and 10th lines of each stanza have in the Latin a peculiar grace of cadence, due to the way in which, though consisting properly of three (accentual) iambic feet of two syllables, they naturally fall into two feet of three syllables (mostly dactylic feet), and so give rise to a counterpoint rhythm. It was not possible to reproduce this grace in the English, owing to the shortness of the words, which are much more often monosyllables than in Latin. In compensation the translator has aimed at another effect: in each 8th and 9th line he seems to have supplied mid-line rhymes, at the 3rd syllable of the line, and by this means also of necessity changed the rhythm, breaking the line up into two shorter lines of three syllables each, and giving at the break or place of meeting the kind of rhythmic effect called *antispastic* or reversing. This may have been suggested to him by the phrasing of the music at this place. Thus he has:

Flésh of *thée*, | máiden bríght, ‖ mánkind *frée* | fór to máke—
Áll man*kín*' | wórth ybóught ‖ Thórough *thí*[ne] | swéet childíng—
Thát I *síth* | hís will is ‖ Maíden *with* | óuten láw—
Whére through *ús* | cáme God wón ‖ Hé bought *ús* | óut of pain—
Ús give *fór* | thínë sáke ‖ Hím so *hére* | fór to sérven.

l. 3. The accentuation *blissfúl tidíng* is no licence. All compound words, even words compounded with suffixes like *ly* or *ness*, have in verse a variable accent in old English. The accent is also often found on the last syllable of words ending in *ing* or *er*.

154 l. 18. [al manken wrth ibout]: This seems to be *All mankind worth ybought*, All mankind becomes bought, comes to be redeemed.

l. 29. [maiden withhuten lawe]: *Withouten law*, contrary to the general law of humanity. The Blessed Virgin was not at this time without law in the sense of being free of a husband, for her so-called espousals were in fact marriage.

l. 38. [war þurw us kam god won]: *Where through us came God won*. Perhaps: Through whom God came to be won over, reconciled, to us,—or, came to be one with us.

l. 42. [Maiden moder makeles]: Although there is a German word *makel*, a spot, and its derivative *makellos*, immaculate, yet *makeles* would seem to be nothing but *matchless*, from *make*, the older form of *mate*.

155 *The Leaden Echo and the Golden Echo*. Autograph in *A* dated 'Stonyhurst. Oct.13 1882'. G. M. H. wrote on it, 'I have marked the stronger stresses, but with the degree of stress so perpetually varying no marking is satisfactory. Do you think all had best be left to the reader?' Text from corrected transcription in *B*, except layout, *Golden Echo*: no comma after 'Somewhere', brackets round 'ah well where!' deleted (l. 6), accent 'prizèd' (l. 8), wimpledwater (one word) (l. 10), maiden manners (two words) (l. 15), semicolons after 'sighs' (l. 17) and 'See' (l. 20), comma after 'Nay' (l. 22), hyphens 'heavy-headed' (l. 24), 'care-killed' (l. 26), l.c. 'do' (l. 30), from *A*. *B* is dated 'Hampstead. 1881'.

Metrical marks in *B*: l. 1 kĕep . . . lâce   l. 4 Dôwn   *Golden Echo*: l. 7 One   l. 9 dĕarly   l. 12 môre   l. 17 sighs.

As early as 5 Sept. 1880 Hopkins wrote to R. B., 'You shall also see *The Leaden Echo* when finished. The reason, I suppose, why you feel it carry the reader along with it is that it is dramatic and meant to be popular. It is a song for St. Winefred's maidens to sing.' He announced in early Oct. 1882 (*L* I, p. 153) that he had finished the poem and was pleased with it. In Nov. 1882 (*L* I, p. 161) he wrote: 'I cannot satisfy myself about the first line. You must know that words like *charm* and *enchantment* will not do: the thought is of beauty as of something that can be physically kept and lost and by physical things only, like keys; then the things must come from the *mundus muliebris*; and thirdly they must not be markedly oldfashioned. You will see that this limits the choice of words very much indeed. However I shall make some changes. *Back* is not pretty, but it gives that feeling of physical constraint which I want.'

When offering to send the poem to Dixon, Hopkins noted of it, 'I never did anything more musical' (*L* II, p. 149). A somewhat similar theme can be seen in 'Morning, Midday, and Evening Sacrifice' (p. 148).

ll. 3–4. *frowning . . . Down*: driving away with disapproval.

rankèd. Ranks of, strongly marked.

l. 8. The wise are the first to realize that physical beauty will fade with age.

Golden Echo:

l. 1. *spare*. A mechanical echo of 'Despair' marking a change in attitude between the two sections of the poem.

l. 10. Fresh, fleeting beauty.

l. 11. *fleece of beauty*. 'The velvetiness of rose-leaves, flesh and other things' (*L* I, p. 215).

156 l. 21. See Matt. 10: 30, quoted by G. M. H. in *S*, p. 252, prose p. 300.

ll. 22–5. See *L* I, p. 159: ' "Nay what we lighthanded" etc. means "Nay more: the seed that we so carelessly and freely flung into the dull furrow, and then forgot it, will have come to ear meantime" etc.'

l. 26. *fagged*: tired, worn out.   *fashed*: (Scot.) anxious.

cogged: perhaps 'vexed', 'blocked'.   *cumbered*: overladen.

ll. 27–30. See *S*, p. 89: God 'takes more interest . . . in a lover's sweetheart than the . . . lover, in a sick man's pain than the sufferer, in our salvation than we ourselves'. The Catholic doctrine of the Resurrection promises believers that their human bodies will be restored for eternity but perfected and consequently the surest way of preserving physical beauty is to dedicate one's earthly body and life to God.

Ribblesdale. Drafts in *C*. Autograph in a letter to Dixon of 25–9 June 1883, printed *L* II, p. 108. Autograph in *A*, which prefaces the poem with the Latin of Rom. 8: 19–20, 'Nam expectatio creaturae . . .'. Below this version G. M. H. wrote, 'It is to be read very pausingly, the voice especially dwelling where the native rhythm overlaps into another line. It is common rhythm counterpointed'. Text is from autograph in *B* dated 'Stonyhurst. 1882' and marked, 'Companion to no. 10' ('In the Valley of the Elwy'), except layout from *A*.

Metrical marks in *B*: l. 1 landscape   l. 10 the heir.

Nam expectatio . . . The Douay version paraphrases Romans 8: 19–21: 'He (St Paul) speaks of the corporeal creation, made for the use and service of man and, by occasion of his sin, made subject to vanity, that is, to a perpetual instability, tending to corruption and other defects: so that by a figure of speech it is here said to groan and be in labour and to long for its deliverance, which is then to come, when sin shall reign no more and God shall raise the bodies and unite them to their souls, never more to separate and to be in everlasting happiness in heaven.'

l. 1. *throng*. See *L* II, p. 109, 'I mean "throng" for an adjective as we use it here in Lancashire' ('thick or crowded', *EDD*). Cf. 'All the herbage enthronged with every fingered or fretted leaf' (*J*, p. 172).

l. 2. *louchèd*. See *L* II, p. 109, ' "louchéd" is a coinage of mine and is to mean much the same as slouched, slouching'.

ll. 6–7. *deal . . . down*. To treat badly, unfairly.

157 l. 11. *selfbent*. Absorbed by his own self-centred desires. Contrast l. 24 of 'The Bugler's First Communion' (p. 146).

ll. 13–14. *this bids . . . concern*. This is why the earth looks so troubled.

'*A Trio of Triolets*'. Printed in the *Stonyhurst Magazine*, vol. i, no. 9 (March 1883), p. 162. They were signed with the pseudonym, BRAN. G. M. H. liked numbers 1 and 3 (*L* I, pp: 190, 317–18), but he considered that no. 2 'was not good, and they spoilt what point it had by changing the title'. The original title is now lost.

No. 1 λέγεταί . . . 'Is there any news today?', Demosthenes, *First Philippic*, 10.

158 *The Blessed Virgin compared*. Autograph in *A* dated 'Stonyhurst. May 1883'. Text from autograph in *B* (similarly dated), except layout, hyphen in 'world-mothering' from *A*. See *L* I, p. 179 (11 May 1883): 'We hang up polyglot poems in honour of the Blessed Virgin this month. I am on one in English in three-foot couplets . . . It is partly a compromise with popular taste, and it is too true that the highest subjects are not those on which it is easy to reach one's highest.' He told Dixon that the poem was 'in the same metre as "Blue in the mists all day" . . .' (*L* II, p. 108).

l. 5. *frailest-flixed*. From 'flix', the fur of various quadrupeds. Hopkins often uses it to describe clouds (see *J*, pp. 153, 192, for example).

ll. 22–33. Papal Bull *Gloriosae Dominae* (1748) compares Mary to 'a heavenly stream through which the flow of all graces and favours reach the soul of every wretched mortal'.

159 l. 37. Mary is called Mother of Mercy. She is seen as Queen of Heaven, possessing the power of dispensing grace but not punishment.

ll. 40–1. 'God has allowed her prayers to dispense his beneficent care.'

ll. 46–8. Christ's statement to John (John 19: 26–7) is often interpreted as a proclamation that Mary is the spiritual mother of the human race.

ll. 53–4. Christian doctrine states that belief in Christ is essential for salvation. (See next note on Christ's mystical body.)

ll. 60–72. Christ's presence on earth after the Resurrection continues within his believers. Contemplation of his divine conception and birth (at Nazareth and Bethlehem respectively) creates new and reaffirms old believers in the Catholic faith. Since Christ's mystical body is his Church of believers on earth, an increase in the number of believers 'more makes' or increases Christ (God's and Mary's Son) both here and in redemption. The dedication to Christ is also made 'more' in each believer.

160 ll. 75–80. See *J*, p. 154 (30 Aug.); *S*, p. 29; prose, p. 275.

l. 103. *god of old*. God of the old dispensation, just but not merciful.

ll. 104–13. Mary is an intermediary between the might of Christ risen and mankind. She is perfect, the only woman completely without sin, and more powerful than all mortals but of lower rank than Christ with whom she intercedes on man's behalf.

161 *'The times are nightfall'*. Text from undated draft in *H* (1885–6?). For the thought, see *L* I, p. 221 (1 September 1885) for example: 'in the life I lead now, which is one of a continually jaded and harassed mind, if in any leisure I try to do anything I make no way—nor with my work, alas! but so it must be.' Hopkins does not appear to have written the sonnet's final three lines.

l. 1. Earlier '. . . nightfall, light of heaven grows less'.

l. 4. Originally 'More make or plainer publish our distress' (cancelled).

l. 5. Originally 'I cannot help', altered to 'Could I but help'.

l. 8. *does dear*: does make (forgetfulness) dear or precious.

ll. 9–11. See 'The Candle Indoors', ll. 9–11 (p. 144).

l. 10. *dragons*. Symbolic of evil; see *S*, pp. 198–9.

St. Winefred's Well. No autographs extant except for Act II, ll. 1–30 in the Dublin notebook. Text of rest from transcription in *A* by Mrs Bridges, but with conventional capitalization. Transcription in *C* is by the poet's mother and was made from *A*.

Metrical marks in *A*: l. 6 ready in his l. 10 Áh l. 11 No man l. 13 mőre times l. 15 Támpering . . . draws l. 16 Whát l. 18 all, all l. 19 nőt l. 24 főol Act II, l. 1 heárt . . . séen l. 3 őff l. 5 Mőnuments . . . rěcords l. 6 whéreas l. 7 Wárned l. 9 Perháps l. 10 mǎkebelieve . . . mőck l. 11 here, here . . . swéats l. 15 cǎn . . . thée ín thy dárk lair; thése drops l. 17 woéful l. 19 rőll l. 23 tó l. 27 lífted, ímmortal . . . immőrtal l. 32 nót l. 34 I . . . dóne víolent l. 35 líke a líon dóne, líonlíke down l. 40 Lóyal . . . own . . . láying . . . own . . . dówn, no láw l. 41 Lőrd . . . hím l. 42 only l. 43 Ŏnly l. 44 líke l. 45 wíll . . . can flésh l. 49 cómfort whére can Í l. 50 time's one rich rose, mý hand l. 52 dáshed down l. 53 nő l. 54 passion-sake. Yes l. 55 nót . . . yét hope ón l. 60 hér neck l. 63 kínd . . . kéen l. 65 Whát do nów . . . Dó l. 66 Déed-bound . . . óne . . . tréads . . . dówn (C) l. 4 thírst l. 5 lípmusic . . . lőst l. 14 rőck l. 15 fráil l. 19 Wǎles l. 21 mőre pilgrims l. 24 cǎme l. 25 nǎme l. 26 bóons . . . háven the héart l. 28 new-dapple l. 29 Amőngst come-back-agǎin . . . thíngs with.

Note that single stress-marks in the text represent double stresses in the MS. See also p. xli.

St Winefred. *c*. 600–60. Winefred was born into a wealthy family in Holywell, Wales. She was guided by her uncle, St Beuno, to dedicate herself to an austere and religious life. Her reputation for wisdom and purity reached the ears of Caradoc, the son of a neighbouring prince. He pressed Winefred to marry him and when she, fearing for her chastity, tried to flee from him, cut off her head. The head rolled down a steep gully and where it came to rest a spring suddenly appeared. St Beuno prayed successfully for Winefred's recovery. He also asked that Caradoc be punished and he was struck dead.

St Winefred became abbess of a convent at Holywell and later at Gwytherin (*Cath. Encycl.* 1910).

G. M. H. was most enthusiastic about St Winefred's well. He wrote to R. B. of her famous spring that it 'fills me with devotion every time I see it and would fill anyone that has eyes with admiration, the flow of ἀγλαὸν ὕδωρ [beautiful water] is so lavish and so beautiful' (*L* I, p. 40, 3–8 April 1877). In 1879 (8 October) he told R. B. that he was writing a drama about the subject in alexandrines. He hoped to be able to send the murder scene and some more soon. 'I mean [it]', he added, 'to be short, say in 3 or even 2 acts; the characters few.' The greatest problems, he found, were not in writing the tragic or stirring scenes, but the 'minor parts' (*L* I, p. 92). The play was not finished although G. M. H. worked at it periodically for seven years. Most of what is extant appears to have been written between October 1884 and April 1885.

Of the rhythm G. M. H. told Dixon, 'It is in an alexandrine verse, which I sometimes expand to 7 or 8 feet, very hard to manage but very effective when well used' (*L* II, p. 143). To R. B. he elaborated, 'I hold that each half line is by nature a dimeter, two bars or four feet, of which commonly one foot is silent or lost at the pause. You will find it sometimes employed in full . . . You will see that as the feeling rises the rhythm becomes freer and more sprung: I think I have written nothing stronger than some of those lines' (*L* I, p. 212).

l. 2. *Caerwys*. G. M. H. notes 'In English pronounced *Caris*, like *heiress*'.

162 l. 9. *Gwenvrewi*. Winefred's original, Welsh name.

163 Act II, l. 17. *Cradock*. This spelling, which gives approximately the Welsh pronunciation and stress of *Caradoc*, appears in the Dublin notebook MS.

166 *The Sonnets of Desolation*. Most of these were found after Hopkins's death, although it is clear from his letters that he had intended to send at least some of them to R. B. Late in May 1885 he wrote, 'I have after long silence written two sonnets, which I am touching: if ever anything was written in blood one of these was' (*L* I, p. 219), and see his letter written in September of the same year (*L* I, p. 221; prose, p. 263). Critics do not all agree about which poems should be covered by the title, which was not chosen by G. M. H., but generally included are: 'To seem the stranger', 'I wake and feel', 'No worst', (Carrion Comfort), 'Patience, hard thing', and 'My own heart'. The sequence is uncertain, since most of them exist only as fair copies on a single sheet of sermon-paper, but they were all most probably written in 1885–6.

'*To seem the stranger*'. Text from autograph in H. Transcription in *A*. Single stresses in the text represent double stresses in the MS; see p. xli.

Metrical marks: l. 7 Ĭ wēar   l. 9 Í . . . thírd   l. 14 únheèded.

ll. 2–3. G. M. H.'s family was not Catholic.

l. 4. See Matt. 10: 34–7.

l. 5. See *L* I, p. 231 (13 October 1886); prose, p. 264.

ll. 6–7. (England) '*would neither hear / Me, were I pleading*', perhaps alluding to his hopes that England would become Catholic again; cf. 'The Wreck of the Deutschland' stanza 35, and 'The Loss of the Eurydice' ll. 97–104, for example.

l. 8 and ll. 11–14. See *S*, p. 262 (prose, pp. 302–3) and *L* I, p. 231 (prose, p. 264) and p. 270 (prose, p. 270), 'It is now years . . .'

ll. 9–10. *third remove.* Perhaps the first remove was the partial estrangement with his family brought about by their holding different religious beliefs (ll. 2–4). The second remove may have been from the English people who were mostly Anglican but whom Hopkins longed to see converted to Catholicism (ll. 5–7). In Ireland, although G. M. H. found himself among Catholics, they were disloyal to England (see *S*, p. 262).

l. 14. *began.* Probably noun, suggesting someone whose early promise has not been fulfilled.

'*I wake and feel*'. Cancelled draft of sestet in *H*. Text from autograph fair copy in *H*. This may well have been the sonnet 'written in blood' (*L* I, p. 219). (N. H. M.)

Metrical mark: l. 2 hõurs we.

l. 1. *fell.* 'blow', but also 'an animal's pelt'; see *J*, p. 174, 'the deep fell of some other animal . . .' See Job 7: 4.

l. 2. *black.* 'Miserable' as well as 'dark'.

l. 7. *dead letters*: undelivered letters.

ll. 9–14 earlier read:

> I am gall and heartburn. God's most deep decree
> Has me taste bitter, and my taste is me.
> My bones build, my flesh fills, blood feeds/ this curse
> Of my selfstuff, by self yeast soured. I see
> The lost are like this, with their loss to be
> Their sweating selves, as I am mine, but worse.

ll. 9–10. *God's most deep decree / Bitter would have me taste.* See *L* II, pp. 108–9 (25 June 1883), 'I see no grounded prospect of my ever doing much not only in poetry but in anything at all. At times I do feel this sadly and bitterly, but it is God's will . . .'

l. 10. Cf. *S*, p. 243, 'sight does not shock like hearing, sounds cannot so disgust as smell, smell is not so bitter as proper bitterness, which is in taste', from G. M. H.'s notes on *The Spiritual Exercises.*

ll. 12–14. *I see . . . but worse.* See *S*, pp. 241–4 (prose, pp. 292–5), which makes it clear that the punishments endured by 'the lost' in hell are directly related to their sins: 'we are our own tormentors, for every sin we then shall have remorse and with remorse torment.'

167 '*Strike, churl*'. Text from a fragment in *H* written on the same page as a cancelled draft of the sestet of 'I wake and feel'. May? 1885. Transcription in *A*. G. M. H. wrote to his mother (17 May 1885), 'The hail today lay long like pailfuls of coarse rice' (*L* III, p. 171). (N. H. M.)

l. 4 *have at*: attack, beat; earlier 'aim at'.

'No worst'. Text from autograph draft in *H*. Transcription in *A*. Metrical marks: l. 6 wórld-s͡orrow; on an̑ l. 8 f͡ell.

l. 1. *Pitched past pitch*. Earlier drafts read: 'grief past pitch of grief', 'Grief past grief', and 'Grief tops grief'.

pitched. Either 'thrown' or 'tuned by stretching a string' as with a violin.

l. 2. *forepangs*: earlier experiences of pain.

l. 3. *Comforter*. The Paraclete; see *S*, pp. 70–1. Christ and the Holy Ghost are Paracletes, comforting and providing encouragement.

l. 6. *world-sorrow*. Perhaps the world-weariness of ennui, perhaps part of the projected amplification of feeling found elsewhere in ll. 5 and 6, or an acknowledgement that all sorrow derives from the Fall.

l. 8. *force*: perforce.

l. 10. Earlier read: 'Frightful, sheer down, not fathomed.'

l. 12. *durance*: endurance.   l. 13. Cf. *King Lear*, III. ii. 60–2.

To what serves Mortal Beauty? Drafts in *H*. Autograph in *D* (printed in *L* II, p. 129, with two small errors: title should be in lower case except for intial 'T', and 'see' (l. 3) should have upper case 'S'). Autograph in *A*, which I take to be later than *D*. This is marked 'common rhythm highly stressed'. Text from amended autograph in *B*, except capitalization in title from *A*, caesural marks editorial. *B* is dated 'Aug. 23 1885'. The ink used suggests that the alterations in *B* are of the same date as those to 'Harry Ploughman' and the copy in *B* of 'St. Alphonsus Rodriguez' (October 1888).

Metrical marks in *B*: l. 1 wha͡t serves   l. 3 See:͡ it ... ke͡eps warm   l. 4 Me͡n's wit ... t͡o what   l. 5 mo͡re may   l. 6 wind͡falls of wa͡r's storm   l. 11 me͡n's selves.

G. M. H. explained in *D* that 'the mark (⌒) over two neighbouring syllables means that, though one has and the other has not the metrical stress, in the recitation-stress they are to be about equal' (*L* II, p. 129).

l. 2. *the O-seal-that-so feature*. One draft reads: 'face feature-perfect'.

ll. 3–4. *keeps warm / Men's wit*. Earlier read: 'keeps warm men's thoughts to what things be'.

ll. 4–5. Earlier versions read: 'One clear glance / May gather more than staring out of countenance' and 'where a glance / Gather more may than gaze me out of countenance'.

ll. 6–8. In the sixth century Pope Gregory saw a group of English boys (Angli) for sale as slaves in a Roman market. The fair-haired beauty of the lads allegedly caused Gregory to compare them to angels (angeli) and he decided that it would be worth trying to convert pagan England to Catholicism. This possibility of salvation is the good fortune that G. M. H. calls 'that day's dear chance'.

ll. 9–11. Earlier read: 'Was man bid love, bid worship, block or barren stone? / Our law is love what are world's loveliest, were all known, / Most worth love, men's selves.'

l. 12. *own*: acknowledge.   l. 13. *home at heart*: deep in your heart (?).

l. 14. *grace* refers to beauty of actions, the result of belief in God.

See *L* I, pp. 95–6 (prose, pp. 240–1) on the different types of beauty, and *S*, pp. 35–6 (prose, p. 277).

168 (*Carrion Comfort*). Text from autograph drafts in *H*. Transcription in *A*. Undated but the earliest draft extant (cancelled) follows a draft of 'To what serves Mortal Beauty' dated 'Aug. 23 '85'. A second draft precedes the only draft of (*The Soldier*), dated 'Clongowes Aug. 1885'. The final draft, which is probably from 1887 since it is written around an early draft of 'Tom's Garland', ends after l. 12. The last two lines are taken from the second draft.

Metrical marks: l. 1 comfort   l. 5 terrible   l. 8 O in . . . tempest . . . there . . . frantic to avoíd   l. 11 rather . . . lo . . . stole   l. 12 The hero whose heaven-handling   l. 13 him . . . which one . . . each one   l. 14 wretch .

l. 1. *carrion comfort*. Despair, the false comfort of abandoning all effort.

l. 4. Originally read: 'Can hold on, hope for comfort; not wish not to be.' Stop from earlier version.

ll. 5–6. An early draft read, 'Yet why, thou terrible, wouldst thou rock rude on me / ~~With~~ Thy wring-earth tread; launch lion-foot on me?' See Job 9: 6, God 'Which shaketh the earth out of her place, and the pillars thereof tremble'. In the final draft G. M. H. wrote 'wring-world' adding as an alternative 'wring-earth' (which he had used in the earlier drafts) before completing the line. See Appendix *C*.

lionlimb. See Job 10: 16, 'Thou huntest me as a fierce lion.'

l. 7. See Job 7: 8, 'thy eyes are upon me, and I shall be no more' (Douay).

fan. to separate the kernel or grain of wheat from the chaff or husk.

l. 8. see Job 9: 17, 'For he breaketh me with a tempest, and multiplieth my wounds without cause.'

l. 9. In Matt. 3: 12 and Luke 3: 17 John the Baptist prophesied that Christ would divide mankind in a similar way. See *S*, pp. 267–8 (prose, pp. 304–5) for G. M. H.'s comments on the second passage.

l. 10. Originally 'because I kissed the rod'. 'Kissed the rod' may mean 'accepted depression, despair as a test set by God'. See *S*, p. 256 (June 1884), 'Take it that weakness, ill-health, every cross is a help' towards greater spirituality (N. H. M., *Reader's Guide*, p. 175).

l. 11. Earlier read, 'Nay from the storm my heart stole joy . . .'

l. 12. An early draft read, 'hero whose heavenforce'.

ll. 13–14. See Genesis 32: 24–30.

(*The Soldier*). R. B.'s title. Text from autograph in *H* dated 'Clongowes Aug. 1885'. Transcription in *A*. Caesura l. 6 from earlier version. Single stress-marks in the text represent double or strong stresses in the MS. See p. xli.

Metrical marks in *H*: l. 1 Whý ... áll, séėing ... sóldier, bléss ... bléss l. 2 rédcoats, óur ... Bóth thêse béing, the gréater párt   l. 3 Here it is: the héart   l. 5 mākesbelieve   l. 6 fáncies, féigns, déems, dêars ... ártist áfter his árt   l. 8 scárlet wêar ... spírit of wár thêre expréss   l. 11 séėing somewhére ... mán   l. 12 lóve he lêans ... néeds ... néck ... fáll ... kíss.

l. 3. G. M. H. had acted as confessor at the Cowley Barracks, Oxford.

l. 6. *dears* ... *after*: values ... as.

l. 9. See *S*, p. 70, 'he (Christ) led the way, went before his troops, was himself the vanguard, was the forlorn hope, bore the brunt of battle alone, died upon the field, on Calvary hill, and bought the victory by his blood. He cried men on ...' Cf. the Ignatian prayer: 'To give and not to count the cost / To fight and not to heed the wounds / To toil and not to seek for rest / To labour and not to ask for any reward / Save knowing I do thy will.'

l. 13. *Christ-done deed*. Both a deed such as Christ would have done, and a deed that maintains Christ's presence on earth.

God-made-flesh. Christ incarnated.

169 '*Thee, God, I come from*'. Text from unfinished draft in *H*. Verso has cancelled drafts of 'To what serves Mortal Beauty?'. Transcription in *A*. A hymn reminiscent of George Herbert.

l. 6. See note to 'The Wreck of the Deutschland', l. 8.

170 '*Patience, hard thing*'. Text from autograph in *H*. Transcription in *A*.

Metrical marks: l. 8 Purple eyes   l. 11 dȯ.

See *S*, p. 204, *Spiritual Exercises* VIII: 'Let him who is in desolation strive to remain in patience, which is the virtue contrary to the troubles which harass him; and let him think that he will shortly be consoled, making diligent efforts against the desolation ...'

ll. 2–4. Patience is a quality that one develops in adversity.

ll. 6–7. Ivy is a vigorous plant that will grow over and hide the ruins of buildings.

l. 8. *Purple eyes*. Berries.

ll. 9–11. See *S*, p. 254 (9 September); prose, p. 301.

l. 10. *dearer*: both more severely, and spiritually more valuable, as in note to ll. 9–11.

l. 14. *combs*: honeycomb.

'*My own heart*'. Text from autograph in *H*, with later revision of ll. 13–14 (*H*). Transcription in *A*.

Metrical marks: l. 13 you; unforseentimes rather.

ll. 6–7 are highly elliptical: comfortless (world) . . . (find) day or thirst can find (ease/drink).

l. 8. *all-in-all*: everything; cf. Coleridge's 'Rime of the Ancient Mariner', ll. 119–22.

l. 9. *Jackself*: everyday, working self. Perhaps an allusion to the saying, 'all work and no play makes Jack a dull boy.'

l. 11. *size*: grow. See 'The May Magnificat', l. 25.

l. 14. *Betweenpie*. Probably a verb, suggesting either that the sky seen between mountains makes a pied or variegated pattern with them, or that sunlight falling into a valley changes the colour of those parts of the mountains it touches.

171 *To his Watch*. Text from undated draft in *H*. Late transcription in *A*.

l. 2. Originally 'With cold beat company, shall you, or I,'.

l. 3. Originally 'Earlier, undone give o'er our work and lie'. See Appendix C.

l. 4. *ruins*. Originally 'wreck'.

l. 5. Originally 'Telling of time our task is; yea ~~not~~ some part,'.

telling time: perhaps, as applied to the poet, it means making time tell— making good use of time (perhaps see Colossians 4: 5).

l. 6. *but we were*. Originally 'time, being both' (cancelled).

l. 7. (We have) *One spell* (on earth) *and well* (must) *that one* (be used).

ah. Earlier 'O'.

l. 8. Earlier alternative, uncancelled, reads 'Is comfort's carol of all or woe's worst smart'. Hopkins inserted 'sweetest' above 'comfort's', then cancelled it and rewrote the line as it now stands in the text.

l. 9. *field-flown, the*. Originally 'O see! The'. Cf. Isaiah 38: 6.

Songs from Shakespeare, in Latin and Greek. Text from autographs in *A*. To R. B., Hopkins said of them, 'Fr. Mat Russell of ours . . . who edits a little half-religious publication the *Irish Monthly*, wrote to me lately for an opinion of some Latin verses furnished him; and this led to . . . my suddenly turning a lot of Shakspere's songs into elegiacs and hendecasyllables (my Latin muse having been wholly mum for years) and sending him one copy (and the rest I believe I can and shall get published in the Trinity *Hermathena* by means of Mr. Tyrrell) . . .' (13 October 1886, *L* I, p. 230). He added on 21 October 1886, 'You will have seen that in one of the pieces were some phrases borrowed from Horace and Virgil. In original composition this is most objectionable, but in translation it is lawful, I think, and may be happy, since there it is a question of matching the best of one language with the best, not the newest, of another.

These verses cannot appear in *Hermathena*, which admits no translations. Mr. Tyrrell said he liked them much, but he did not himself approve of my Catullian rhythms. I employ them of choice, taking Catullus for my warrant only, not my standard, for metrically Catullus was very unsure' (*L* I, p. 232).

(i) Text is from *A*, which is less heavily punctuated than that printed in the *Irish Monthly* (vol. xv (1887), p. 92).

l. 10. *Nos*. The *Irish Monthly* reads 'Et'.

l. 11. *Adlatrant* [ed.]. *Irish Monthly* uses the spelling 'Allatrent'.

occinit, occinat. *Irish Monthly* reads 'occinit. Occinat'.

(ii) Text from *A* (printed in the *Irish Monthly*, vol. xiv (1886), p. 628).

l. 7. W. H. G. said that 'Phorcys' was metrically indefensible placed last and changed the line to 'Exsequias Phorcys, quod tu miraberis, illi'. N. H. M. and Dr Ross Kilpatrick suggest that Hopkins may have been following Catullus 66 ll. 3, 41, 61.

(iv) Text from *A*. At least one line has been cut out of the MS after l. 10.

(v) Text from *A²*. Below the title of *A¹* G. M. H. noted: 'Greek: Dorian rhythm, freely syncopated, as in drama.' He wrote to R. B., 'I added two metrical schemes to my Greek verses for you . . . one is fuller than the other.' The one for (v) fits only the first version (*A¹*) and is therefore omitted. For G. M. H.'s discovery about Dorian rhythm see *L* I, pp. 233–4. In the second version Hopkins added a second line to the first two strophes and altered a few words.

l. 3. R. B. amended καρδίαν, which belonged to an earlier version, to καρδίας.

I am grateful to N. H. M. for pointing out that the layout in (v) and (vii) needed emendation.

174 (vii) Text from *A*, where G. M. H. noted: 'Dorian rhythm, syncopated, and with triplets in resolution'. The metrical scheme which Hopkins sent to R. B. has an extra line:

σχῆμα

⏕ is placed over Ὀρφέως . . . δενδρεσὶν (l. 1), νιφοκτύπων. .δαμείσαις (l. 2), κελαδοῦντι . . . ἁλίου (l. 3), χιόνος . . . πόντιον

(l. 5), ἐριβρόμων . . . γαλάνᾳ (l. 6), κιθάρᾳ . . . παυσίλυπον
(l. 7). ⌐⌐ is placed over οὐρανίου (l. 4), ἀδύνατον (l. 8).
In his letter of 21 October Hopkins added, 'The Dorian rhythm . . . arises
from the Dorian measure or bar. The Dorian bar is originally *a march step in
three-time executed in four steps to the bar*. Out of this simple combination of
numbers, three and four, simple to state but a good deal more complicated
than any rhythm we have, arose the structure of most of Pindar's odes and
most of the choral odes in the drama. In strict rhythm every bar must have
four steps. Now since four were to be taken to three-time, say three
crotchets, (1) one crotchet had to be resolved, (2) only one at a time, and that
(3) never the last. Hence the two legitimate figures of the Dorian bar were
these: ∪∪—— (the rising Ionic) and —∪∪— (the choriambus).' He
admits certain irregularities: ∪∪—∪ (third paeon); —∪—— (second
epitrite); resolution of the first long into three shorts instead of two, 'exactly
as we employ triplets in music'; —∪—∪ (double trochee). 'When the
measure is more loosely used two new licences appear—syncopation, by
which syllables are lengthened so that three fill a bar and so that the last of
one bar becomes the first of the next; and triple resolution, so that a bar can
have five syllables. By means of syncopation the measure can be made
dactylic and practically brought into common time. The strict Dorian can
only be found in odes meant to be marched to.' Of his peculiar stressing he
says: 'Naturally the strongest place in the Dorian bar is the second crotchet,
not the first, and I have so marked it in the schemes I sent, but perhaps it
would be best to mark the first as strongest. . . .'
For a discussion of the effect of these metres on G. M. H.'s sprung and
outriding rhythms see Gardner, *Study*, ii, chapter 2.

Robert Bridges's 'In all things beautiful, I cannot see'. Text from *L* I, p. 242. Sent
to R. B. in a letter of 31 October 1886, headed 'first draught'. Below l. 12
G. M. H. wrote 'Here follow the first three lines of the sextet, which I do not
correctly remember: please send me them'. Under l. 16 he noted, 'They are
not satisfactory, I feel.' On 26 November he wrote, 'I have improved the
Latin of "In all things beautiful". I asked you to send me three missing lines
and you have not done so'.

l. 7. MS 'fabebor'.

R. B. later revised the poem for the 1889 edition of *The Growth of Love*, and
again for *Poetical Works* (1914), where it is Sonnet 31. The early version
which G. M. H. seems to have used, reads:

> In all things beautiful, I cannot see
>   Her sit or stand but love is stirred anew:
> 'Tis joy the foldings of her dress to view,
> And all she doth is past expectancy.
>
>   If she be silent, silence let it be;
> He who would bid her speak would sit and sue
> The deep-browed Phidian Jove to be untrue
> To his two thousand years' solemnity.

> Ah! but her launchèd passion when she sings
> Wins on the hearing like a shapen prow
> Borne by the mastery of its urgent wings.
>
> Or, if she deign her wisdom, she doth show
> She hath the intelligence of heavenly things
> Unsullied by man's mortal overthrow.

175 *Spelt from Sibyl's Leaves.* Drafts in the Dublin Notebook. Autograph of the first two lines in *C.* Autograph in *A* marked 'sprung rhythm: a rest of one stress in the first line'. Text from autograph in *B*, except caesura from *A*.

Metrical marks in *B*: l. 10 O̅ our.

Hopkins wrote to R. B., 'I have at last completed but not quite finished the longest sonnet ever made and no doubt the longest making. It is in 8-foot lines and essays effects almost musical' (*L* I, p. 245). When he sent it to R. B. on 11 December 1886, he added that it should be read with 'loud, leisurely, poetical (not rhetorical) recitation, with long rests, long dwells on the rhyme and other marked syllables and so on. This sonnet shd. be almost sung: it is most carefully timed in *tempo rubato*' (*L* I, p. 246). See *J*, p. 200 (prose, pp. 202–3): 'This busy working . . .'.

Title. An allusion to the opening verse of the 'Dies irae' in the Catholic Mass for the Dead: 'Day of wrath, that day when the world is consumed to ash as David and the Sibyl testify.' Cf. also *Aeneid* vi. 11–12, 268–72, 539–43.

l. 1. *earnest*: solemn.   *stupendous*: terrifying.

l. 2. The evening darkens into night that seems to envelop everything from the creation of the earth to its end. See Matt. 24: 29.

l. 3. *hornlight.* The yellow glow like the light emitted through a lantern's horn window.

hoarlight. See *J*, p. 199: 'the burnished or embossed forehead of sky over the sundown; of beautiful "clear" / Perhaps the zodiacal light'.

l. 4. *waste*: fade away.

l. 6. *throughther.* I am grateful to Mr Myrddin Jones for pointing out that this dialect word, 'through-other', means confused, disorderly. [See *EDD*.]

pashed: crushed by blows.   l. 7. *disremembering*: forgetting.

dismembering. See *J*, p. 236: 'darkness and despair. In fact being unwell I was quite downcast: nature in all her parcels and faculties gaped and fell apart, *fatiscebat*, like a clod cleaving and holding only by strings of root.'

round: rebuke.

l. 9. *dragonish.* See *S*, pp. 199 and 243.

damask: make a pattern like that on ornate metal swords.

ll. 11–12. See Matt. 25: 31–3.

ll. 13–14. *ware of.* Remember that there will be a 'world' in the Day of Judgement when the behaviour of all men will be judged to be either right or wrong and those condemned to hell will suffer; see *S*, p. 243, 'the worm of conscience, which is the mind gnawing and feeding on its own most miserable self' (Meditation on Hell).

176 *On the Portrait, &c.* Drafts in *H*. Text from autograph fair copy in *A* dated 'Monasterevan, Co. Kildare. Christmas 1886'. G. M. H. wrote to Dixon, 'I was at Xmas and New Year down with some kind people in Co. Kildare, where I happened to see the portrait of two beautiful young persons, a brother and sister, living in the neighbourhood. It so much struck me that I began an elegy in Gray's metre, but being back here I cannot go on with it' (*L* II, p. 150).

l. 3. Earlier read 'Rich runs the juice in violets and fresh leaves'.

ll. 3–4. In the portrait the children's faces are surrounded by a garland of flowers, vines, and fruit.

l. 7. *time's aftercast.* Earlier read 'fate's afterthrow'.

ll. 7–8. Perhaps 'as for the future, the children will be (for their parents) a source of effort, hope, risk, and interest.'

l. 8. Earlier read 'Things all of care, heft, hazard, interest.'

l. 12. *Barrow*: The river that runs through Monasterevan (see *L* I, p. 306).

ll. 23–4. See Matt. 19: 16–24.

Before l. 25 one draft read:

> Who yet was inward-lovely, bravèd well
> That world-breath's ransack nor wrestling nor stealth
> The least foil. How then? Rise he would not; fell
> Rather; he wore that millstone you wear, wealth.
>
> Ah, life, what's like it?—Booth at Fairlop Fair;
> Men/ boys brought in to have each our shy there, one
> Shot, mark or miss, no more. I miss; and 'There!—
> Another time I' . . . 'Time' says Death 'is done'.

l. 29. *eye*: (verb) 'see'. l. 30. *banes*: cause of trouble (OED).

ll. 29–31. Earlier read:

>   feast of
> your ~~lovely~~ youth and that most earnest air,
> They do but call your banes to more carouse;
> Worst ~~will~~ batten on best: (cancelled)

177 *Harry Ploughman.* Autograph in *A*, dated 'Dromore Sept. 1887' and sent to R. B. on 11 October 1887, has a plethora of metrical marks. There is a facsimile of it in *L* I, opposite p. 262. Fair copy (*D*) in BL (see *L* II, p. 153). Text from amended autograph in *B* (1888).

Metrical marks in *B*: l. 1 hurdle͡ arms l. 4 shŏuldér . . . shánk l. 6 stånd l. 8 Sŏared ŏr sánk l. 9 Though as a͡ . . . finds his, as at a

l. 10 ín ... whát ... he each ... do͡   l. 11 His͡   l. 12 to it, Harry͡   l. 13 In him ... quâil ... wallowing o' the ... crîmsons   l. 14 wînd lifteͅd wind-laced   l. 16 Chūrlsgrace ... chîld ... Amansstrength ... hângs or hūrls   l. 17 Them—broad ín ... híde ... fró̆wning féet ... ráced   l. 18 cold fūrls.

On *A* G. M. H. wrote 'Marks used:

(1) ∧ strong stress; which does not differ much from

(2) ⌒ pause or dwell on a syllable, which need not however have the metrical stress;

(3) ´ the metrical stress, marked in doubtful cases only;

(4) ∼ quiver or circumflexion, making one syllable nearly two, most used with diphthongs and liquids;

(5) ⌢ between syllables slurs them into one;

(6) ——— over three or more syllables gives them the time of one half foot

(7) ——— the outride; under one or more syllables makes them extra-metrical: a slight pause follows as if the voice were silently making its way back to the highroad of the verse'.

Hopkins wrote to R. B. on 28 September 1887 that he had done the whole of one sonnet ('Harry Ploughman') and most of another ('Tom's Garland'). 'The one finished is', he said, 'a direct picture of a ploughman, without afterthought.' On 11 October 1887 G. M. H. wrote to R. B. that he would enclose the sonnet on Harry Ploughman 'in which burden-lines [indented] (they might be recited by a chorus) are freely used: there is in this very heavily loaded sprung rhythm a call for their employment. The rhythm of this sonnet, which is altogether for recital, not for perusal (as by nature verse should be) is very highly studied' (*L* I, p. 263). He remarked that he wanted the ploughman 'to be a vivid figure before the mind's eye; if he is not that the sonnet fails' (*L* I, p. 265).

l. 1. *hurdle*: willow twigs twisted together.

flue. A woolly or downy substance (OED), here 'hair'.

l. 2. *rack*: (technical) ribcage (N. H. M.).

l. 3. *knee-nave*. Earlier read 'knee-bank' = knee-cap.

l. 7. *curded*: bulged.

l. 9. *beechbole*. Beech trees are known for their strength and straightness. Beech trunks were used for ship's masts.

l. 14. *crossbridle*: tangle.   l. 15. See p. 414.

l. 16. *Churlsgrace*: hardy, peasant gracefulness.

ll. 16–17. *Churlsgrace ... how it hangs or hurls them*. Grammatically: his churlsgrace controls his frowning feet broad-lashed in bluff hide as they race against the clods of earth turned by the plough.

frowning. Moving forward aggressively? W. H. G. suggested that it referred to wrinkles in the leather of Harry's boots.

l. 17. Hopkins wrote to R. B. (*L* I, p. 265, 6 November 1887), 'dividing a compound word by a clause sandwiched into it was a desperate deed, I feel, and I do not feel that it was an unquestionable success.'

l. 18. *cold furls*: earlier 'flame-furls'.

l. 19. Earlier read:

With-a-wet-sheen-shot furls.
and: With-a-wet-fire-flushed furls.

(*Ashboughs*). R. B.'s title. Text from the second version in *H*. 1887? (N. H. M.). Written first as a curtal sonnet. A fair copy of this is written above drafts of 'To seem the stranger' and 'I wake and feel'. A second version, written on another sheet below the final lines of 'My own heart', has an additional revision that expands the poem to thirteen lines. On MS *A* of 'Tom's Garland' G. M. H. noted, 'Heavy stresses marked double, thus ″ and stresses of sense, independent of the natural stress of the verse, thus ` '.

Single stress-marks in the text represent double stresses in the MS. See p. xli.

Metrical marks in MS: l. 1 Nót óf . . . wàndering   l. 2 ánything . . . mílk . . . mínd . . . só sìghs dèep.

l. 3 Poetry tó it, as a . . . brèak   l. 4 Say it is áshboughs: whether on a December   l. 5 ór . . . láshtender cŏmbs   l. 13 Heaven . . . chìlds things by.

See *L* III, p. 202 for G. M. H.'s early enthusiasm for ash trees.

178 ll. 5–6. See *J*, pp. 154 and 177 (prose, pp. 192 and 195); *J*, p. 222, 'The ashes thrive and the combs are not wiry and straight but rich and beautifully curved'; *J*, p. 23, '. . . in wind. Noticed also frequent partings of ash-boughs'.

l. 7 earlier read 'They touch: their wild weather-swung talons sweep'.

tabour: play on it as on a small drum.

l. 9 *mells*: mixes.

ll. 9–11. Earlier:

. . . a thousand fing-
Ers: then they are old Earth groping towards that steep
Heaven once Earth childed by.

Tom's Garland. Draft of the first ten lines in *H*. Autograph in *C* sent to Dixon 22 December 1887; autograph in BL. Autograph in *B* was probably written in December 1887 or January 1888 (see *L* I, p. 270). Text from autograph in *A*, written 12 January 1888, except hyphen 'no-one' (l. 16) editorial. All autographs are dated 'Dromore Sept. 1887'.

A has numerous metrical marks. Hopkins noted that it is a 'sonnet common rhythm, but with hurried feet: two codas . . . Heavy stresses marked double, thus ″ . . .'

Metrical marks in *B*: l. 3 By him and   l. 5 lustily he his   l. 7 Seldomer

l. 10 honour enough in   l. 12 nő   l. 14 nó   l. 15 beyond   l. 16

glory, earth's.

G. M. H. asked R. B. how to construct codas, 'it is', he remarked, 'the only time I have felt forced to exceed the beaten bounds'. (*L* I, p. 263). Neither Dixon nor R. B. found the poem's meaning easy to untangle and on 10 February 1888 G. M. H. sent R. B. a crib: 'it means . . . that . . . the commonwealth or well ordered human society is like one man; a body with many members and each its function; some higher, some lower, but all honourable, from the honour which belongs to the whole. The head is the sovereign, who has no superior but God and from heaven receives his or her authority: we must then imagine this head as bare . . . and covered, so to say, only with the sun and stars, of which the crown is a symbol. . . . The foot is the daylabourer, and this is armed with hobnail boots, because it has to wear and be worn by the ground; which again is symbolical; for it is navvies or day-labourers who . . . in gangs and millions, mainly trench, tunnel, blast, and in other ways disfigure, "mammock" the earth and . . . stamp it with their footprints. And the "garlands" of nails they wear are therefore the visible badge of the place they fill, the lowest in the commonwealth. But this place still shares the common honour, and if it wants [lacks] one advantage, glory or public fame, makes up for it by another, ease of mind, absence of care; and these things are symbolized by the gold and the iron garlands. . . . Therefore the scene of the poem is laid at evening, when they are giving over work and one after another pile their picks, with which they earn their living, and swing off home, knocking sparks out of mother earth not now by labour and of choice but by the mere footing, being strongshod and making no hardship of hardness, taking all easy. And so to supper and bed . . . the labourer—surveys his lot, low but free from care; then by a sudden strong act . . . tosses it away as a light matter. The witnessing of which lightheartedness makes me indignant with the fools of Radical Levellers. But presently I remember that this is all very well for those who are in, however low in, the Commonwealth and share in any way the Common weal [goods]; but that the curse of our times is that many do not share it, that they are outcasts from it and have neither security nor splendour; that they share care with the high and obscurity with the low, but wealth or comfort with neither. And this state of things, I say, is the origin of Loafers, Tramps, Cornerboys, Roughs, Socialists and other pests of society' (*L* I, pp. 272–4).

Hopkins remarked that the poem resembled 'Harry Ploughman', the result of their having been conceived at the same time (*L* I, p. 271).

l. 2. *fallowbootfellow*. Draft read: 'The fallow booted navvy'.

fallow. No longer at work once the day's toil is over.

bootfellow: workmate.

179 *Epithalamion*. Draft in *C*; text from unfinished drafts in *H*. Hopkins intended the poem for his brother, Everard, who married in April 1888 (*L* I,

p. 277). The drafts are written on examination paper of the Royal University of Ireland, suggesting that G. M. H. began it while invigilating. However, he found himself without the inspiration to complete it.

Metrical mark: l. 37 hoār-huskèd.

l. 37. *selfquainèd*. See 'quoin'—an architectural term for an exterior angle or interior corner in a building.

180 l. 38. *shivès*. Splinters; see 'That Nature is a Heraclitean Fire', l. 4.

'The sea took pity'. Text from an undated fragment in *H* written in pencil.

That Nature is a Heraclitean Fire, &c. Text from autograph in *A* dated 'July 26 1888 / Co. Dublin' and marked 'sprung rhythm, with many outrides and hurried feet: sonnet with two codas'. It actually has three codas. Hopkins wrote that the sonnet was 'provisional only'.

The single stress marked in the text is a double stress in the MS. See p. xli.

Metrical marks: l. 1 Cloud-puffball . . . pillows l. 2 thoroughfare: heaven-roysterers, in gay-gangs . . . marches l. 3 roughcast . . . whitewash l. 4 Shivelights . . . ín . . . lashes l. 5 boisterous l. 6 creases l. 7 crúst . . . stánches, stárches l. 9 Foótfretted in it. Million-fuelèd . . . bonfire l. 10 to her, her l. 11 Mán . . . fást . . . firedint l. 12 Bóth are in an únfáthomable, áll is in an enórmous dárk l. 13 indignation l. 15 ány of him at áll so stárk l. 16 the Resurrection l. 17 héart's-clarion! Awáy . . . gásping l. 20 Fáll to the resíduary l. 23 Thís . . . jóke . . . pótsherd.

Heraclitus: a Greek philosopher (*c.*500 BC). It is difficult to establish the text and interpretation of Heraclitus' work since it has survived only in fragments and quotations. However, he seems to have suggested that the underlying substance of the universe is fire, understood not as flame but as a hot wind that was thought to exist in a pure form outside the earth's atmosphere, where it composed the sun and stars, and in a less pure, moist form close to the earth. (Fire is the material part of the world order or Logos.) Similar qualities were accorded the soul. Heraclitus suggested that the overall quantites of earth, water and 'fire' were constant but that a continuous, cyclic process operated whereby some water evaporated to form 'fire', and other bits dried leaving behind earth while elsewhere earth and 'fire' dissolved or condensed to water. I am indebted to W. K. C. Guthrie, *A History of Greek Philosophy* (Cambridge, 1969) and G. S. Kirk, *Heraclitus: The Cosmic Fragments* (Cambridge, 1954).

ll. 1–2. Clouds show the world's constant flux, as 'fire' becomes moist and water dries. See *J*, pp. 203–4, 207 (prose, pp. 203–4 and 206–7).

181 l. 3. *roughcast*: a mixture of lime and gravel used to coat walls.

l. 4. *shivelights*: splinters of light.

shadowtackle. Perhaps the shadows form a net-like pattern on the wall.

l. 5. *bright wind*. A term used by Diogenes Laertius to describe 'fire', the air rising to form the celestial bodies (Fr. IX. 9).

l. 7. *squeezed dough, crust, dust*. Stages in the drying-out of the mud to dust.

stanches: stops the flow.

starches: stiffens. See 'Fragments of *Richard*', iii, ll. 2–3.

l. 9. *Million-fuelèd, nature's bonfire burns on*. Fr. 30: the cosmos is compared to a huge bonfire, part of it already 'fire' while earth and water are potentially 'fire'.

l. 10. According to Macrobius, Heraclitus called the soul 'a spark of the substance of the stars' (Guthrie, p. 481).

ll. 13–14. *that shone / Sheer off, disseveral*. Individual and unlike any other.

ll. 14. Heraclitus spoke of the soul as being like 'fire', which could be destroyed by sin as 'fire' by water.

l. 19. Perhaps influenced by Fr. 96 (Heraclitus): 'Corpses are more fit to be cast out than dung.'

ll. 21–2. By allowing himself to be incarnated and crucified Christ won for believers eternal life in a perfect body; see 1 Cor. 15: 52.

G. M. H. wrote to Dixon, 'What a preposterous summer! It is raining now: when is it not? However there was one windy bright day between floods last week: fearing for my eyes, with my other rain of papers, I put work aside and went out for the day, and conceived a sonnet. Otherwise my muse has long put down her carriage and now for years "takes in washing" . . . ' (*L* II, p. 157, 29–30 July 1888). On 25 September Hopkins wrote to R. B., 'lately I sent you a sonnet, on the Heraclitean Fire, in which a great deal of early Greek philosophical thought was distilled; but the liquor of distillation did not taste very Greek, did it?' (*L* I, p. 291).

'What shall I do for the land that bred me?' Text from MS letter to Bridges (printed *L* I, p. 292 and *J*, pp. 491–2). R. B. dated it August 1885, but G. M. H.'s letters suggest that most, if not all, of it must have been written in 1888. On 8 September 1888, for instance, he wrote to R. B., 'I had in my mind the first verse of a patriotic song for soldiers, the words I mean: heaven knows it is needed. I hope to make some 5 verses, but 3 would do for singing: perhaps you will contribute a verse. In the Park I hit on a tune, very flowing and spirited . . . I find I have made 4 verses, rough at present, but I send them . . . I hope you may approve what I have done, for it is worth doing and yet it is a task of great delicacy and hazard to write a patriotic song that shall breathe true feeling without spoon or brag' (*L* I, p. 283). He sent a revised version on 25 September 1888. See *L* I, pp. 289–90, 292, and 301–2. An accompaniment was written by W. S. Rockstro.

182 *St. Alphonsus Rodriguez*. Drafts in *C* and *B*. Text from final version in *C* (see Appendix C).

Metrical marks in *C*: l. 1 Glory is   l. 13 by of   l. 14 Majorca Alfonso.

G. M. H. sent a draft to R. B. on 3 October 1888, saying 'I ask your opinion of a sonnet written to order on the occasion of the first feast since his canonisation proper of St. Alphonsus Rodriguez, a laybrother of our Order, who for 40 years acted as hall-porter to the College of Palma in Majorca: he was, it is believed, much favoured by God with heavenly lights and much persecuted by evil spirits. The sonnet (I say it snorting) aims at being intelligible.' The penultimate version (MS *B*) read:

> Honour is flashed off exploit, so we say;
> And those strokes once that gashed flesh or galled shield
> Should tongue that time now, trumpet now that field,
> And, on the fighter, forge his glorious day.
> On Christ they do and on the martyr may;
> But be the war within, the brand we wield
> Unseen, the heroic breast not outward-steeled,
> Earth hears no hurtle then from fiercest fray.
> Yet God (that hews mountain and continent,
> Earth, all, out; who, with trickling increment,
> Veins violets and tall trees makes more and more)
> Could crowd career with conquest while there went
> Those years and years by of world without event
> That in Majorca Alfonso watched the door.

l. 1. *so we say*. 'I mean "This is what we commonly say, but we are wrong" ' (*L* I, p. 297).

l. 10. To an objection made by R. B., Hopkins conceded, 'it is true continents are partly made by "trickling increment"; but', he added, 'what is on the whole truest and most strikes us about them and mountains is that they are made what now we see them by trickling *de*crements, by detrition, weathering and the like . . . And at any rate naturally said to be hewn, and to *shape*, itself, means in old English to hew and the Hebrew *bara*/ to create, even, properly means to hew. But life and living things are not naturally said to be hewn: they grow, and their growth is by trickling increment' (*L* I, pp. 296–7).

183 *'Thou art indeed just'*. Drafts in *H*. Autograph pasted into *B*. Text from autograph fair copy in *A*. One draft (*H*), *B*, and *A* all dated 'March 17 1889'.

Metrical marks in *A*: l. 3 Why do sinners'   l. 4 Disappointment   l. 7 me? Oh   l. 10 they are.

Hopkins wrote to R. B. that 'it must be read *adagio molto* and with great stress' (*L* I, p. 303, 21 March 1889).

l. 9. *brakes*: thickets.

l. 11. *fretty chervil*. Cow parsley, whose flowers are delicate, white, and lacy; G. M. H. recorded its flowering in March 1868 (*J*, p. 162).

ll. 12–13. See *S*, p. 262 (prose, p. 303) and *L* I, p. 270 (prose, p. 270).

'*The shepherd's brow*'. Drafts and fair copy in *H*, dated 'April 3 1889'.

Metrical marks: l. 3 they are.

l. 1 earlier read: 'The shepherd fronting heaven's fork-lightning owns' and 'The shepherd's eye, fronting forked lightning owns'.

ll. 1–2. G. M. H. describes the 'forked' appearance of lightning in *J*, p. 212 (prose, p. 209) and its effect in *J*, p. 221 (19 July; prose, p. 211) and *J*, pp. 233–4 (22 July; prose, p. 216). The shepherd has been variously identified as Christ, Moses, or even Michael; all acknowledge (*own*) the power and glory of God, represented by lightning (cf. 'The Wreck of the Deutschland', stanzas 2 and 9).

l. 3. *towers*. See Milton, *Paradise Lost*, v. 907, where the rebellious angels are described as 'those proud towers to swift destruction doomed'.

l. 8 earlier read: 'What bass can his lute bear to tragic tones?'

viol. The translation sometimes used for *nēbel*, a kind of harp or lyre probably used to supply the bass when played with other instruments. It was played by David (*New Bible Dict*. ed. J. D. Douglas). Milton in 'The Passion' declared 'softer strings / of lute, or viol still, more apt for mournful things' (ll. 27–8) were appropriate to his poem on Christ's crucifixion ('still' = quiet).

l. 11. Cf. *King Lear*, III. iv. 109–10, 'unaccommodated man is no more but such a poor, bare, forked animal'.

ll. 13–14. Earlier read:

> In spoons have seen my masque played and how tame
> My tempest and my spitfire freaks how fussy.

184   *To R. B.* Drafts in *H*. Text from autograph in *A*. Both dated 'April 22 1889'.

Metrical marks: l. 4 mother of   l. 7 widow of   l. 10 rapture of.

l. 2. *blowpipe flame*. Cf. *S*, p. 242, 'You have seen a glassblower breathe on a flame; at once it darts out into a jet taper as a lance head and as piercing too'.

l. 5. *nine years*. Horace, in *The Art of Poetry* (ll. 386–9), suggested that a poem should be kept for nine years before publication so that it could be revised. (N. H. M.) See *L* II, p. 135, when in talking about music G. M. H. remarked to Dixon 'the disproportion is wonderful between the momentary conception of an air and the long long gestation of its setting'.

l. 6. *combs*: straightens out, grooms for presentation.

l. 12. Drafts read:
> 'The fire, the fall, the courage, the creation'
> and 'The wild wing, waft, cry, carol, and creation'.

ll. 13–14. Draft reads:
> 'Believe my withered world knows no such bliss
> Rebuke no more, but read my explanation.'

PROSE

EARLY DIARIES

185 *Flick*. . . Notes such as this are common in G. M. H.'s early diaries. He was keenly interested in etymology, the interrelation of words, and dialect all his life, speculating even on Egyptian words during his last few years (see *L* III, pp. 284–5, for example) and contributing to the *English Dialect Dictionary* (see N. H. MacKenzie, 'Hopkins, Yeats and Dublin in the Eighties' in *Myth and Reality in Irish Literature* (ed. J. Ronsley, 1977), pp. 89–90). A discussion of Hopkins's philology can be found in *J*, Appendix III; see especially pp. 504–5 on 'Flick . . .'. James Milroy provides a much more detailed study in *The Language of Gerard Manley Hopkins* (London, 1977).

Whitby Abbey. G. M. H.'s interest in architecture is evident throughout his early diaries and journals, the notes often accompanied by drawings of detail (see *J*, figures). He possessed John H. Parker's *Introduction to Gothic Architecture* (1849, a gift of 1857) and Parker's *Glossary of Terms Used in Grecian, Roman, Italian and Gothic Architecture* (1845, a gift of 1864).

The poetical language . . . These are notes towards an essay for the Hexameron Society. See *L* III, pp. 215–20 (10 September 1864) and Introduction, p. xvi.

186 *Enoch Arden's island*. In Tennyson's poem 'Enoch Arden' (1864) Enoch is shipwrecked. His sojourn on a tropical island is described in ll. 546–646 of the poem.

The prose scenes from 'Floris in Italy'. They were written before soliloquy (ii), see pp. 186–7 and 188–9.

189 *For Lent*. See Introduction, p. xviii.

JOURNAL

191 *July 17 (1866)*. G. M. H. did not keep his resolution to say nothing (see *J*, p. 147, 24 July 1866, and *L* III, p. 397) and, for reasons given in a letter to his father (*L* III, pp. 91–5; prose, pp. 223–6), joined the Catholic Church on 21 October 1866.

Aug. 22 (1866). Hopkins was fond of trees, especially the ash (see *L* III, p. 202), and made numerous detailed descriptions and drawings of them.

192 διὰ . . . 'through growth' (*J*, p. 371).

July 6, 1866. G. M. H. writes 1866 by mistake. It should be 1867.

Aug. 23 (1867). H. H. argues persuasively that this refers to a resolution to discontinue the writing of poetry and destroy what he had written if it became clear that it would interfere with his dedication to God. At this date G. M. H. does not appear to have decided whether or not to join a religious order (see *J*, pp. 537–9).

Aug. 30 (1867). A clear example of observations noted and later used in

Hopkins's poetry. See 'The Blessed Virgin compared to the Air we Breathe', ll. 75–8.

May 2 (1868). See notes to Aug. 23 1867 and May 11 1868.

193 *May 5 and 7 (1868).* G. M. H. spent Holy Week and Easter 1867 at the Benedictine House, Belmont Abbey and liked it very much. In April 1868 he was allowed a private retreat at the Jesuit Novitiate (Roehampton) to help him towards deciding whether he had a religious vocation.

May 11 (1868). H. H. suggests that this refers to Hopkins's burning of many of the copies he possessed of his poems. See Aug. 23 1867 above.

July 11 (1868). This was written during G. M. H.'s walking holiday in Switzerland with Edward Bond (see Introduction, p. xx).

pagharee. 'a fashion imported from India, a silk scarf worn round the head or hat and falling down behind as a shade' (*J*, p. 393).

195 *July 18–19 (1868).* See Introduction, p. xx. G. M. H.'s drawing of the Baths of Rosenlaui can be seen in *J*, plate 28.

196 *July 26 (1868). Tyndal.* John Tyndall (1820–93), whose research on glaciers had led to his becoming a well-known mountaineer. Tyndall was making a third attempt to climb the Matterhorn and one of his guides insisted on attending mass before climbing on Sunday. The expedition was successful.

197 τύχη . . . ' "[by] chance that loves art" based on a quotation from Agathon', *J*, p. 394.

198 *Jan. 24 (1869). penance.* The cause of the penance is unknown but it may, in part, have been intended to increase the spirituality of Hopkins's response to physical beauty.

199 *Dec. 23 (1869). A Secretis.* Slang at Manresa for the novice responsible for cleaning the toilets (*J*, p. 406). See Introduction, p. xxi.

brought into no scaping. Originally '. . . inscaping', but G. M. H. cancelled 'in'.

201 *Caesar's Camp.* An earthwork south of Wimbledon Common.

202 *April 4 (1870).* Entry cancelled, probably because it is out of sequence.

206 *April 15 (1871).* '*The young lambs bound . . .*' from Wordsworth, 'Ode: Intimations of Immortality', ll. 20–1.

April 16 (1871). drawing. See *J*, p. 205.

209 *July 8 (1871). χάλκεον . . .* 'Brazen heaven'; MS omits first accent. See note to 'My prayers must meet a brazen heaven', p. 321.

210 *gutturs.* Thus in MS.

211 *Aug. 3 (1872). Baddely Library.* Misspelt. A collection bequeathed by Edward Badeley to Stonyhurst in 1868. It contained Scotus's *Scriptum Oxoniense super Sententiis*, 2 vols., Venice, 1514 (*J*, p. 417). See Introduction, p. xxiii.

213 *Oct. 5 (1872). cieling.* G. M. H.'s spelling.

Cyril Hopkins. He became a partner with his father in the firm Manley Hopkins, Son and Cookes. (*J*, p. 310).

214 *Oct. 27 (1872)*. *Blandyke*. The word used at Stonyhurst for a monthly day's holiday. The 'old complaint' was piles, which necessitated an operation at Christmas and continued to cause problems (*L* I, p. 84).

215 *June 15 (1873)*. For G. M. H.'s explanation of the significance of the Corpus Christi procession see *L* I, p. 149.

Aug. 16 (1873). *villa*: I am grateful to the Revd J. Brayley, SJ, who tells me that 'villa' is a Jesuit term for the 'annual holiday' (usually 14 days). Like 'Blandyke', the usage has a Continental origin, and stems from the country house or villa in which the holidays were spent.

218 *Aug. 27 (1873)*. *Roehampton*. Manresa House, London, where G. M. H. had been a novice.

Provincial. Father Peter Gallwey (1820–1906), who was appointed Provincial in 1873 and was kind to Hopkins on a number of occasions. He became Superior of St Beuno's in 1876 while Hopkins was studying theology there.

Sept. 18 (1873). *Kensington Museum*. Now the Victoria and Albert Museum. The bronze gilt doors of the Baptistry at Florence were by Ghiberti, who had been trained as a goldsmith. The Victoria and Albert Museum's reproduction of these was obtained in 1866. The cartoons were Raphael's designs for tapestries. William Mulready (1786–1863) was an Irish artist who specialized in painting contemporary incidents.

Fétis. Probably François Joseph Fétis's *Instruments de Musique* (1868).

219 $Φυμός$: the soul, essence.

May 23 (1874). *Briton Rivière* (1840–1920) frequently painted classical scenes that included animals.

220 *Millais*, Sir John Everett (1829–96), whose work G. M. H. had keenly admired when at Oxford (see *L* III, p. 201).

'*Bared ruined choirs*': Shakespeare, Sonnet 73, 'That time of year thou mayst in me behold'.

July 14 (1874). The debate concerned the suspension of the Endowed Schools Commission which had been established by W. E. Y. Forster when Gladstone had been Prime Minister.

221 *Oct. 8 (1874)*. St Winefred's Well. See pp. 102 and 161.

LETTERS

223 *16 Oct. 1866*. See Introduction, p. xviii.

Master. The Revd Robert Scott (1854–70), who evidently marked some of G. M. H.'s essays.

Alfred William *Garrett* (1844–1929). G. M. H. believed that his own conversion had prompted both Garrett and Wood to follow suit. In 1868

Garrett entered the Indian Education Service, Bengal and in 1884 he became an Inspector of Schools in Tasmania, his birthplace.

William *Addis* (1844–1917). One of Hopkins's closest friends at Oxford, he became a Catholic about a fortnight before G. M. H. In 1868 he joined the Oratory in London. In 1888 he renounced Catholicism, married and moved to Australia. G. M. H. was much upset (see *L* I, p. 298). Addis returned to England and the Anglican Church in 1901, later becoming Vicar of All Saints, Ennismore Gardens in 1910.

Alexander *Wood* (1845–1912). After living in Rome and America he settled in England, writing a number of pamphlets of interest to Catholics.

224 *Monsignor Eyre* (1817–1902). Archbishop for the Western District and Delegate Apostolic for Scotland.

225 *Liddon*; *Pusey*. Tractarians. See Introduction, pp. xvii–xviii.

If Dr. Pusey is in Oxford . . . Pusey wrote to G. M. H. refusing to do this (10 (actually 20th?) October 1866).

226 *Bridges*. G. M. H. had stayed with Robert Bridges and his family while waiting to see Dr (later Cardinal) Newman. See *L* I, pp. 5–7.

Arthur Hopkins (1848–1930). A younger brother of Gerard's who became an artist.

Christmas Eve 1875. Kate Hopkins (1856–1933). G. M. H.'s second sister. She had a keen sense of humour (see *L* III, pp. 164 and 240) and a gift for drawing. She remained unmarried and after her mother's death, lived on at the family home with her brother, Lionel, and sister, Grace.

Grace Hopkins (1857–1945). Hopkins's youngest sister who was musical and set accompaniments to G. M. H.'s tunes for Bridges's and Dixon's poems. She was engaged in 1882 but, after the death of her fiancé, never married.

Nicholas Breakspear: probably Eustace John Breakspeare who later wrote a book on Mozart in the Musical Masters series.

The wreck of the *Deutschland*. There were numerous newspaper accounts of the accident and subsequent inquiry. Extracts of a couple of these from *The Times* and the *Illustrated London News* can be found in *L* III, pp. 439–43.

227 *comet*. It 'turned out to be a wellknown nebula of great size, Praesepe it is called, in Cancer' (*L* III, p. 137).

Aug. 21 1877. if it were it. Thus in MS.

Harry Ellis *Wooldridge* (1845–1917), a close friend of R. B.'s who was very knowledgeable about Renaissance art. He was elected to the Slade Chair of Fine Art at Oxford three times. He was a painter and musician, and wrote the first two volumes of the *Oxford History of Music*. R. B. shared a house with him in London during part of his medical training.

Bremen stanza. Stanza 12 of 'The Wreck of the Deutschland'.

molossic. It should be 'amphibrachic'.

228 *I have mingled the two systems.* 'The Sea and the Skylark' and 'In the Valley of the Elwy', for example.

229 *April 2 1878. cure me.* G. M. H.'s previous letter had said, 'Life here is as dank as ditch-water and has some of the other qualities of ditch-water: at least I know that I am reduced to great weakness by diarrhoea, which lasts too, as if I were poisoned.'

Mrs. Molesworth. Robert Bridges's mother, who had married the Revd Dr Molesworth after the death of R. B.'s father.

Pater. Walter Pater (1839–94). See Introduction, p. xvi.

Violet. A mistake for 'The Daisy'.

230 *May 13 1878.* MS dated '1877' in error.

Marzials. Theophilus Julius Henry Marzials, who wrote the libretto to *Esmeralda* and a book of poetry.

231 *'If it were done . . .'* Macbeth, I vii.

critic . . . Athenaeum . . . writes very long reviews. C. C. Abbott suggests that it was Theodore Watts-Dunton.

232 *May 30 1878. 'hearts of oak' etc.* 'The Loss of the Eurydice', see p. 135.

233 *Lower Isis.* The Thames below Oxford.

No trace has been found of either G. M. H.'s ode on the Vale of Clwyd or R. B.'s ode on the *Eurydice*.

234 *Postcard, June 9 1878.* R. B.'s volumes of poetry of 1878 and 1879 show his experiments with sprung rhythm. In the 1879 volume R. B. included a note about the new metre which ended: 'The author disavows any claim to originality for the novelty: this is almost entirely due to a friend, whose poems remain, he regrets to say, in manuscript.'

Feb. 15 1879. 'Silver Jubilee', see p. 119.

your album. MS *A*; the first collection R. B. made of G. M. H.'s poems.

235 *preliminary.* Thus in MS.

'winding the eyes'. 'The Lantern out of Doors', see p. 134.

That is how . . . The rest of this letter is missing.

236 *May 26 1879. your brother's book. Wet Days*, poetry written by R. B.'s older brother, John.

The poem you send. 'The Voice of Nature', virtually unaltered in *Poetical Works* (1952).

Coventry *Patmore* (1823–96) worked as an assistant in the Printed Books Room of the British Museum and was well-known for his volumes of poetry. He became a Catholic in 1864. G. M. H. met him in 1883 and corresponded with him for the rest of his life. See Introduction, p. xxxii.

I wrote a little . . . 'Winter with the Gulf Stream', see p. 15.

the Purcell sonnet. 'Henry Purcell', see p. 143.

237 *Aug. 14 1879. Picnic verses*. They are no longer extant.

Lionel *Muirhead* (1845–1925). One of R. B.'s closest friends, he withdrew from Oxford after losing the sight in one eye, travelled extensively in the Middle East, and became an artist and gentleman farmer in Oxfordshire.

music you have. See Introduction, p. xxiii.

238 *Candle sonnet*. 'The Candle Indoors', p. 144.

the Handsome Heart. See p. 144.

little scene . . . Mount St. Mary's. 'Brothers', p. 151.

finer thing. 'The Bugler's First Communion', p. 146.

a little song. 'The furl of fresh-leaved dogrose' (N. H. M.).

Lastly I enclose a sonnet. 'Andromeda', p. 148.

239 *birthday lines*. 'A Complaint', p. 76.

Gosse's paper. See the *Academy*, 26 July 1879, pp. 60–1.

'Linden Ore'. It should be 'Lindenore'.

prose of Michelangelo's sonnets. Unknown, but nos. 35 and 64 of 'The Growth of Love' as printed in *Poetical Works* were based on madrigals by the artist.

Non ha l'ottimo. Michelangelo, translated by J. A. Symonds under the title, 'The Lover and the Sculptor'.

240 *'Tis joy the falling*. From 'In all things beautiful', no. 31 in 'The Growth of Love', *Poetical Works* (1952). G. M. H. mentioned this line several times, not always accurately. In 1886 he began a translation of the poem into Latin. See p. 174.

244 *22 Dec. 1880. little piece . . . in walking from Lydiate*. 'Spring and Fall', p. 152.

Bridges' book. MS *A*, see p. xl.

April 3 1881. play. 'St. Winefred's Well', p. 161.

245 *Sept. 16 1881. You did run well . . .* Gal. 5:7, 'Ye did run well; who did hinder you that ye should not obey the truth?'

Canon. Canon Richard Watson Dixon (1833–1900), a contemporary at Oxford of Burne-Jones and William Morris; he later became vicar of Hayton (Humberside) and then of Warkworth (Northumberland). He published several volumes of poetry and a chronicle history of the Church of England in five volumes.

vacare Deo: to be available for God.

ode on Edmund Campion. Nothing of this is known. Campion (1540–81) was a fellow of St John's College, Oxford. He was a novice at Brno in Moravia and was ordained in Prague, where he was sent to teach. In 1580 he entered England secretly on a mission to promote Catholicism in the country. He was captured, tortured, and executed at Tyburn with Ralph Sherwin and Alexander Briant on 1 December 1581.

Alexander's Feast. An ode by John Dryden.

246 *I have loved flowers. . .* A poem by R. B.: *Poetical Works*, p. 263.

247 *29 Oct. 1881. Non ha l'ottimo.* See above, p. 239.

248 τοὺς περὶ: 'followers of'.

'outriding' feet. See 'Harry Ploughman', notes, p. 382, for example.

249 *My sister.* Grace, see note, p. 392. *have.* Preceded by 'may', cancelled.

the waste of time the very compositions. . . See Introduction, pp. xxxii and xxxiv.

a tragedy. 'St Winefred's Well', see p. 161.

251 *Dec. 1 1881. Stanislaus* Kostka (1550–68) was a Jesuit only during the last year of his life. He was canonized in 1726.

Louis *Bourdaloue* (1632–1704), a French Jesuit who preached in Paris for thirty-four years, during which his command of oratorical technique was said to have improved steadily to perfection.

Francisco *Suarez* (1548–1617). It was the thought of Aquinas as commented on by Suarez that G. M. H. would have been taught at St Mary's Hall.

Louis de *Molina* (1535–1600), a Spanish theologian considered to be one of the pre-eminent moral theologians of the sixteenth century, whose reconciliation of the ideas of man's free will and God's omniscience is still accepted today.

Canon Dixon's book was a *History of the Church of England from the Abolition of the Roman Jurisdiction.* (London, 1878).

252 *my sister.* Grace, see above, p. 392.

William *Morris* (1834–96), poet, artist, and designer. He was an extremely productive man who wrote a number of long poems and romances, made translations from Greek and Icelandic as well as designing patterns for wallpaper and textiles and founding the Kelmscott Press. He had strong socialist sympathies evident not only in his numerous lectures but also in such romances as *News from Nowhere* and *The Dream of John Ball.* Dixon had known many of the Pre-Raphaelites at Oxford and his poetry has, as G. M. H. remarked, a Pre-Raphaelite flavour.

254 *Oct. 18 1882. 'Pete'.* 'Come up From the Fields Father' in *Leaves of Grass.*

review by Saintsbury. 10 October, 1874, pp. 398–400, printed in *L* I, note P.

Constable. 'The Hay Wain', 'A View near London', and 'The Lock on the Stour' were exhibited at the Paris Salon in 1824.

or a handkerchief. See Saintsbury's review (*L* I, note P).

256 *C. D.* Dixon.

257 *Jan. 4 1883. Prometheus the Firegiver.* Written by R. B. in 1881–2.

the sonnet a fine work. Sonnet 42 in *The Growth of Love* (1889).

my sonnet. 'God's Grandeur'.

Joseph *Noel Paton* (1821–1901), a popular Scottish artist who frequently chose subjects from fairy-tales or history. He exhibited a number of pictures at the Royal Academy.

sonnet on Purcell. 'Henry Purcell', p. 143.

259 *Vernon Lee.* 'Impersonality and Evolution in Music'.

roking the pot. See *EDD*, 'rauk', to stir, poke about; search.

March 7 1884. Dublin. See Introduction, p. xxxiii.

260 *'Wild air, world-mothering air'.* 'The Blessed Virgin compared to the Air we Breathe'. In MS *B*, R. B. later copied in the final section in pencil. G. M. H. then strengthened this in ink, erasing the pencil. Dishearteningly, it was this incomplete poem that Coventry Patmore chose as his favourite (*L* I, p. 192).

a poem of Tennyson's. 'Early Spring'. *two triolets.* See p. 157.

Aug. 21 1884. MS book. MS *B*, see p. xl. *wedding.* R. B. married M. Monica Waterhouse, daughter of the architect, on Sept. 3 1884.

261 *Non omnium . . .* The basic meaning is 'not everything sets/fades'.

I yield, you do come . . . 'Peace', p. 149.

Margaret. 'Spring and Fall', p. 152. *Arrah . . .* See note to 'Flick', p. 389.

Dec. 9 1884. Kate, see above, p. 392.

Alexander *Baillie* (1843–1921). A close friend of G. M. H.'s at Oxford, he also obtained a double first. He became a draftsman and conveyancer until 1874, when he went to Egypt because of ill-health and there developed his interests in Egyptian language and archaeology. See *L* III, pp. 199–294, for his correspondence with G. M. H.

263 *Sept. 1 1885. compromising.* Fr. Goldie, SJ, recounts that G. M. H. would not even stay to tea or dinner if he did not have the specific permission of his superiors. When the Revd Mr Wade protested on one occasion that he would take the whole responsibility himself if Hopkins stayed for dinner, G. M. H. refused, quipping, 'You may be weighed (Wade), but I should be found wanting' (*Letters and Novices* (1890), reprinted in *Hopkins Quarterly*, Summer 1981, p. 58).

Mrs. Bridges' disappointment. A miscarriage.

five or more. See p. 372, n. to p. 166.

Four. Written above 'Three', cancelled.

William *Barnes* (1801–86). A graduate of the University of Cambridge, he became rector of Came (Dorset) and wrote three volumes of *Poems of Rural Life in the Dorset Dialect.*

264 *Oct. 13 1886. true.* MS 'necessary', cancelled.

τῇ ποιήσει . . .: to be energetic in writing poetry.

Let your light . . . Matt. 5:16.

265 *Oct. 28 1886. writer in a late Academy.* H. C. Beeching, a relative of R. B.'s. The article appeared on 16 October 1886.

interchange. 'No——of the means employed, wh. is physical and sh^d have been magical' (R. B.).

more than yours. '(but does not make chemistry of)' (R. B.).

266 *Feb. 17 1887. Canon's book.* R. W. Dixon's *Lyrical Poems* (Oxford, 1887), dedicated to G. M. H.

my judgment . . . G. M. H. wrote an appreciative note on Dixon for *A Manual of English Literature Historical and Critical* (ed. Tom Arnold, 5th edn., Dublin, 1885), reprinted in *L* II, note K.

piece being a fragment. 'On a Portrait of Two Beautiful Young People', p. 176.

Miss Tynan. Katherine Tynan (1861–1931), a popular poet at the time and good friend of W. B. Yeats.

267 *Feast of Bacchus.* One of R. B.'s eight plays. G. M. H.'s praise of them contributed to R. B.'s decision to have them published.

William J. *Walsh* (1841–1921), appointed Archbishop of Dublin in 1885 despite the protests of the English Government. He had supported Gladstone's Land Bill in 1881 and was a nationalist.

Thomas W. *Croke* (1824–1902), Archbishop of Cashel from 1874 and an active supporter of Home Rule.

268 *July 30 1887. Classical Review etc.* Nothing by G. M. H. was published.

269 *Grand Old Mischiefmaker.* The Prime Minister, W. E. Gladstone.

Nero. Another of R. B.'s plays.

Ignatius *Persico* (1823–96). In March 1887 he had been appointed Archbishop of Tamiatha, commissioned to report on the involvement of the clergy in Irish politics. However, the Holy See condemned the Plan of Campaign, nullifying his report before it could be published. By the Plan of Campaign (1886–7), tenants whose landlords refused to reduce their rents were to pay what they considered fair rent to a political leader who was to act as intermediary, retaining the money till the landlord agreed to the new rent.

270 *Jan. 12 1888. Monasterevan.* A country house in the village of Monastervan belonging to Miss Cassidy where Hopkins spent a number of recuperative holidays. See letter of 29 April 1889, p. 273.

Wooldridge. See note on p. 392 above.

Orpheus with his lute. Henry VIII, III i. G. M. H. wrote Greek and Latin metrical versions of the song, see pp. 173 and 174.

Sir Frederick Arthur Gore *Ouseley* (1825–1889), prolific composer and canon residentiary of Hereford Cathedral. His three books on musical theory were: *A Treatise on Harmony* (Oxford, 1868), *A Treatise on Counterpoint, Canon and Fugue: based upon that of Cherubini* (Oxford, 1869), and *A*

Treatise on Musical Form and General Composition (Oxford, 1875). It is presumably to the second of these that G. M. H. was referring.

James *Higgs, Fugue* (London, 1878).

271 *lady of the strachey. Twelfth Night*, II. v; Malvolio, contemplating the possibility of marrying the Lady Olivia, says 'There is example for't: the lady of Strachy married the yeoman of the wardrobe' (ll. 38–9).

April 29 1889. Charles Henry *Daniel* (1836–1919), provost of Worcester College, who established his own press and published a number of R. B.'s early volumes, generally in small impressions.

Edward *Dowden* (1843–1913), Professor of English at Trinity College, Dublin. G. M. H. left him several of R. B.'s volumes. As requested, these were not acknowledged, but Dowden later mentioned the gifts with gratitude.

272 *a new sonnet.* G. M. H.'s last sonnet, 'To R. B.', p. 184.

Swinburne's new volume was *Poems and Ballads*, third series (1889).

my song. 'What shall I do for the land that bred me?'

273 *The two beautiful* . . . See p. 76.

SERMONS

275 *Oct. 5 1879.* See 'Ad Matrem Virginem' and 'The Blessed Virgin compared to the Air we Breathe', pp. 95 and 158.

Et erat . . . (Luke 2:33): 'And his father and his mother marvelled at those things which were spoken of him.'

276 *Nov. 23 1879. hero.* See 'The Loss of the Eurydice', l. 112.

277 *beautiful in mould* . . . Ps. 44: 3 (Douay).

278 *Plato* . . . *foretold of him. Republic*, 361–362A, 'the just man will have to endure the lash, the rack, chains, the branding-iron in his eyes, and finally, after every extremity of suffering, he will be' fixed on a stake, impaled (Loeb, translated by Paul Shorey).

279 *Oct. 25 1880. Not a sparrow.* Matt. 10: 29.

280 *guardian angel.* See 'The Bugler's First Communion', stanza 5.

SPIRITUAL WRITINGS

280 *'Homo creatus est'.* See *S*, pp. 338–51, and Introduction, p. xxxi.

282 *Aug. 7 1882.* This sort of thought occurs time and again in the poems; see for example, 'As kingfishers catch fire', 'Morning, Midday, and Evening Sacrifice', 'The Leaden Echo and the Golden Echo'.

283 *Notes on Suarez, De Mysteriis.* For Hopkins's opinion of Suarez, see p. 251. See 'The Wreck of the Deutschland', stanza 10.

NOTES TO PAGES 285–304 399

285 *Marie Lataste* (1822–47) grew up in Gascony and had numerous religious visions which she was persuaded to record. She became a lay-sister in the Society of the Sacré Cœur in 1844. In 1862 her writings were published posthumously. Hopkins evidently used the 1877 edition in the Beaumont Library (*S*, p. 325).

286 *God / in man / knowing his own truth* . . . See 'As kingfishers catch fire'.

The will is surrounded . . . See 'The Handsome Heart'.

κεῖται ἐν τῷ πονηρῷ: lies in a bad state.

287 *quickening.* See 'The Wreck of the Deutschland', stanza 2.

elevating, which lifts. See 'The Wreck', stanzas 1 and 3.

subito probas eum: from Job 7: 17, 18, 'What is man that thou shouldst . . . try him every moment'.

Veniam et curabo eum (Matt. 8: 7): 'I will come and heal him'.

Volo, mundare (Matt. 8: 3): Jesus said to the leper, 'I will: be thou clean'.

ramener à la route: bring back to the way.

288 *Nov. 8 1881. Ipsius enim* . . . *ambulemus* (Eph. 2: 10): 'For we are his workmanship, created in Jesus Christ unto good works, which God hath before ordained that we should walk in them.'

omnia enim . . . *Dei* (from I Cor. 3: 22, 23): 'Whether Paul, or Apollos, or Cephas, or the world, or life, or death, or things present, or things to come; all are yours; And ye are Christ's; and Christ is God's.'

289 *The snake or serpent a symbol of the Devil.* See 'Spelt from Sibyl's leaves'.

290 *The Principle or Foundation* (1882). See p. 281 and 'As kingfishers catch fire'.

291 *Domine* . . . *angelis*: from Ps. 8, 'O Lord our Lord, how excellent is thy name . . . What is man, that thou art mindful of him? . . . For thou hast made him a little lower than the angels, and hast crowned him with glory and honour.'

302 *Jan. 1 1889.* MS reads 1888 in error.

304 *exiit sermo a Caesare* . . . (Luke 2: 1): 'And it came to pass . . . that there went out a decree from Caesar Augustus, that all the world should be taxed.' G. M. H. had tried in vain to refuse the appointment.

ego quidem aqua . . . (from Luke 3: 16, 17): John the Baptist said, 'I indeed bapize you with water; but one mightier than I cometh . . . he shall baptize you with the Holy Ghost and with fire. Whose fan is in his hand . . .'.

FURTHER READING

BIBLIOGRAPHY

Cohen, Edward H., *Works and Criticism of Gerard Manley Hopkins: a Comprehensive Bibliography* (Washington, 1969).

Dunne, Tom, *Gerard Manley Hopkins: A Comprehensive Bibliography* (Oxford, 1976). This covers work up to 1970–1.

MAJOR EDITIONS

The Oxford Authors Hopkins is intended to replace the fourth edition of *The Poems of Gerard Manley Hopkins*, edited by W. H. Gardner and N. H. MacKenzie (Oxford, 1967, 1984), which has been the most accurate and comprehensive edition of Hopkins's poetry available. N. H. MacKenzie's forthcoming edition in the Oxford English Texts series (along with 2 vols. of *Poetic Facsimiles* published by Garland, vol. 1, 1989) will provide scholars with far more information than is included in either of these smaller books.

The Journals and Papers were edited by Humphry House (London, 1959, reprinted 1966) and completed after his death by Graham Storey. Christopher Devlin, SJ prepared the edition of Hopkins's *Sermons and Devotional Writings* (London, 1959, reprinted 1967) while C. C. Abbott produced three volumes of letters; those of *Hopkins to Robert Bridges*, second edition revised (London, 1955, repr. 1970); *The Correspondence of Hopkins to R. W. Dixon*, second edition revised (London, 1955, repr. 1970); and *Further Letters*, second edition revised and enlarged (London, 1956, repr. 1970). A number of additional letters were published in *The Hopkins Research Bulletin*.

BIOGRAPHY

The two most recent biographical books on Hopkins are by Bernard Bergonzi (London, 1977) and Paddy Kitchen (London, 1978). Norman White is preparing a much longer biography for Oxford University Press.

CRITICAL BOOKS

Ball, Patricia M., *The Science of Aspects: The Changing Role of Fact in the Work of Coleridge, Ruskin and Hopkins* (London, 1971).

Bender, Todd K., *Gerard Manley Hopkins: The Classical Background and Critical Reception of his Work* (Baltimore, 1966).

Bottrall, Margaret (ed.), *Gerard Manley Hopkins: Poems—A Casebook* (London, 1975).

Boyle, Robert, SJ, *Metaphor in Hopkins* (Chapel Hill, 1960, 1961).

Cotter, James Finn, *Inscape: the Christology and Poetry of Gerard Manley Hopkins* (Pittsburg, 1972).

Downes, David A., *Gerard Manley Hopkins: A Study of his Ignatian Spirit* (London, 1960).

—— *Victorian Portraits: Hopkins and Pater* (New York, 1965).

Fulweiler, Howard, *Letters from the Darkling Plain: Language and the Grounds of Knowledge in the Poetry of Arnold and Hopkins* (Columbia, 1972).

Gardner, W. H., *Gerard Manley Hopkins, 1844–89: a Study of Poetic Idiosyncrasy in Relation to Poetic Tradition*, 2 vols. (London, 1944, 1949, repr. London, 1966).

Hartman, Geoffrey, *The Unmediated Vision: an Interpretation of Wordsworth, Hopkins, Rilke, and Valéry* (New Haven, 1954, repr. New York, 1966).

—— (ed.), *Hopkins: A Collection of Critical Essays* (Englewood Cliffs, 1966).

Heuser, Alan, *The Shaping Vision of Gerard Manley Hopkins* (London, 1958, repr. New York, 1968).

Holloway, Sister Marcella M., *The Prosodic Theory of Gerard Manley Hopkins* (Washington, 1947, repr., 1964).

Johnson, Wendell Stacy, *Gerard Manley Hopkins: the Poet as Victorian* (Ithaca, 1968).

Keating, John E., SJ, *'The Wreck of the Deutschland': an Essay and Commentary* (Kent, Ohio, 1963).

Kenyon Critics, The, *Gerard Manley Hopkins* (New York, 1945, repr. London, 1975).

Lees, Francis N., *Gerard Manley Hopkins* (New York, 1966).

McChesney, Donald, *A Hopkins Commentary* (London, 1968).

MacKenzie, Norman H., *A Reader's Guide to Gerard Manley Hopkins* (London, 1981).

—— *Hopkins*, Writers and Critics series (Edinburgh, 1968).

Mariana, Paul, *Commentary on the Complete Poems of Gerard Manley Hopkins* (London, 1970).

Milroy, James, *The Language of Gerard Manley Hopkins* (London, 1977).

Milward, Peter, SJ, *Commentary on G. M. Hopkins' 'The Wreck of the Deutschland'* (Tokyo, 1968).

—— *A Commentary on the Sonnets of G. M. Hopkins* (Tokyo, 1969).

—— and Raymond V. Schoder, SJ, *Landscape and Inscape: Vision and Inspiration in Hopkins's Poetry* (London, 1975).

—— and Raymond V. Schoder, SJ (eds.), *Readings of 'The Wreck': Essays in Commemoration of the Centenary of G. M. Hopkins' 'The Wreck of the Deutschland'* (Chicago, 1976).

Peters, W. A. M., SJ, *Gerard Manley Hopkins: a Critical Essay towards the Understanding of his Poetry* (London, 1948, repr. Oxford, 1970).

Phare, E. E., *Gerard Manley Hopkins* (Cambridge, 1933, reprinted New York, 1967).

Pick, John, *Gerard Manley Hopkins: Priest and Poet*, second edition (London, 1966).

—— (ed.), *Gerard Manley Hopkins: 'The Windhover'* (Columbus, 1969).

Plotkin, Cary H., *The Tenth Muse: Victorian Philology and the Genesis of the Poetic Language of Gerard Manley Hopkins* (Carbondale, 1989).

Ritz, Jean-Georges, *Robert Bridges and Gerard Manley Hopkins, 1863–1889: a Literary Friendship* (London, 1960).

—— *Le Poète Gérard Manley Hopkins, S.J.: L'homme et l'œuvre* (Paris, 1963).

Schneider, Elisabeth, *The Dragon in the Gate: Studies in the Poetry of G. M. Hopkins* (Berkeley and Los Angeles, 1968).

Sulloway, Alison G., *Gerard Manley Hopkins and the Victorian Temper* (London, 1972).

Thomas, Alfred, SJ, *Hopkins the Jesuit: the Years of Training* (London, 1969).

Thornton, R. K. R., *All My Eyes See: the Visual World of Gerard Manley Hopkins* (Sunderland, 1975).

—— *Gerard Manley Hopkins: the Poems* (London, 1973).

Weyand, Norman, SJ, and Raymond V. Schoder, SJ (eds.), *Immortal Diamond: Studies in Gerard Manley Hopkins* (London, 1949).

Zanniello, Tom, *Hopkins in the Age of Darwin* (Iowa City, 1988).

APPENDIX A

The Convent Threshold[1]

By CHRISTINA ROSSETTI

THERE'S blood between us, love, my love,
There's father's blood, there's brother's blood;
And blood's a bar I cannot pass:
I choose the stairs that mount above,
Stair after golden skyward stair,
To city and to sea of glass.
My lily feet are soiled with mud,
With scarlet mud which tells a tale
Of hope that was, of guilt that was,
Of love that shall not yet avail;
Alas, my heart, if I could bare
My heart, this selfsame stain is there:
I seek the sea of glass and fire
To wash the spot, to burn the snare;
Lo, stairs are meant to lift us higher:
Mount with me, mount the kindled stair.

Your eyes look earthward, mine look up.
I see the far-off city grand,
Beyond the hills a watered land,
Beyond the gulf a gleaming strand
Of mansions where the righteous sup;
Who sleep at ease among their trees,
Or wake to sing a cadenced hymn
With Cherubim and Seraphim;
They bore the Cross, they drained the cup,
Racked, roasted, crushed, wrenched limb from limb,
They the offscouring of the world:
The heaven of starry heavens unfurled,
The sun before their face is dim.

You looking earthward, what see you?
Milk-white, wine-flushed among the vines,
Up and down leaping, to and fro,
Most glad, most full, made strong with wines,

---

[1] The text is that of the first edition, published in *Goblin Market and other Poems*, 1862.

Blooming as peaches pearled with dew,
Their golden windy hair afloat,
Love-music warbling in their throat,
Young men and women come and go.

You linger, yet the time is short:
Flee for your life, gird up your strength
To flee; the shadows stretched at length
Show that day wanes, that night draws nigh;
Flee to the mountain, tarry not.
Is this a time for smile and sigh,
For songs among the secret trees
Where sudden blue birds nest and sport?
The time is short and yet you stay:
To-day, while it is called to-day
Kneel, wrestle, knock, do violence, pray;
To-day is short, to-morrow nigh:
Why will you die? why will you die?

You sinned with me a pleasant sin:
Repent with me, for I repent.
Woe's me the lore I must unlearn!
Woe's me that easy way we went,
So rugged when I would return!
How long until my sleep begin,
How long shall stretch these nights and days?
Surely, clean Angels cry, she prays;
She laves her soul with tedious tears:
How long must stretch these years and years?

I turn from you my cheeks and eyes,
My hair which you shall see no more—
Alas for joy that went before,
For joy that dies, for love that dies.
Only my lips still turn to you,
My livid lips that cry, Repent.
Oh weary life, Oh weary Lent,
Oh weary time whose stars are few.

How should I rest in Paradise,
Or sit on steps of heaven alone?
If Saints and Angels spoke of love,
Should I not answer from my throne:
Have pity upon me, ye my friends,
For I have heard the sound thereof:

Should I not turn with yearning eyes,
Turn earthwards with a pitiful pang?
Oh save me from a pang in heaven.
By all the gifts we took and gave,
Repent, repent, and be forgiven:
This life is long, but yet it ends;
Repent and purge your soul and save:
No gladder song the morning stars
Upon their birthday morning sang
Than Angels sing when one repents.

I tell you what I dreamed last night:
A spirit with transfigured face
Fire-footed clomb an infinite space.
I heard his hundred pinions clang,
Heaven-bells rejoicing rang and rang,
Heaven-air was thrilled with subtle scents,
Worlds spun upon their rushing cars:
He mounted shrieking: 'Give me light.'
Still light was pour'd on him, more light;
Angels, Archangels he outstripped
Exultant in exceeding might,
And trod the skirts of Cherubim.
Still 'Give me light,' he shrieked; and dipped
His thirsty face, and drank a sea,
Athirst with thirst it could not slake.
I saw him, drunk with knowledge, take
From aching brows the aureole crown—
His locks writhed like a cloven snake—
He left his throne to grovel down
And lick the dust of Seraphs' feet
For what is knowledge duly weighed?
Knowledge is strong, but love is sweet;
Yea all the progress he had made
Was but to learn that all is small
Save love, for love is all in all.

I tell you what I dreamed last night:
It was not dark, it was not light,
Cold dews had drenched my plenteous hair
Through clay; you came to seek me there.
And 'Do you dream of me?' you said.
My heart was dust that used to leap
To you; I answered half asleep;
'My pillow is damp, my sheets are red,

There's a leaden tester to my bed:
Find you a warmer playfellow,
A warmer pillow for your head,
A kinder love to love than mine.'
You wrung your hands; while I like lead
Crushed downwards through the sodden earth:
You smote your hands but not in mirth,
And reeled but were not drunk with wine.

For all night long I dreamed of you:
I woke and prayed against my will,
Then slept to dream of you again.
At length I rose and knelt and prayed:
I cannot write the words I said,
My words were slow, my tears were few;
But through the dark my silence spoke
Like thunder. When this morning broke,
My face was pinched, my hair was grey,
And frozen blood was on the sill
Where stifling in my struggle I lay.

If now you saw me you would say:
Where is the face I used to love?
And I would answer: Gone before;
It tarries veiled in Paradise.
When once the morning star shall rise,
When earth with shadow flees away
And we stand safe within the door,
Then you shall lift the veil thereof.
Look up, rise up: for far above
Our palms are grown, our place is set;
There we shall meet as once we met
And love with old familiar love.

APPENDIX B

The Nix

By RICHARD GARNETT

THE crafty Nix, more false than fair,
Whose haunt in arrowy Iser lies,
She envied me my golden hair,
She envied me my azure eyes.

The moon with silvery cyphers traced
The leaves, and on the waters play'd;
She rose, she caught me round the waist
She said, Come down with me, fair maid.

She led me to her crystal grot,
She set me in her coral chair,
She waved her hand, and I had not
Or azure eyes, or golden hair.

Her locks of jet, her eyes of flame
Were mine, and hers my semblance fair;
'O make me, Nix, again the same,
O give me back my golden hair!'

She smiles in scorn, she disappears,
And here I sit and see no sun,
My eyes of fire are quenched in tears,
And all my darksome locks undone.

APPENDIX C

Alterations made to the 1984 reprint of the fourth edition of *The Poems*, edited by W. H. Gardner and N. H. MacKenzie, include:

The Escorial. l. 9 MS comma added after 'palace'; l. 106 'continually' now 'continuously' (see note to poem); l. 112 comma changed to MS dash; l. 122 'full' changed to 'null' MS.

Promêtheus Desmotês. MS accents added in title; l. 5 comma after 'you' deleted to follow MS.

A Vision of the Mermaids. l. 136 'Ocean', MS 'ocean'.

Pilate. Stanzas reordered and numbered as in the MS; l. 8 'and' restored (MS).

Richard. l. 8 'one drunk' changed to the later reading, 'drinking'; l. 9 'In' similarly to 'True'.

A Soliloquy. Stanzas reordered by N. H. M. to follow MS; comma added after l. 7 'sand'; exclamation mark inserted in l. 11; l. 15 comma changed to MS stop; l. 25 stop replaces comma after 'stand'; l. 39 comma changed to MS stop; l. 43 initial capital inserted in 'Who made'; ll. 53–4 later reading taken so that 'And bring your offerings to a grateful god, / And fear no iron rod' is now 'Bring wheat-ears from the loamy stintless sod, / To a more grateful god'.

New Readings. Comma deleted at the end of l. 6.

Heaven-Haven. Subtitle: MS brackets added; l.c. 'a'.

'Why should their foolish bands'. l. 8 later reading preferred so that 'Far from its head' is now 'If so it be,'; editorial changes to punctuation in ll. 5 and 7.

'Why if it be so'. 'Bala' moved from the text to the notes.

'Or else their cooings' and 'It was a hard thing' and 'Think of an opening page'. 'Maentwrog' moved from the text to the notes.

'Glimmer'd along'. l. 2 stop, MS semicolon.

'Miss Story's character'. ll. 19–20 later version taken so that '. . . but less than female tact, / Sees the right thing to do, and does not act;' is now '. . . if she would make it known / And charms—but they should be more freely shewn.' ll. 21–2 placed earlier as indicated in MS.

'Did Helen'. l. 3 later reading taken so that 'She is too plain' is now 'But she's too plain'.

Seven Epigrams: (v). Colon deleted after 'lark'.

Io. l. 15 later reading used so that 'The knot of feathery locks' now reads 'The feathery knot of locks'.

Floris (iii). l. 21 MS comma added after 'sight'; l. 25 accent deleted from 'learnèd'; later alternatives taken throughout.

'No, they are come'. l. 10 MS comma added; l. 13 MS 'etc' added.

A Voice from the World. l. 36 final semicolon deleted; l. 117 comma added.

Richard (iv). l. 9 second hyphen added (MS); l. 15 comma added (MS).

The Queen's Crowning. l. 120 hyphen added in 'true-love'.

Easter Communion l. 10 semicolon after 'gladness' changed to MS comma.

To Oxford. Layout now follows MS.

'Confirmed beauty'. l. 6 comma added (MS); l. 11 'it.' altered to MS 'it; what'.

The Beginning of the End. Layout now follows autograph; (i) l. 5 'Winter' now l.c.; l. 8 final comma deleted.

'Myself unholy'. l. 7 comma deleted after 'trust'; l. 9 final semicolon now a comma; l. 13 final comma added (MS).

To Oxford. Layout altered to follow MS.

'See how Spring opens'. Layout changed (MS).

Continuation of 'Nix'. l. 6 hyphen added.

'Mothers are doubtless'. l. 4 'born in' now 'hold'.

Castara Victrix. (ii) l. 4 'best begun' now 'well begun'; (iii) l. 21 'upon' now 'on'.

'My prayers must meet'. l. 10 comma added after 'Yea'.

'Let me be to Thee'. Layout changed (MS).

The Half-way House. l. 3 colon replaced by MS semicolon.

A Complaint. Text follows MS A leading to these changes: l. 2 comma added; l. 3 'And if you write at last,' now 'And now if at last you write'; l. 5 'Me—neglectful that you . . . not.' now 'me, Neglectful you . . . not:'; l. 7 'could' now 'can'; l. 9 'a' added; l. 11 'bright; . . . charms' now 'bright, . . . charms,'; l. 12 'It is' now 'It's'; l. 13 hyphen deleted; l. 15 italics added; l. 17 'never: you' now 'never. You'; l. 18 colon now semicolon; l. 19 colon deleted before brackets; l. 20 quotation marks added.

'Moonless darkness'. l. 3 hyphen added.

'The earth and heaven'. l. 7 comma added (MS); l. 28 dots deleted.

The Nightingale. Initial quotation marks deleted except from l. 1; l. 53 indented.

Summa. Stops replace commas in ll. 2 and 12.

Jesu Dulcis Memoria. Stanzas reordered and added to; ll. 5–6 later version taken so that 'Song never was so sweet in ear, / Word never was such news to hear' is now 'No music so can touch the ear, / No news is heard of such sweet cheer'.

Equis binas. l. 4 later variant chosen so that. 'I in an instant would be gone' is now 'In instant time I would be gone'.

'Alget honos'. Not previously printed. (N.H.M.)

Horace: 'Persicos odi'. Layout altered (MS).

Horace: 'Odi profanum'. l. 4 later version used so that 'bid' now reads 'make'; l. 21 dash deleted before 'sleep'.

Rosa Mystica. Layout changed; stanzas numbered (MS); l. 1 dash replaced by comma; l. 2 dash added at end; l. 16 commas after 'birth' and 'bloom' deleted; l. 45 initial capital 'Breath'.

'Mirror surgentem.' l. 10 'putes' changed to 'putas'.

S. Thomae Aquinatis. ll. 1–2 altered to later version; l. 25 'veilèd' taken as final choice.

The Wreck of the Deutschland. Subtitle follows MS *B*; accents from *B*; l. 38 and l. 55 'though' for 'tho' ' since G. M. H. generally used 'though' while R. B. frequently wrote 'tho' '; l. 71 semicolon replaces colon; l. 90 indented (MS); l. 137 final comma deleted to follow *B*.

Moonrise. l. 1 l.c. 'midsummer'.

The Woodlark. Fr. Bliss's lines deleted; layout, punctuation, and sequence of lines now follows MS.

Penmaen Pool. l. 5 dash added at end; l. 9 dash added; l. 30 initial capital 'Month'.

God's Grandeur. l. 12 'eastwards' from all autographs.

'As kingfishers catch fire. l. 7 'itself' changed to MS 'its self'.

Spring. l. 11 comma after 'get' deleted. G. M. H. checked the punctuation in this poem carefully.

The Windhover. l. 5 'ecstacy' (MS).

The Caged Skylark. Ink and pen-tracks suggest: l. 1 final comma added by G. M. H.; l. 3 comma changed to semicolon; l. 5 final comma by R. B. deleted; l. 12 comma added after 'flesh-bound' by G. M. H.

Lantern out of Doors. Ink and pen-tracks suggest l. 5 comma added by G. M. H. as a result of cancelling 'So' from the line's beginning and consequently also deleting the elision 'me whom'.

May Magnificat. l. 45 'ecstacy' (MS).

'Denis'. Layout (MS).

Candle Indoors. Ink and pen-tracks suggest l. 10 colon changed to semi-colon by G. M. H.

Handsome Heart. R. B. preferred the early versions of this poem and transcribed a composite of these into MS *B*. In 1883–4 G. M. H. cancelled this version and wrote the one used here for text.

'How all is one way wrought' (On a Piece of Music). Title changed by N. H. M.; I have reordered the stanzas (MS); stanzas 1 and 3, later versions used.

Bugler's First Communion. l. 30 ink and pen-tracks suggest G. M. H. changed colon to dash; l. 43 comma added by G. M. H.

Morning, Midday, and Evening Sacrifice. l. 17 G. M. H. revised the line and omitted the final comma.

At the Wedding March. Layout altered (MS).

Brothers. Following MS *B* leads to a number of changes: l. 8 colon replaces semicolon; l. 11 stop replaces semicolon; l. 13 stop replaces colon; l. 14 comma after 'called' deleted; l. 16 hyphen inserted; l. 19 comma replaces semicolon; l. 22 final comma deleted; l. 33 semicolon replaces exclamation mark; l. 35 'But' replaced by 'Oh,'; comma added after 'Harry'; l. 37 dash added; l. 38 indentation removed; l. 39 exclamation mark replaces semicolon; l. 40 comma after 'Nature' deleted.

Spring and Fall. Capitals inserted in subtitle (MS).

Leaden Echo and Golden Echo. Golden Echo l. 2 semicolon replaces comma; l. 19 comma after 'God' deleted; l. 24 hyphen inserted in 'heavy-headed'.

Ribblesdale. l. 3 comma deleted after 'To'; l. 10 dash deleted before 'Ah'.

Blessed Virgin compared. l. 124 hyphen added to 'World-mothering': see *L* I, pp. 190, 195. When G. M. H. copied in ink R. B.'s pencilled transcription of the last two-thirds of the poem, he did not notice the missing hyphen.

'The times are nightfall'. Layout changed (MS); dots at end (not in MS) deleted.

'To seem the stranger'. l. 7 accent added; l. 8 initial capital added.

To what serves Mortal Beauty. Using *B* instead of the earlier *D* as text leads to: l. 4 'wit' instead of 'wits'; 'to' inserted before 'what'; initial capitals inserted in ll. 2 and 8; l. 9 'once' instead of 'needs'; l. 10 virgule replaces colon; l.c. verb 'love'.

(Carrion Comfort). l. 6 the final alternative, also used in all earlier drafts extant, 'earth', replaces 'world'.

(The Soldier). Caesuras inserted (MS); l. 7 editorial comma deleted.

'Thee, God, I come from'. Two lines and a final verse added.

'Patience, hard thing'. l. 2 'patience' l.c. (MS); l. 6 hyphen inserted 'No-where'; l. 11 accent added.

'My own heart: l. 6 editorial comma after 'comfortless' deleted; l. 13 initial capital added; 'unforseentimes' made one word (MS).

To his Watch. l. 1 final comma added (MS); l. 3 'force' retained, it seems to me more appropriate than 'forge' to the mood and meaning of the poem; l. 8 later alternative taken so that 'Is comfort's carol of all or woe's worst smart' is now 'Is sweetest comfort's carol or worst woe's smart'; l. 11 dots deleted.

Songs from Shakespeare. (i) l. 1 comma moved to far side of 'has'; final comma deleted; l. 11 'occinit. Occinat' replaced by 'occinit, occinat' (*A*).

(ii) l. 7 'Phorcys' moved to end of line as in MS.

(v) and (vii). Layout changed by N. H. M.

Latin version of 'In all things beautiful'. l. 7 l.c. 'tacet'.

Spelt from Sibyl's Leaves. Accents from *B*; l. 6 initial capital 'Tray' (MS).

Harry Ploughman. l. 15 later version taken so that 'See his wind-lilylocks-laced' becomes 'Wind-lilylocks-laced'.

Ashboughs. Layout altered (MS).

Tom's Garland. l. 16 hyphen inserted in 'no-one'.

Epithalamion. Breaks between ll. 13–14, 18–19, 45–6 deleted (not in MS); l. 28 'No', accent added; l. 33 'fingerteasing', hyphen deleted; l. 45 dash deleted (MS); l. 47 stop deleted; two extra lines added after l. 47.

'What shall I do'. l. 2 dash added; l. 5 'we' is editorial; l. 8 'There' is underlined in MS.

St. Alphonsus Rodriguez. Ink and paper, when compared with those used in letters written by G. M. H. to R. B. at the time, suggest that the version in *C* is the final one. This is also the only copy to include the revision to l. 1 mentioned *L* I, p. 297 (19 October 1888). Resultant changes in wording occur in ll. 1, 2, 4–7, 9–11.

*   *   *

Some Corrections to the Second Impression

I am most grateful to Professor N. H. MacKenzie for pointing out that on:

p. 28 'Why should their foolish bands' the stop l. 5 'away.' though cited was omitted.

p. 321 'My prayers must meet a brazen heaven' has imagery influenced by Deut. 28: 23 and Lev. 26: 19 rather than Homer.

pp. 350 and 351 Lady Pooley's manuscripts are owned by the University of Texas.

p. 355 l. 5 of 'The Lantern out of Doors' opens with a spondee rather than counterpoint.

p. 356 note to p. 138 l. 94 blown = mature, rather than 'lost'.

p. 361 l. 1 'which' should be followed by a comma, not a semi-colon.

p. 372 Caerwys rhymes with 'heiress', not 'herres'.

Mrs E. E. Duncan-Jones has convincingly argued that 'Belleisle' in 'To Oxford' refers to Balliol, not Oxford.

Dr Norman White has helpfully mentioned that:

p. 381 The portrait identified in *Landscape and Inscape* as being of 'two beautiful young persons' is unlikely to be of them.

p. 395 Hopkins was taught Suarez at St Mary's Hall, not Mount St Mary's.

pp. 428–9 There were errors in the index: Stevenson was misspelt Stephenson and under Mrs Thomas Marsland, Hopkins's surname was omitted so that the entry appeared under Marsland instead of Hopkins.

The most important of the other changes are:

p. 191 the notes for Aug. 22 belong to 1867, not 1866.

p. 391 Cyril Hopkins became an average adjuster, not a linguist.

p. 392 I am now practically certain that Dr Pusey's letter refusing to see G. M. H.
       was written on the 20th, not the 10th, of October and that Hopkins was
       being perfectly honest in telling his father (and subsequently his mother)
       that he had not yet approached Pusey.

INDEX OF SHORT TITLES AND
FIRST LINES

(Titles are set in italic, first lines in roman)

INDEX TO PROSE